EMERGENT PHENOMENA IN ATOMIC NUCLEI FROM LARGE-SCALE MODELING

A Symmetry-Guided Perspective

EMERGENT PHENOMENA IN ATOMIC NUCLEI FROM LARGE-SCALE MODELING

A Symmetry-Guided Perspective

Kristina D Launey
Louisiana State University, USA

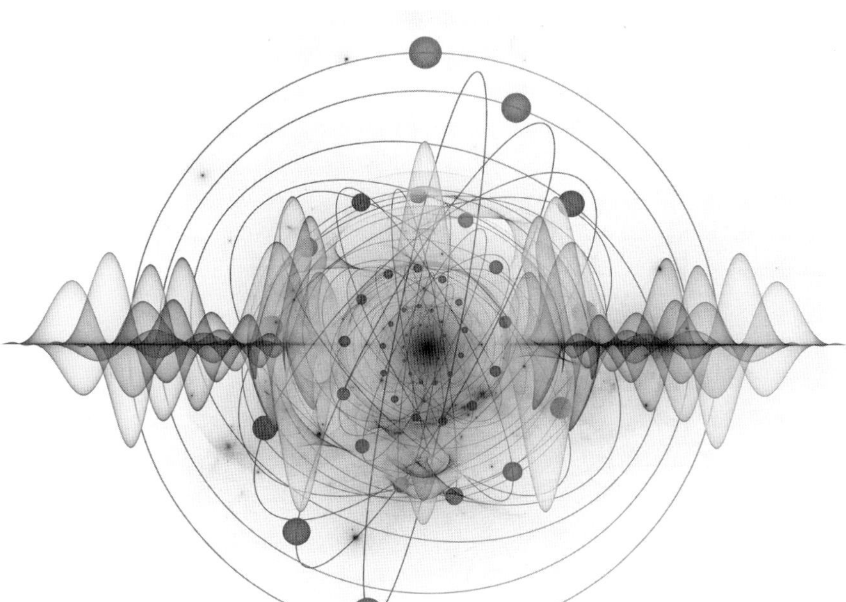

World Scientific

NEW JERSEY • LONDON • SINGAPORE • BEIJING • SHANGHAI • HONG KONG • TAIPEI • CHENNAI • TOKYO

Published by

World Scientific Publishing Co. Pte. Ltd.
5 Toh Tuck Link, Singapore 596224
USA office: 27 Warren Street, Suite 401-402, Hackensack, NJ 07601
UK office: 57 Shelton Street, Covent Garden, London WC2H 9HE

Library of Congress Cataloging-in-Publication Data
Names: Launey, Kristina D., 1976– editor.
Title: Emergent phenomena in atomic nuclei from large-scale modeling :
a symmetry-guided perspective / editor, Kristina D. Launey, Louisiana State University, USA.
Description: Singapore ; Hackensack, NJ : World Scientific, [2017]
Identifiers: LCCN 2017019358| ISBN 9789813146044 (hardcover ; alk. paper) |
ISBN 9813146044 (hardcover ; alk. paper)
Subjects: LCSH: Nuclear physics. | Symmetry (Physics) | Nuclear collective models.
Classification: LCC QC776 .E54 2017 | DDC 539.7/5--dc23
LC record available at https://lccn.loc.gov/2017019358

British Library Cataloguing-in-Publication Data
A catalogue record for this book is available from the British Library.

Printed in Singapore

Preface

This book does what might be impossible for a single-authored book to do. As a unique collection of reviews that share a common topic, *Emergent Phenomena in Atomic Nuclei*, the book offers different perspectives of the multifaceted nature of the subject and encompasses different scales, from quarks through few-particle nuclear systems to heavy nuclei. It tells an amazing story of a decades-long journey of trials and successes, up to the present, with the aim to understand the vast array of experimental data and to determine the fundamentals of strongly interacting fermions. The emphasis is on discovering emergent orderly patterns amidst the overarching complexity of many-particle quantum-mechanical systems. Recent findings are discussed within an interesting framework: a combination of nuclear theory and experiment, of group theory and computational science, and of pivotal models of astonishing simplicity and state-of-the-art models empowered by supercomputers.

A special theme resonates throughout the book: the important role of symmetries, both exact and approximate, in exposing emergent features and guiding large-scale nuclear modeling. World-renowned experts offer their unique perspective on symmetries in the world of quarks and gluons, and that of protons and neutrons — from chiral symmetry, through spin–isospin and quasi-spin symmetries, to symplectic symmetry — as well as on the emergent nature of nuclear collectivity, deformation, clustering, and pairing, as viewed from first principles, microscopic considerations, and analysis of experimental data. The book provides an excellent foundation that allows researchers and graduate students in physics and applied mathematics to review the current status of the subject, and to further explore the research literature through exhaustive sets of references (about 900 references are cited in the book) that also point to studies underpinned by similar techniques in condensed matter and atomic physics along with quantum information and astronomy.

The book chapters have been carefully selected as insightful representatives, *albeit* not an exhaustive list, of modern approaches that have been remarkably successful in modeling the challenging nuclear dynamics. Starting with brief overviews of our current perception of the nucleus shaped by the vast body of experimental data and pivotal nuclear models, the book then unveils the intricacy of the many-particle self-bound nuclear system and the remarkable simplicity found within. It encompasses widely used microscopic structure approaches, such as the shell model, pairing and collective models, mean-field models, and calculations on the lattice, as well as the cluster model that enhances structure modeling with the ability to describe cluster formations and nuclear reactions. Chapters can be read independently but often intersect, as indicative by many cross-references I have included throughout the book.

Why is the atomic nucleus so intriguing? Why now? The foremost reason is, perhaps, that now is the time when the enigmatic complexity of nuclei is beginning to be unraveled — as a result of advances being made in experiment and theory; of massively parallel computers facilitating large-scale simulations of nuclei; and of the link that is being built from the underlying physics of quarks and gluons to systems of a few nucleons. This book speaks up for such a major progress, addressing topics of utmost importance:

- the challenging Hoyle state, formation of alpha clusters, shape coexistence, as well as nuclear moments and their experimental status;
- giant resonances and soft dipole modes, nuclear and neutron matter;
- the elusive neutron electric dipole moment and physics beyond the standard model;

- ◦ electroweak scattering and light-ion reactions of interest to nuclear physics, astrophysics, and neutrino physics;
- ◦ level densities and pairing gaps in medium-mass and heavy nuclei;
- ◦ extended Hohenberg–Kohn theorem for density functional theory and its mathematical proof, along with new insights into symmetry patterns with efficacious numerical algorithms; and so much more.

I am deeply indebted to the authors of the book chapters who devoted considerable time and effort in compiling years of experience onto the pages of this book.

K. D. Launey

Contents

Part 4: Symmetries and Emergent Features in Medium-mass and Heavy Nuclei 265

PART 1

Symmetries in Nuclei: From Experimental Signatures to Pivotal Nuclear Models (Overview)

Chapter 1

Nuclear Collectivity — Its Emergent Nature Viewed from Phenomenology and Spectroscopy

John L. Wood
School of Physics, Georgia Institute of Technology
Atlanta, GA 30332, USA

The collective nature of the nucleus was formalized 63 years ago in the work of Aage Bohr. In other arenas of physics it took 63 years to progress from the "quantum" to the "quark". Progress in elucidating the details of nuclear collectivity appears to have almost stood still by comparison. The Author gives a personal perspective on nuclear collectivity, using phenomenology and examples from spectroscopy, based on his 50 years as an active nuclear physicist.

Contents

1. Introduction

Nuclear collectivity is, by definition, a many-body phenomenon. Unlike the physics that one encounters when dealing with a quantum or a quark, or an independent nucleon in a shell-model or Nilsson-model state, nuclear collectivity is orders-of-magnitude more complicated. This difference in complexity is proven by experience: we do not possess a universally accepted, unified model of nuclear collectivity (even after 63 years of trying to achieve this). The present exposition is a personal view by the Author that addresses this issue and may help to understand the lack of progress. (It is worth noting that we are in the excellent company of high-temperature superconductivity and the fractional quantum Hall effect.)

The atomic nucleus is a finite many-body quantum system. Extraordinary advances have been made in probing nuclear structure by way of experimental ingenuity and sophisticated model building. The shell model has been a reassuring guide to understanding a significant number of nuclei, at least their low-lying states, through the key concepts of energy shells and independent particle degrees of freedom. The pairing model provides remarkable insight into structures that fall under the heading of "seniority" degrees of freedom (seniority = number of unpaired nucleons). The rotor model is realized in some nuclei at the level of sub-one-percent precision in low-energy excitations and selected electromagnetic properties. Yet, the majority of nuclei do not come close to matching these ideal model examples; and there is seemingly little insight to what is missing, despite vast amounts of data.

Confronted with lack of insight regarding a unified model of nuclear collectivity, from the perspective of the Author, most recent activity has turned to shorter-term exploration: discovering a new isotope, adding a few excited states, determining a mass or a mean-square charge radius, measuring an electromagnetic transition probability. There are rallies to the collapses of closed shells, but few concerted efforts to probe nuclear collectivity in a manner that will move nuclear structure towards a unified view that is universally accepted. The present exposition looks at facets of nuclear structure where there is a basis for agreement, and takes a personal view of where much work still needs to be done.

2. Basics of the Nuclear Many-Body Problem

A number of philosophical criteria are important to keep in mind:

> "*Everything should be made as simple as possible, but not simpler*"
>
> — Albert Einstein;

> "*We are to admit no more causes of natural things than such as are both true and sufficient to explain their appearances. Therefore, to the same natural effects we must, so far as possible, assign the same causes*"
>
> — Isaac Newton (William of Ockham).

These two adages define useful bounds for model building; unfortunately many (most) models of nuclear collectivity overstep these bounds. But, without models we would be powerless to organize nuclear data and make any sense thereof.

The array of nuclear data is, after 63 years of having ideas from Bohr's collective model in hand (66 years for the shell model, 57 years for the pairing model), vast. It should be noted that were it not for the Evaluated Nuclear Structure Data File (ENSDF), maintained by the International Network of Nuclear Structure and Decay Data Evaluators, the Community would be buried in largely meaningless numbers. Consultation of ENSDF (and their link to the Experimental Unevaluated Nuclear Data List, XUNDL) provides a remarkable "jump-start" to any nuclear structure investigation, experimental or theoretical. However, what is lacking is "horizontal" perspective.

Occasionally a review is made of some specific feature of nuclear structure, e.g., the electric quadrupole transition probability, $B(E2)$, from the first-excited 2^+ state in a doubly even nucleus [1]; but such work requires great experience and enormous amounts of time.

At a more subtle level, despite organized databases and useful models for organizing the data, new phenomena emerge. Philip Anderson underlined the operative meaning of "emergent" [2]. In the face of emergent behavior, all that models can do is fail — this is of great value, because when a model fails we have learned something. Note, the failure of the Standard Model of particles and fields is being sought, avidly. It is from data that emergent behavior is recognized; but data generally have to be organized to realize this, recall the above point regarding horizontal reviews. However, the first signs of new emergent data may be subtle. (A very good example of this was the first indication that nuclei might be deformed, which emerged from optical hyperfine structure and the suspicion that there may be the presence of a quadrupole electric field [3].)

2.1. *Nuclear Monopole Moments*

Nuclear monopole moments, or nuclear mean-square charge radii, form one of the two fundamental and broad horizontal views of nuclear structure (the other is nuclear masses). It is formalized in the equation, $R = R_0 A^{1/3}$, where R is the root-mean-square radius, $R_0 = 1.2\,\text{fm}$, and A is the mass number. It reflects a remarkable property of atomic nuclei — they all possess nearly the same (central, there is a surface diffuseness) density. What is not so obvious is that this can be regarded as a scale invariance possessed by nuclei. This translates into similar behavior over a wide range of masses (A values). Thus, for example, ^{133}Sn and ^{209}Pb have similar excitation patterns [4].

2.2. *Nuclear Quadrupole Moments*

Nuclear quadrupole moments form a horizontal view of nuclear structure which, if not regarded as fundamental, ought to be so regarded. They are the basis on which the collective model of the nucleus was introduced. Thus, nuclear collective models are dominated by quadrupole degrees of freedom. Possibly, the lack of fundamental status for nuclear quadrupole moments is because "some nuclei do not possess them", i.e., spherical nuclei; but a spherical nucleus has a quadrupole moment, it equals zero. This point is made because nuclei are often introduced as if their normal manifestation is spherical, ... but some show deformation. It is becoming evident that non-zero deformation is far more widespread than zero deformation and that even nuclei that are spherical (in their ground states) have deformed excited states [5]. Thus, it is more useful to regard all nuclei as possessing a range of possible deformations that includes zero, and that sometimes a zero-deformation state is the ground state.

There is a serious lack of quadrupole moment data. One fundamental difficulty is that 25% of all nuclei (even–even nuclei) have spin-zero ground states and so, for quantum mechanical reasons, their ground-state quadrupole moments are zero. In order to "see" the deformation of such nuclei, either a model-dependent feature such as a rotational band of excited states is needed, or a model-independent feature such as a quadrupole moment of the 2_1^+ state is needed. The former is very easy to observe, the latter is extremely difficult to measure.

There is an additional challenge involving models of nuclear collectivity: they are often formulated in terms of shape variables. However, these variables are in the intrinsic frame and all observations are made in the laboratory frame. Thus, there is a fundamental quantum "fuzziness" to deduced nuclear shapes. This has been circumvented, to a limited degree, by the so-called Kumar–Cline sum rules [6, 7]. Essentially, these are sums of products of (measurable) $E2$ matrix elements that are rotationally invariant; but they involve excited-state quadrupole moments (diagonal $E2$ matrix elements).

2.3. *Simple Nuclear Models*

2.3.1. *The shell model*

The shell model is an extreme independent-particle description of the nucleons that constitute a nucleus. It led to the understanding of the emergent magic numbers, or closed shells, manifested in the nuclear data accumulated in the late 1940s [8, 9]. It achieved this by way of invoking a single-particle central potential with the addition of a very strong spin-orbit interaction. Figure 1(a) shows spectroscopic data for ^{207}Pb "reverse-engineered", by way of removing the spin-orbit contributions, to reveal the pattern of angular momentum eigenstates possessed by the central potential for this nucleus. One immediately infers that a harmonic-oscillator potential (which would possess angular momentum degeneracy) is inappropriate for this nucleus. However, an empirical correction to the oscillator potential is provided by a one-body term of the form $-D\mathbf{l}^2$, sometimes called the "well-flattening" term. Sven-Gösta Nilsson recognized the addition of such a term as being a most effective way to retain the harmonic-oscillator radial dependence in the modeling of the shell-model central potential [10].

The direct applicability of the shell model to nuclear spectra is limited to doubly closed shell nuclei plus or minus one nucleon. Carrying out the process embodied in Fig. 1(a), for nuclei adjacent to closed shells, is a most informative exercise provided that one works with single-particle energy *centroids* constructed using spectroscopic factors determined from one-nucleon transfer reactions. However, the shell model offers a much more important contribution to nuclear structure: it provides a powerful *basis of representation* for many-nucleon configurations.

The process by which the shell model leads to a basis for the description of a many-nucleon configuration is achieved (in simple terms) by taking A-copies of independent-particle eigenstates in an antisymmetrized product. If, in this process, harmonic-oscillator radial wave functions are retained, the properties of such a basis lead to remarkable mathematical simplicity. For example, the kinematics of the center-of-mass of the many-body system is separable. At a deeper level, such a many-body basis possesses elegant and powerful Lie algebraic properties, as first put forward by Elliott [11].

There is another aspect of the shell model that emerges from the nuclear properties sketched in the foregoing discourse: the "scaling" provided by the monopole properties of the nucleus leads directly to a

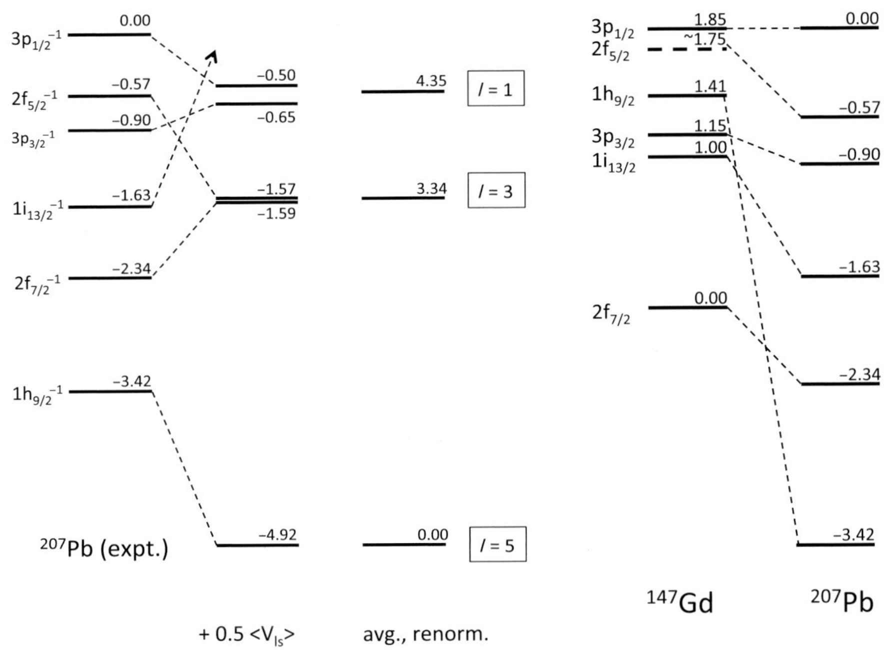

(a) (b)

Fig. 1. (a) The low-energy states of ^{207}Pb with the effect of removing the spin-orbit coupling energy using an empirical strength of 0.5 MeV. This reveals the non-degeneracy of the different angular momentum eigenstates in the central potential describing the independent-particle mean field. The data are taken from ENSDF. (b) The low-energy states of ^{147}Gd and ^{207}Pb, plotting the energies of ^{207}Pb inverted with respect to excitation in recognition of their hole character. Note the enormous energy shift of the $1h_{9/2}$ state relative to the other states. This is due to the filling of the $1h_{11/2}$ proton subshell between ^{147}Gd and ^{207}Pb and a two-body proton–neutron monopole residual interaction. The data are taken from ENSDF.

self-consistency condition between the oscillator frequency possessed by a given nucleus and the nuclear mean-square charge radius. This is often termed the "volume-conservation constraint". In simple terms, nucleons in an unconstrained oscillator potential would minimize their energy by organization into oscillator states with very low oscillator frequencies. This self consistency is the reason that one can recognize, e.g., in the excitation spectra of ^{133}Sn and ^{209}Pb [4]. the emergence of two manifestations of the same underlying many-body physics, despite the fact that they differ by about a factor of two in the number of particles.

A less well recognized feature of "shell-model" nuclei, i.e., nuclei adjacent to doubly closed shells, is manifested in Fig. 1(b). This shows that even where extreme independent-particle behavior is expected to dominate, other important factors play a major role. In this instance it reflects the fact that there are many-body effects, in particular a two-body "monopole" interaction between the $1h_{9/2}$ neutrons and $1h_{11/2}$ protons. It is impossible to incorporate this into a universal one-body potential for the $82 < N < 126$ shell. Similar "monopole" shifts occur in all shells. Such behavior has led to the currently popular tensor-force developments [12].

While the extreme independent-particle shell model is limited in the ways defined above, qualitatively one observes the survival of states that look like independent-particle states in many nuclei in the vicinity of closed shells. An example is depicted in Fig. 2:

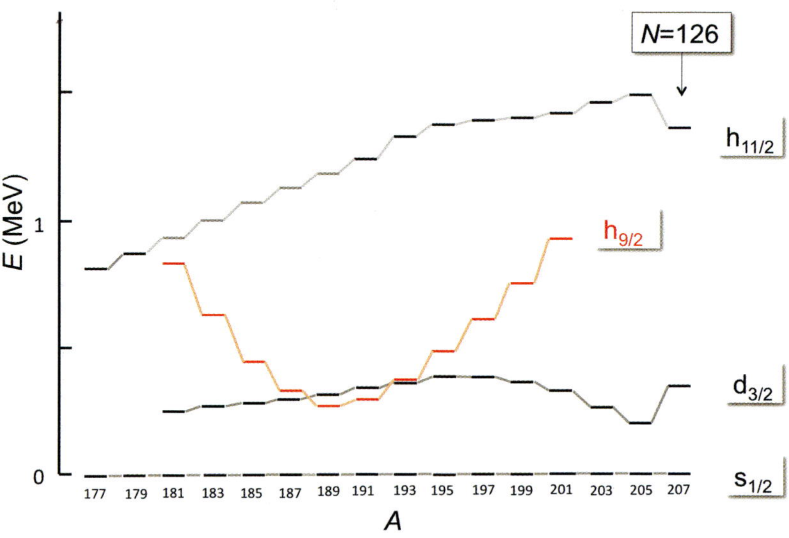

Fig. 2. Selected low-energy states of the odd-mass Tl isotopes. The lowest states with spin-parity $1/2^+$, $3/2^+$, $9/2^-$, and the candidate $11/2^-$ state for the $1h_{11/2}$ configuration are shown. The labeling indicates the dominant shell-model configurations. The $9/2^-$ state is discussed in the text. The data are taken from ENSDF.

the structure of the odd-mass Tl isotopes requires much more than simple shell-model ideas for an explanation; but to the degree that the shell model is influencing these isotopes, it is exhibiting remarkable stability. Indeed, when one notes that the three lightest Tl isotopes shown are unbound with respect to proton emission, this pattern seems extraordinary. The behavior of the state labeled as the $1h_{9/2}$ orbital is discussed later.

An in-depth treatment of the nuclear shell model is presented by Kris Heyde in a monograph [13].

2.3.2. *The pairing model*

The pairing model refers to a family of models that have as a common feature a short-ranged two-body force acting on independent-particle configurations in such a manner as to produce so-called Cooper pairs. Cooper pairs are fermion pairs that are strongly correlated in time-reversed orbitals. They can be regarded as forming independent-pair configurations. They are manifestly not independent-particle configurations: they cannot be factorized into products of independent-particle states. In basic quantum mechanical terms — the pairs are entangled.

Cooper pairs have properties that simplify low-energy nuclear structure in distinct ways. For example, subshell structure (due to energy separations of orbitals with different angular momentum in an oscillator shell, cf. Fig. 1(a)) is "washed out": in practical terms, this leads to smooth behavior across entire shells of quantities such as mean-square charge radii and masses (and their differences: isotope shifts and nucleon separation energies). It also "stabilizes" the configurations occupied by unpaired nucleons, as noted by Maria Goeppert-Mayer [14]. The Cooper-pair "environment" of an unpaired nucleon is the basis of a simplified description of many-fermion systems, which possess pairing forces, using independent quasi-particles by way of the Bogolyubov–Valatin transformation and Bardeen–Cooper–Schrieffer (BCS) theory [15–17]. Yet, further simplification of such many-fermion systems is achievable because Cooper pairs possess an SU(2) algebraic structure called quasi-spin, first put forward by Arthur Kerman [18].

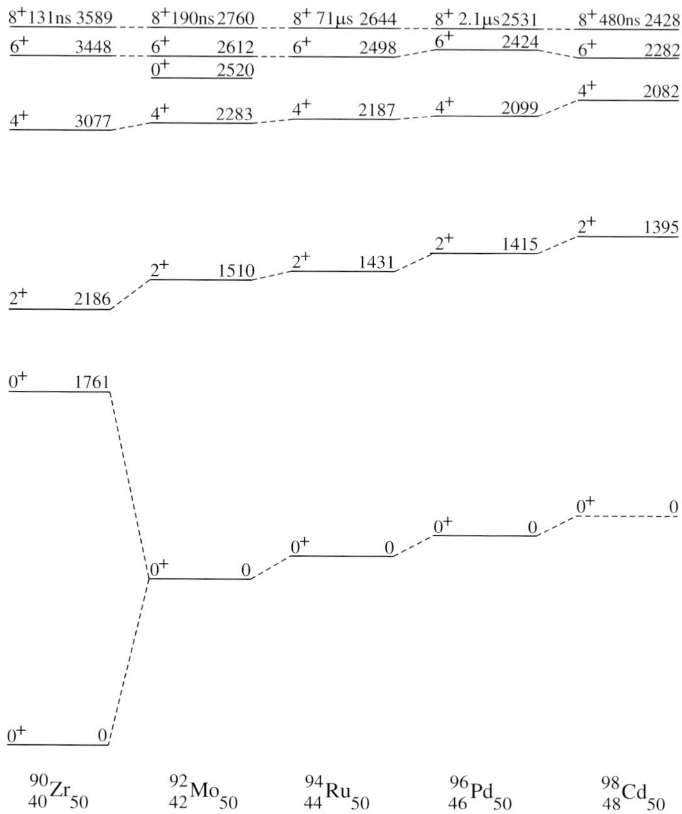

Fig. 3. The systematics of the low-energy states in the $N = 50$ isotones that reveal the $(\pi 1g_{9/2})^n$ seniority-dominated structure (energies in keV). The reference energy is set on the 8^+ state to dramatize the nature of the excited 0^+ state in ^{90}Zr as a mixed seniority-zero structure involving the $(\pi 2p_{1/2})^2$ and $(\pi 1g_{9/2})^2$ configurations (see, e.g., [19]). The data are taken from ENSDF.

Quasi-spin structure leads to very simple patterns of excitation energy and electromagnetic properties in singly closed shell nuclei. An example of an energy pattern is shown in Fig. 3 and of patterns of $B(E2)$ values in Figs. 4(a) and 4(b). Behind these patterns are simple SU(2)-algebraic properties. These extend to a Wigner–Eckart theorem for one-body transition operators through their bilinear combinations of a fermion creation and an annihilation operator, and the property of single-fermion operators that they are quasi-spin

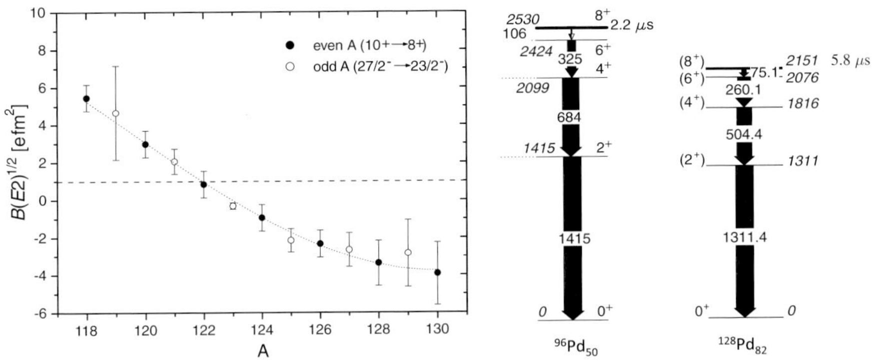

(a) (b)

Fig. 4. (a) The systematics of the $\sqrt{B(E2)}$ values for the seniority-dominated structures in the Sn isotopes with $A = 118 - 130$. The transitions involve the $10^+ \to 8^+$ seniority-2 and $27/2^- \to 23/2^-$ seniority-3 $(\nu 1h_{11/2})^n$ structures. The odd-A values are multiplied by 0.514, from a Clebsch–Gordan coefficient reflecting the SU(2) quasi-spin dynamical symmetry possessed by the seniority structures. The dotted curve is to guide the eye. The figure is taken from Ref. [20]. (b) The $(\pi 1g_{9/2})^n$ seniority-2 dominated structures in ^{96}Pd and ^{128}Pd (energies in keV). These structures are strong proof of the robustness of the $N = 82$ shell closure in the extremely neutron-rich isotopes below the $Z = 50$ closed shell at ^{132}Sn. The figure is adapted from figures presented in Refs. [21, 22].

tensors of rank-1/2. (For a detailed development of these issues, see the monograph by David Rowe and the Author [17].)

2.3.3. *The rotor model*

The rotor model refers to a family of models that treat the nucleus as an inert deformed rigid body. A deformed quantum object can exhibit quantized rotations. The prototype emerged from molecular spectroscopy. Indeed, the migration of the idea manifested in Aage Bohr's original paper [23] may have been influenced by Bohr's time spent at Columbia University where Isidor Rabi, one of the first persons to quantize the rotor [24], was a member of the physics faculty.

The rotor model is manifestly not a many-body quantum mechanical theory. One of the biggest challenges to nuclear structure is the completion of this unification (which appears to be close at

Table 1. Comparison of ground-state band transition energies for ^{242}Pu and ^{174}Yb.

I_i	$E(^{242}\text{Pu})$ (keV)	$E(^{174}\text{Yb})$ (keV) $\times$ 0.5824	$E(^{174}\text{Yb})$ (keV)
2	44.54^2	44.54 [norm.]	76.471^1
4	102.8^1	102.9	176.645^2
6	159.0^1	158.9	272.918^6
8	211.7^4	211.8	363.64^5
10	260.5^6	260.4^6	447.2^{10}
12	305.8^8	305.4^8	524.4^{13}
14	347.3^{10}	347.1^{10}	595.9^{17}
16	385.0^{11}	384.4^{11}	660^2
18	419.3^{12}	418.7^{17}	719^3
20	450.2^{13}	450.8^{29}	774^5

Notes: Uncertainties are quoted as superscripts, viz. $44.54^2 = 44.54 \pm 0.02$. The scale factor of 0.5824 is arbitrary and is chosen to match $E(2^+)$. Data are taken from ENSDF.

hand). The single most important thing about the rotor model, as applied to nuclei, is that in some nuclei it is realized at the level of precision of a fraction of a percent. To achieve this, all that is needed is the necessary lowest-order corrections for centrifugal and Coriolis effects (which are inevitably present in a rotating system). Table 1 illustrates a remarkable scaling between rotational energies in the rare-earth and actinide regions. Figure 5 gives some details of a very high precision realization of rotor $B(E2)$ values [25].

Using model-dependent relationships, $E2$ properties of "rotational" nuclei can be converted to intrinsic quadrupole moments. Figure 6 shows such moment data for the Yb isotopes. The plot suggests that intrinsic quadrupole moments are a smoothly varying property for sequences of isotopes (and isotones). Figure 7(a) presents evidence that the intrinsic quadrupole moment for a given nucleus does not change with spin, i.e., there is no centrifugal stretching. Thus, it is surprising that moments of inertia appear to exhibit a dependence on spin, as shown in Fig. 7(b) for the same nuclei as presented in Fig. 7(a). The most conservative statement that can be made about nuclear moments of inertia is that we do not understand them.

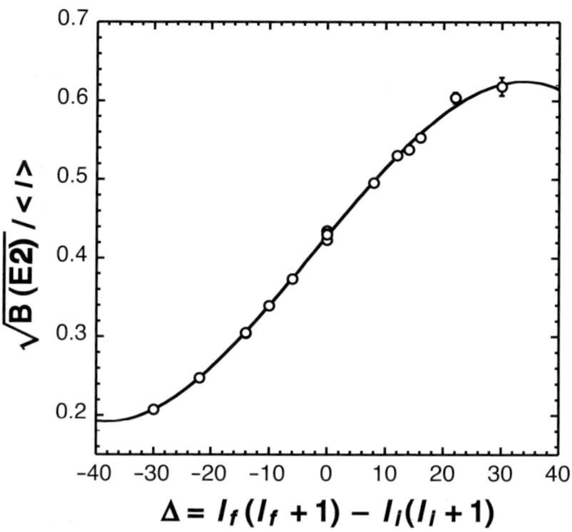

Fig. 5. Precision test of the rotor model for $B(E2)$ values between the lowest $K = 2$ band and the $K = 0$ ground-state band in ^{166}Er. The uncertainties in the data points are smaller than the circles for all but a single value for which the error bar is shown. The curve is a least-squares-fitted cubic polynomial. The figure is taken from Ref. [25].

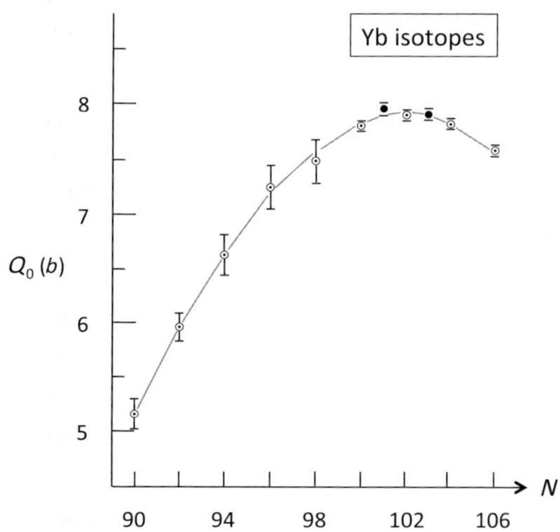

Fig. 6. Intrinsic quadrupole moments, Q_0, for the Yb isotopes extracted using rotor-model relationships for $E2$ matrix elements. The data are taken from ENSDF and Raman *et al.* [1]. (The values for ^{170,171,172,173,174,176}Yb are taken from muonic X-ray measurements of Ref. [26]: moments deduced from muonic X-ray data, which are of the highest precision available, are generally omitted from ENSDF, without explanation.)

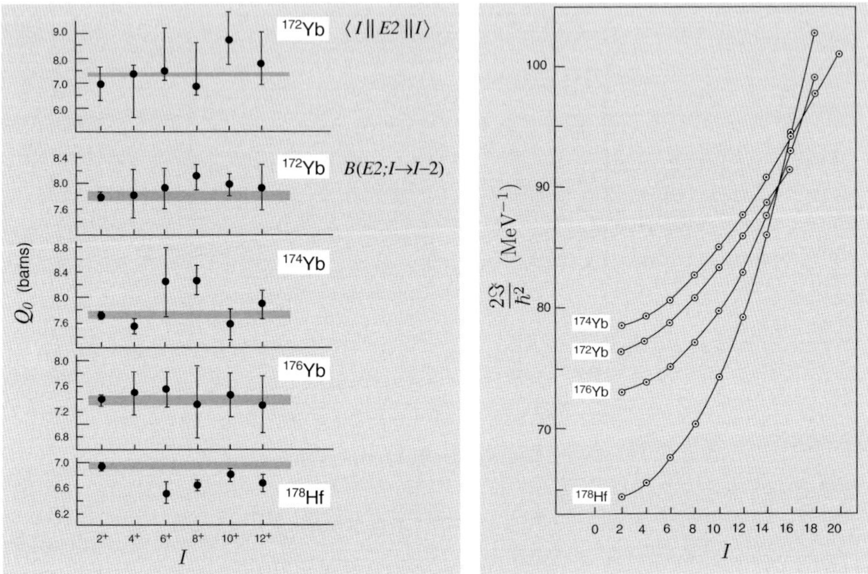

(a) (b)

Fig. 7. (a) Intrinsic quadrupole moments, Q_0, for ^{172,174,176}Yb and ^{178}Hf extracted using rotor-model relationships for *E2* matrix elements. The figure is taken from Ref. [17]. (b) Moments of inertia for ^{172,174,176}Yb and ^{178}Hf extracted using rotor-model relationships for the energies. The figure is taken from Ref. [17].

3. Nuclear Collectivity — A Current View

Essentially, all of the concepts outlined in Sec. 2 were in place by 1960. Since then there has been a vast expansion of detailed nuclear data; but this has not been matched by a unified understanding of nuclear collectivity. Within the scope of the Bohr model, collective vibrations of nuclei are possible. Spherical nuclei were expected to exhibit low-energy quadrupole vibrations if they were not too "rigid", i.e., in nuclei adjacent to closed shells. Deformed nuclei were expected to exhibit low-energy, normal-mode, vibrations of the quadrupole-deformed shape, i.e., the so-called β-vibrations and γ-vibrations. There was also the possibility that some deformed nuclei might be axially asymmetric and exhibit excitations characteristic of an axially asymmetric rotor. None of these simple collective features have been established beyond all doubt in nuclei. In what follows, the Author provides a personal view of the current situation.

Two major realizations, regarding nuclear collectivity, have expanded the view provided by the foregoing. The first is that shape coexistence occurs in many nuclei across the entire mass surface [5]; it probably occurs in nearly all nuclei. The second is that, with the inclusion of Coriolis and centrifugal effects, the rotor model can be applied to nuclei with quite small deformations [27]. These two realizations also provided critical points of confrontation with respect to low-energy collective quadrupole vibrations and their manifestation in nuclei. Low-energy quadrupole vibrations give rise to excited 0^+ states in double even nuclei, so does shape coexistence. The coupling of an unpaired particle to quadrupole collective core excitations gives rise to distinct patterns of spin sequences in excited states of odd-mass nuclei. These patterns are very similar for (anharmonic) quadrupole vibrations of the core and for rotations of weakly deformed cores (see, e.g., [28]).

To address the questions, defined above, regarding nuclear collectivity, the present narrative takes a view that focuses on a few specific sets of nuclei:

- The Cd isotopes have been regarded traditionally as spherical and "soft" with respect to quadrupole collectivity, i.e., they have been regarded as a best-case region for observing and studying the manifestation of low-energy quadrupole vibrations [29].
- The $N = 90$ isotones have been widely regarded as lying in a region of sudden onset of deformation and to be moderately deformed, i.e., they have been regarded as a best-case region for observing and studying the manifestation of β vibrations [30].
- The Os and Pt isotopes have been particularly a focus for testing models with axial asymmetry and γ "softness" [31].

3.1. *The* Cd *Isotopes*

The Cd isotopes possess both low-energy weakly collective excitations built on their ground states and strongly deformed "intruder states". Figure 8 shows this pattern for ^{110,112,114,116}Cd. One observes

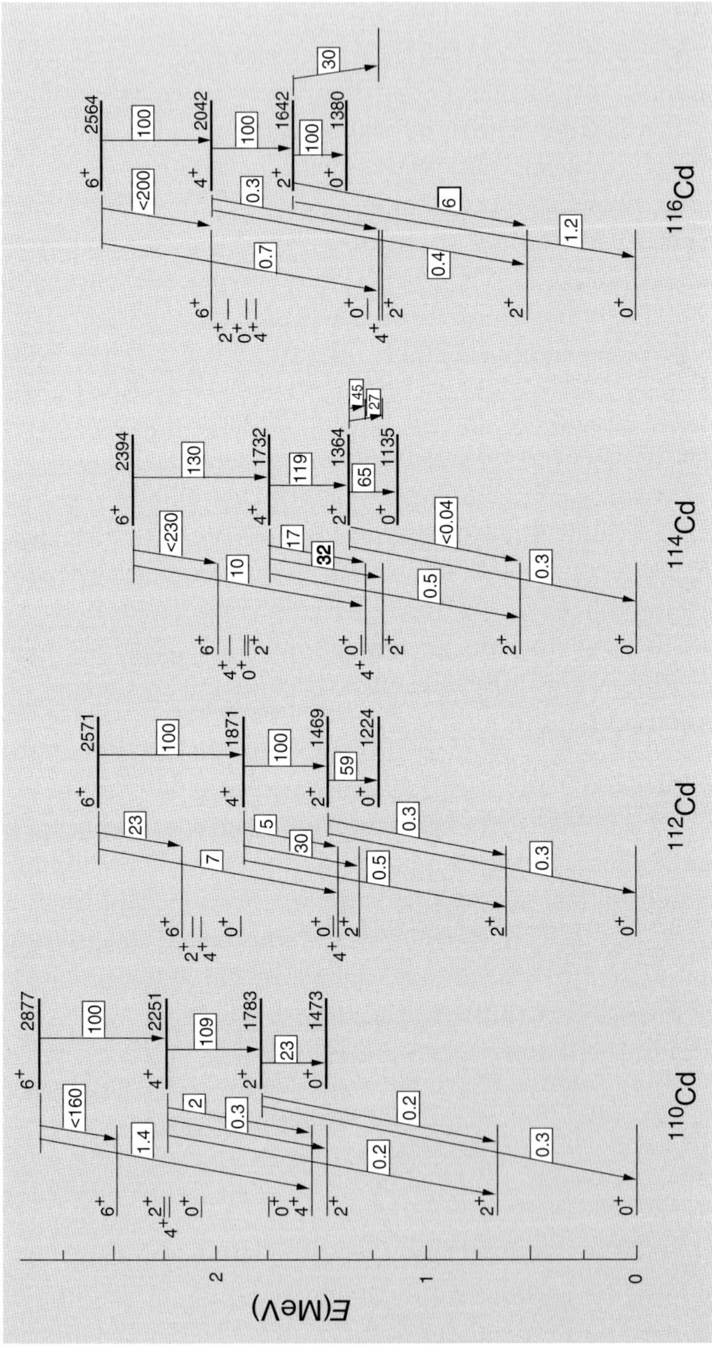

Fig. 8. Deformed intruder bands in ^{110,112,114,116}Cd. The figure is taken from Ref. [17].

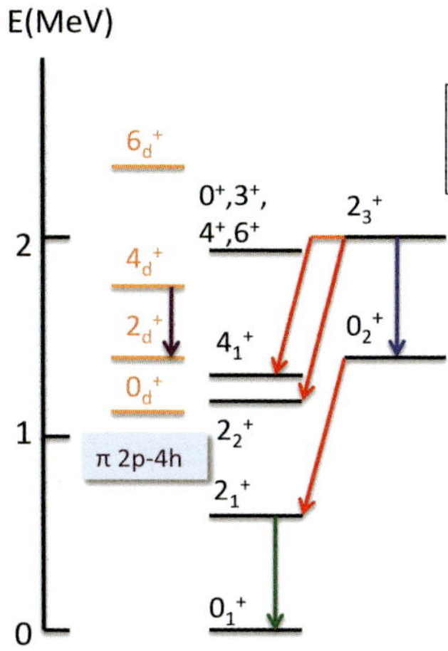

trans.	^{110}Cd	^{112}Cd	^{114}Cd	^{116}Cd	harm. vib.
4_d2_d	115^{35}		119^{12}		
2_30_2	24^{2}	27^{8}	17^{5}	35^{10}	42
2_34_1	< 5	< 0.4	< 0.3	< 7	31
2_32_2	$< 0.7^{6}$	< 2	2.8^{4}	2.0^{6}	17
0_22_1	< 7.9	0.0099^{44}	0.0026^{4}	0.55^{4}	60
2_10_1	27	30	31	34	30 (norm)

Fig. 9. $B(E2)$ values (in W.u.) for ^{110,112,114,116}Cd that are the basis for the refutation of low-energy quadrupole vibrations in these nuclei. The data are taken from ENSDF and Ref. [32].

a "triplet" of states with spin 0^+, 2^+, 4^+ at approximately twice the excitation energy of the 2_1^+ states, that are distinct from the deformed band states. This triplet of states, which occurs through all of the Cd isotopes, centered on the mid shell at $N = 66$, were universally regarded as forming a two-phonon multiplet characteristic of a near-harmonic quadrupole vibrator. However, with detailed measurements of lifetimes of low-spin states using $(n, n'\gamma)$ spectroscopy, it has become evident that the pattern of $B(E2)$ values involving these states is not characteristic of a quadrupole vibrator [32]. The relevant data are shown in Fig. 9. The nature of the 0^+ state, that was the candidate two-phonon state, remains to be determined: a possible interpretation in terms of the role of $j = \frac{1}{2}$ subshells is given in Ref. [33].

The emerging pattern of collectivity in the even-mass Cd isotopes, reflected in the $B(E2)$ values, also points to the possibility of asymmetric rotor behavior built on the ground state. Rotational behavior would be consistent with the large $Q(2_1^+)$ values for these isotopes (which were previously rationalized as resulting from large anharmonicities present in the quadrupole vibrational degree of

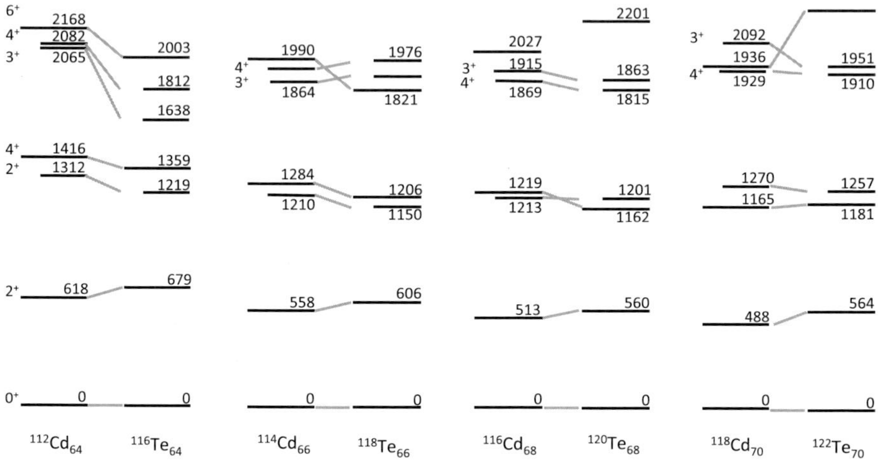

Fig. 10. Similarity between selected low-energy excited states in the Cd and Te isotones. The energies (in keV) and $B(E2)$ values of these states suggest quasi-rotational behavior with axial asymmetry. The data are taken from ENSDF.

freedom). Figure 10 shows the relevant data for the Cd isotopes and the Te isotones: previously it had been a puzzle why the Cd isotopes (two proton holes) and the Te isotopes (two proton particles), adjacent to the Sn isotopes (closed proton shell), were so different in their collective structure. This difference is with respect to their excited 0^+ states.

An interpretation of the low-energy collectivity of the even-mass Cd isotopes in terms of an axially symmetric rotor would solve a long-standing and important problem. Over 30 years ago, detailed spectroscopic study of the odd-mass Cd isotopes made a convincing case that these nuclei exhibit rotational bands [34]. The excited states of ^{111}Cd, organized into rotational bands, and its isotones are shown in Fig. 11. What is lacking for the odd-mass Cd isotopes, if this unified view is to be adopted, are detailed $B(E2)$ data.

3.2. *The $N = 90$ Isotones*

The $N = 90$ isotones, ^{150}Nd, ^{152}Sm, ^{154}Gd, and ^{156}Dy, exhibit very low-energy excited 0^+ states. These have been regarded as quintessential examples of β-vibrations ever since their discovery in

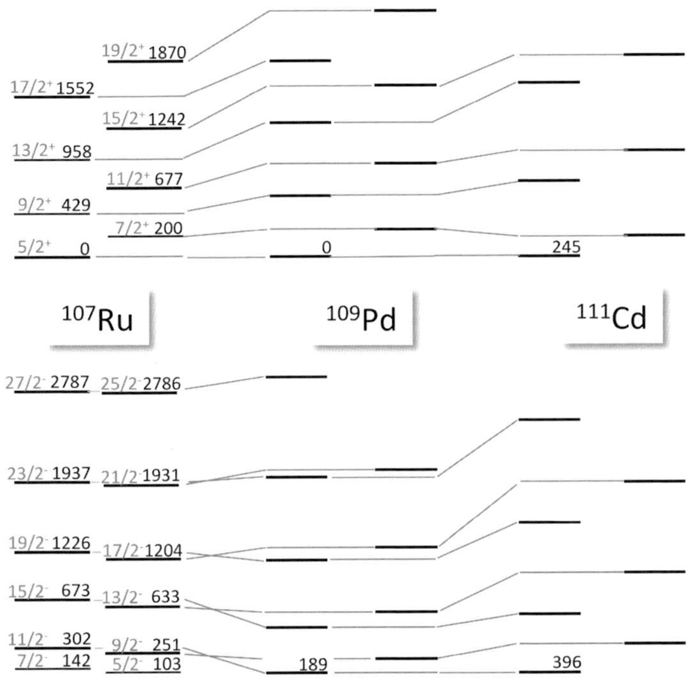

Fig. 11. Excitation energies (in keV) in ^{111}Cd and the $N = 63$ isotones, ^{109}Pd and ^{107}Ru, organized into rotational bands. The data are taken from ENSDF.

the 1960s. However, Mottelson [35] noted that the γ-ray branching ratios for the decay out of the bands built on these states in ^{152}Sm and ^{154}Gd seriously violate the decay branching rules, the so-called Alaga rules (modified by rotor-model $\Delta K = 0$ mixing), for β-vibrational bands.

Very recently, the $N = 90$ isotones have been studied by detailed and multiple nuclear spectroscopic techniques [36–41]. These studies reveal that the $N = 90$ isotones possess near-degenerate coexisting configurations that strongly mix. Further, the possibility that the low-energy excited 0^+ state in ^{152}Sm is a β-vibration is refuted [40]. The key data are shown in Figs. 12 and 13, and in Table 2. Besides the pattern of $B(E2)$ values, that of $E0$ transitions and radii are crucial for understanding this interpretation.

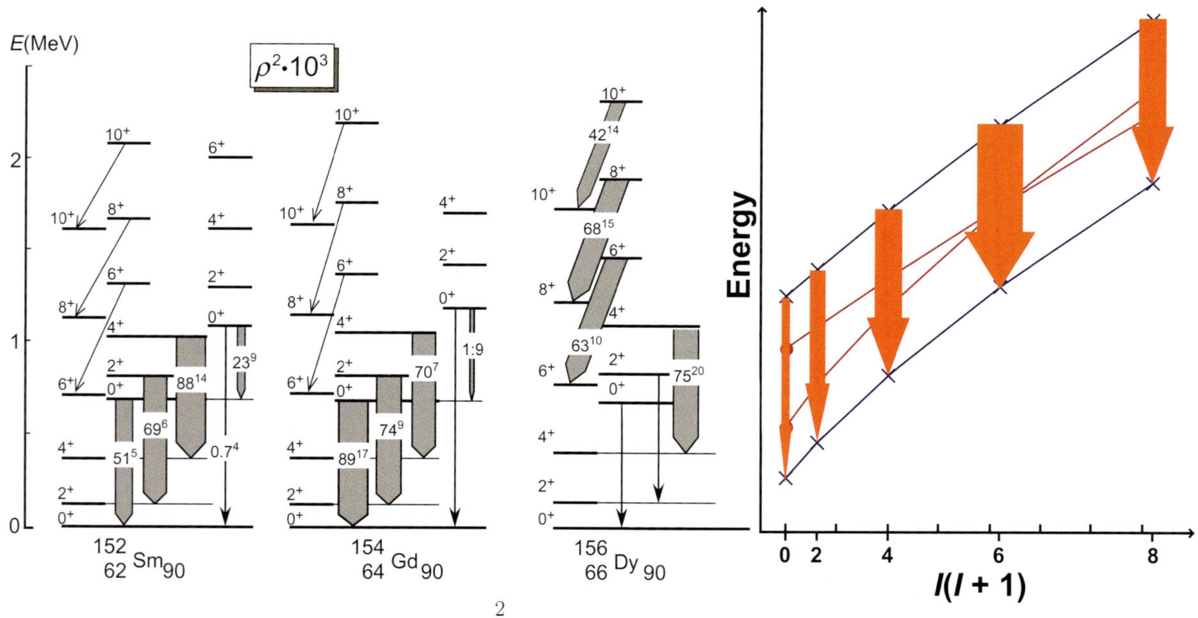

Fig. 12. $N = 90$ band mixing and $E0$ transition strengths. The right-hand side schematically illustrates unmixed rotational bands (red) and the effect of a constant interaction that mixes and repels configurations with the same spin to produce identical rotational bands (blue). The resulting $E0$ transition strengths are shown as orange vertical arrows with widths reflecting their strengths. The crossing point for the unmixed bands is arbitrary: the experimental data for ^{152}Sm are consistent with this point being at $J > 2$ and for ^{154}Gd at $J = 0$. The figure is adapted from one in Ref. [5].

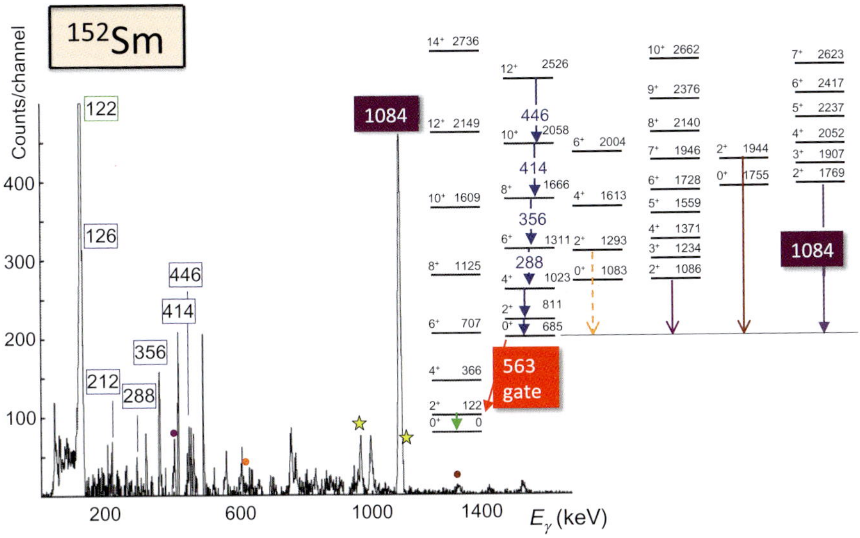

Fig. 13. Response of the first excited 0^+ state in the nucleus ^{152}Sm to multi-step Coulomb excitation. The only state that is observed with strong population is the head of a $K^\pi = 2^+$ band at 1769 keV. (The population of the rotational band built on the first excited 0^+ state is masked by a decay branching ratio of 99.7% out of the band at 811 keV.) The figure is taken from Ref. [33].

The above developments suggest that β vibrations may not exist in deformed nuclei. Indeed, this is the conclusion reached in a global assessment for the rare-earth isotopes using a simple adiabatic limit of the algebraic collective model (ACM) [44].

The above developments also raise the question of the nature of the low-lying 0^+ states in the $N = 90$ isotones. A strong argument can be made for the identification of neutron pairing isomers in ^{152}Sm and ^{154}Gd [36,38]. This leaves open the nature of the first excited 0^+ state in these nuclei. Studies of the $N = 91$ nucleus, ^{155}Gd, suggest, from particle-core coupling patterns, that the unpaired neutron is blocking the neutron pairing correlations [45]. Further, a study of the ^{150}Nd(^{3}He, n)^{152}Sm reveals that the 0_2^+ state in ^{152}Sm is weakly populated, which is also consistent with different neutron-pair distributions for the 0_1^+ and 0_2^+ states in ^{152}Sm [46]. However, with the 0_3^+ state as a strong candidate for a neutron pairing isomer in ^{152}Sm, the situation appears to contain contradictions.

Table 2. Comparison of experimental data for the lowest two $K = 0$ bands in ^{152}Sm with values calculated using a two-band mixing model (for details see W. D. Kulp *et al.*, Ref. [40]).

	Level energy (keV)			
	Expt. [A]		Calc.	
J^π	Band 1	Band 2	Band 1	Band 2
0^+	0	685	4	692
2^+	122	810	140	792
4^+	366	1023	393	1023
6^+	707	1311	730	1350
8^+	1125	1666	1131	1758
	$10^3 \times \rho^2(E0)$			
Transition	Expt. [B, C]		Calc.	
$0_2^+ \rightarrow 0_1^+$	51 (5)		72	
$2_2^+ \rightarrow 2_1^+$	69 (6)		77	
$4_2^+ \rightarrow 4_1^+$	88 (14)^a		84	
$6_2^+ \rightarrow 6_1^+$			87	
$8_2^+ \rightarrow 8_1^+$			85	
	$B(E2)$(W.u.)			
Transition	Expt. [A]		Calc.	
$2_1^+ \rightarrow 0_1^+$	145.0 (16)		133	
$4_1^+ \rightarrow 2_1^+$	209.5 (22)		197	
$6_1^+ \rightarrow 4_1^+$	240 (4)		230	
$2_2^+ \rightarrow 0_2^+$	170 (12)		170	
$4_2^+ \rightarrow 2_2^+$	250 (40)		233	
$0_2^+ \rightarrow 2_1^+$	33.3 (12)		31	
$2_2^+ \rightarrow 0_1^+$	0.94 (6)		1	
$2_2^+ \rightarrow 2_1^+$	5.7 (4)		5	
$2_2^+ \rightarrow 4_1^+$	18.0 (12)		23	
$4_2^+ \rightarrow 2_1^+$	0.74 (12)		1	
$4_2^+ \rightarrow 4_1^+$	5.0 (10/7)		5	
$4_2^+ \rightarrow 6_1^+$	17 (3)		21	

[A] M. J. Martin (2013), Ref. [42].
[B] T. Kibédi and R. H. Spear (2005), Ref. [43].
[C] Ref. [19].
^{a}Calculated using data from [A].

The resolution of the above is in two properties that appear to be possessed by the $N = 90$ isotones. The first is that the structures underlying the 0_1^+ and 0_2^+ states are configurations with different deformations: this will result in energy shifts in particle-core coupling that can mimic blocking. The second is that the coexisting structures strongly mix in the $N = 90$ isotones: this will produce near-identical ground-state structures in ^{150}Nd and ^{152}Sm, with the consequence that the two-proton transfer strength is concentrated in the ground-state-to-ground-state transition.

3.3. *The* Os *and* Pt *Isotopes*

The isotopes ^{186,188,190,192}Os and ^{192,194,196}Pt exhibit low-lying excitation patterns characteristic of axially asymmetric rotors, possibly with some γ softness. The definitive evidence is multi-step Coulomb excitation [31]. These data have been analyzed [47] using a version of the rigid axially asymmetric rotor model with independent inertia and electric quadrupole tensors [48]. This model reveals a number of basic properties of rigid axially asymmetric rotor descriptions:

- The ratios of the components of the rotational inertia tensor are very different from those that would result from irrotational flow (such as was used by all previous formulations of axially asymmetric rotor models) [48].
- It resolves the long-standing puzzle of the P_3 interference-term sign anomaly [49].
- The model possesses a destructive interference property between the inertia and the electric quadrupole tensors [50]. This feature is the reason that an irrotational inertia tensor results in vanishing quadrupole moments at maximal triaxiality.
- The destructive interference property explains the vanishingly small value for $B(E2; 2_2^+ \to 0_1^+)$ in ^{196}Pt [50]. This $B(E2)$ value in ^{196}Pt has been taken as evidence for an O(6) boson symmetry [51], despite the fact that such an interpretation of this nucleus is seriously contradicted by a large value of $Q(2_1^+)$ — the quadrupole

moments are zero for all states in a nucleus with O(6) dynamical symmetry.

4. A Forward-Looking View of Nuclear Quadrupole Collectivity

The foregoing sketch of the current emerging situation with respect to low-energy quadrupole collectivity in nuclei suggests that nuclear deformation is fairly rigid, that nuclei widely exhibit coexistence of states with different deformations, and that nuclei formerly thought to be too weakly deformed to be amenable to a rotational description can be described using rotor models.

Caution is needed when looking for the failure of simple degrees of freedom in the nucleus, when those degrees of freedom are widely substantiated. It has become popular to herald the downfall of the shell model in exotic isotopes such as in neutron-rich nuclei with $N = 20$ and 28 [52]. However, all of the data cited as indicating the failure of the shell model are consistent with the broad understanding of the occurrence of shape coexistence in nuclei [5]. For example, in Fig. 2 the "$1h_{9/2}$" state in the Tl isotopes appears to severely violate shell-model expectations for an independent-particle state. The answer is that this state is not an independent-particle state with the characteristics of the $1h_{9/2}$ shell-model configuration; it is a proton 1p-2h configuration which, through pairing correlations and quadrupole correlations between the proton pair and the multiple neutron pairs, produces a deformed structure [5].

A leading question is: "what prospects are there for a unified view of the various phenomena (shape coexistence, dominance of rotational behavior) detailed in the foregoing narrative?" It appears that the symplectic shell model of Rosensteel and Rowe [53] (detailed in Chapter 3) exhibits all of these phenomena. Indeed, these phenomena would exhaust the manifestations of collective behavior at low energy possessed by the symplectic shell model. Very recent work has been directed to the exploration of this realization. In particular, the apparent near-universal manifestation of ground-state band $B(E2)$ values, but distinctly not a universal manifestation of ground-state band energies, shown in Fig. 14, can be explained

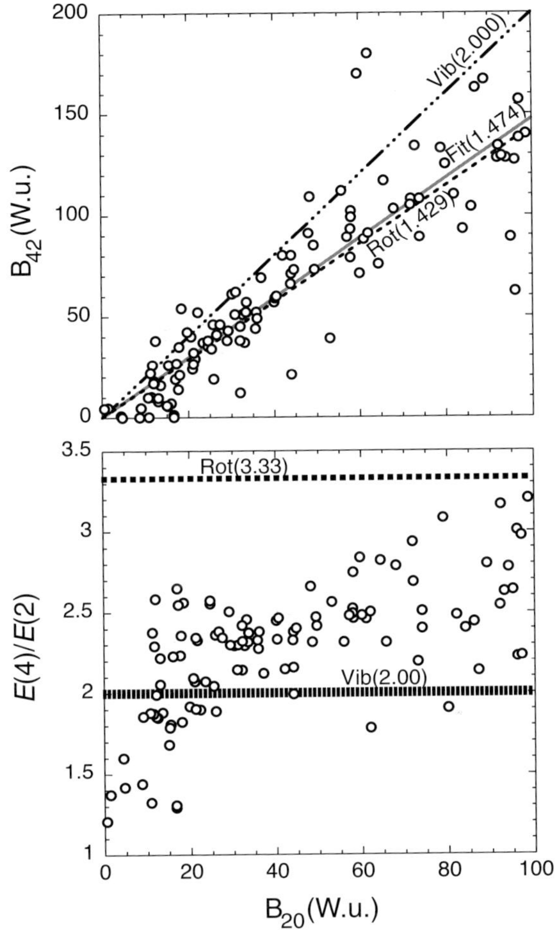

Fig. 14. The top figure shows values of $B_{42} = B(E2; 4_1^+ \to 2_1^+)$ in comparison with corresponding values of $B_{20} = B(E2; 2_1^+ \to 0_1^+)$ for all doubly even nuclei with $A > 56$, which are considered "non-rotational", i.e., $B_{20} < 100$ W.u. and $E(4)/E(2) < 3$. The three lines indicate the slopes corresponding to the harmonic vibrator, the axially symmetric rigid rotor and a least-square fit to the data. The lower figure shows ratios of the excitation energies of the 4_1^+ and 2_1^+, $E(4)/E(2)$, vs. the corresponding B_{20} values. The ratios given by vibrational and axially symmetric rigid-rotor models are indicated by horizontal lines. The figure is taken from Ref. [54].

by simple mixing of SU(3) irreps within a symplectic shell-model framework [54].

In its simplest depiction, the symplectic shell model is based on nucleons occupying harmonic-oscillator shells with important correlations within each shell and between shells differing by $\pm 2\hbar\omega$. The in-shell correlations are dominated by interactions of the quadrupole–quadrupole type, as first introduced by Phil Elliott [11]. The inter-shell correlations are of the giant monopole and giant quadrupole type. The inter-shell correlations enhance the electric quadrupole collectivity in such a manner as to obviate the need for effective $E2$ charges. It is found that in many strongly deformed nuclei the SU(3) quantum numbers, λ and μ, familiar from the Elliott model, possess very large values, e.g., $\lambda \sim 100$ and $\mu \sim 10$ [55]. This naturally leads to contraction of the SU(3) algebra to a rotor algebra [56]. The algebra of the symplectic shell model, Sp(3, $\mathbb{R}$), possesses SU(3) as a subalgebra, and also possesses the microscopic version of the Bohr model [56] (cf. Chapter 3) with ties to the ACM referenced above [57, 58].

Symplectic shell-model structures naturally contain low-energy shape coexisting excitations and they do not exhibit low-energy collective quadrupole vibrational excitations. This points to the critical need for focused programs of nuclear spectroscopy that systematically map $E2$ and $E0$ matrix elements, diagonal and off-diagonal, for low-lying states in nuclei, across the entire mass surface. Further, the occurrence of subshell gaps needs to be identified: monopole shifts (cf. Fig. 2) may be producing new subshell gaps that become obscured by the mixing of the associated shape coexisting configurations [5]. Finally, a clear answer is needed to the question of whether or not nuclei are exhibiting triaxial-rotor degrees of freedom.

5. Conclusions

This short synopsis of the status of low-energy (quadrupole) collectivity in nuclei illustrates a number of key criteria with respect to the study of the nuclear many-body problem: the first is that some patterns of excitation energies, while they may appear to be near-perfect examples of simple collective dynamics in nuclei, can be very misleading. Detailed maps of $E2$ matrix elements are vital to the exploration of low-energy quadrupole collectivity in nuclei. The second is that models can be misleading. Some features of a model may appear to be realized nearly perfectly, but other aspects of the model are manifestly not exhibited by nuclei. Often, what has happened is that with limited, apparently near-perfect realizations, no further work has been carried out to test the model. Indeed, when inconsistencies have appeared, they are rationalized with explanations such as "anharmonicities" or "mixing with non-collective degrees of freedom".

Acknowledgments

I would like to acknowledge collaborations, some spanning 40 years, with Mitch Allmond, Paul Garrett, Kris Heyde, David Kulp, David Rowe, Martin Venhart, Steve Yates, and Ed Zganjar, which substantially influenced my ideas, expressed herein.

References

[1] S. Raman, C. W. Nestor, Jr., and P. Tikkanen, *At. Data Nucl. Data Tables.* **78**, 1 (2001).
[2] P. W. Anderson, *Science*, **177**, 393 (1972).
[3] J. A. S. Smith, *Chem. Soc. Rev.* **15**, 225 (1986).
[4] J. M. Allmond *et al.*, *Phys. Rev. Lett.* **112**, 172701 (2014).
[5] K. Heyde and J. L. Wood, *Rev. Mod. Phys.* **83**, 1467 (2011).
[6] K. Kumar, *Phys. Rev. Lett.* **28**, 249 (1972).
[7] D. Cline, *Ann. Rev. Nucl. Part. Sci.* **36**, 683 (1986).
[8] M. Goeppert-Mayer, *Phys. Rev.* **75**, 1969 (1949).
[9] O. Haxel, J. H. D. Jensen, and H. E. Suess, *Phys. Rev.* **75**, 1766 (1949).
[10] S. G. Nilsson, *Mat. Fys. Medd. Dan. Vid. Selsk.* **29**(16) (1955).
[11] J. P. Elliott, *Proc. Roy. Soc. A.* **245**, 128 (1958); *ibid.*, 562 (1958).
[12] T. Otsuka, T. Suzuki, R. Fujimoto, H. Grawe, and Y. Akaishi, *Phys. Rev. Lett.* **95**, 232502 (2005).
[13] K. L. G. Heyde, *The Nuclear Shell Model.* Springer-Verlag, Berlin (1990).
[14] M. Goeppert-Mayer, *Phys. Rev.* **78**, 16 (1950); *ibid.*, 22 (1950).
[15] S. T. Belyaev, *Mat. Fys. Medd. Dan. Vid. Selsk.* **31**(11) (1959).
[16] D. M. Brink and R. A. Broglia, *Nuclear Superfluidity: Pairing in finite systems.* Cambridge University Press, Cambridge (2005).

[17] D. J. Rowe and J. L. Wood, *Fundamentals of nuclear models: Foundational models.* Chapter 6, World Scientific, Singapore (2010).
[18] A. K. Kerman, *Ann. Phys.* **12**, 300 (1961).
[19] J. L. Wood, E. F. Zganjar, C. De Coster, and K. Heyde, *Nucl. Phys. A.* **651**, 323 (1999).
[20] R. L. Lozeva *et al.*, *Phys. Rev. C.* **77**, 064313 (2008).
[21] M. Palacz *et al.*, *Phys. Rev. C.* **86**, 014318 (2012).
[22] H. Watanabe *et al.*, *Phys. Rev. Lett.* **111**, 152501 (2013).
[23] A. Bohr, *Mat. Fys. Medd. Dan. Vid. Selsk.* **26**(14) (1952).
[24] R. de L. Kronig and I. I. Rabi, *Phys. Rev.* **29**, 262 (1927).
[25] W. D. Kulp *et al.*, *Phys. Rev. C.* **73**, 014308 (2006).
[26] A. Zehnder *et al.*, *Nucl. Phys. A.* **254**, 315 (1975).
[27] F. S. Stephens, *Rev. Mod. Phys.* **47**, 43 (1975).
[28] P. Kemnitz *et al.*, *Nucl. Phys. A.* **293**, 314 (1977).
[29] J. Kern, P. E. Garrett, J. Jolie, and H. Lehmann, *Nucl. Phys. A.* **593**, 21 (1995).
[30] P. E. Garrett, *J. Phys. G: Nucl. Part. Phys.* **27**, R1 (2001).
[31] C. Y. Wu *et al.*, *Nucl. Phys. A.* **607**, 178 (1996).
[32] P. E. Garrett and J. L. Wood, *J. Phys. G.* **37**, 064028 (2010); **37**, 069701 (2010).
[33] J. L. Wood, *J. Phys. Conf. Ser.* **403**, 012011 (2012).
[34] P. C. Simms, F. A. Rickey, and R. K. Popli, *Nucl. Phys. A.* **347**, 205 (1980).
[35] B. R. Mottelson. In *Proc. Int. Conf. on Nuclear Structure*, Tokyo (1967).
[36] W. D. Kulp *et al.*, *Phys. Rev. Lett.* **91**, 102501 (2003).
[37] W. D. Kulp *et al.*, *Phys. Rev. C.* **69**, 064309 (2004).
[38] W. D. Kulp *et al.*, *Phys. Rev. C.* **71**, 041303(R) (2005).
[39] W. D. Kulp *et al.*, *Phys. Rev. C.* **76**, 034319 (2007).
[40] W. D. Kulp *et al.*, *Phys. Rev. C.* **77**, 061301(R) (2008). arXiv: 0706.4129.
[41] P. E. Garrett *et al.*, *Phys. Rev. Lett.* **103**, 062501 (2009).
[42] M. J. Martin, *Nucl. Data Sheets.* **114**, 1497 (2013).
[43] T. Kibédi and R. H. Spear, *At. Data Nucl. Data Tables.* **89**, 77 (2005).
[44] D. J. Rowe and J. L. Wood, *Fundamentals of Nuclear Models: Foundational Models.* Chapter 2, World Scientific, Singapore (2010).
[45] J. F. Sharpey-Schafer *et al.*, *Eur. Phys. J. A.* **47**, 6 (2011).
[46] J. F. Sharpey-Schafer, priv. comm. (2014).
[47] J. M. Allmond, R. Zaballa, A. M. Oros-Peusquens, W. D. Kulp, and J. L. Wood, *Phys. Rev. C.* **78**, 014312 (2008).
[48] J. L. Wood, A. M. Oros-Peusquens, R. Zaballa, J. M. Allmond, and W. D. Kulp, *Phys. Rev. C.* **70**, 024308 (2004).
[49] J. M. Allmond, J. L. Wood, and W. D. Kulp, *Phys. Rev. C.* **80**, 021303(R) (2010).
[50] J. M. Allmond, J. L. Wood, and W. D. Kulp, *Phys. Rev. C.* **81**, 051305(R) (2010).

[51] F. Iachello and A. Arima, *The Interacting Boson Model.* Cambridge University Press, Cambridge (1987).
[52] O. Sorlin and M. G. Porquet, *Progr. Part. Nucl. Phys.* **61**, 602 (2008).
[53] G. Rosensteel and D. J. Rowe, *Phys. Rev. Lett.* **38**, 10 (1977).
[54] G. Thiamova, D. J. Rowe, and J. L. Wood, *Nucl. Phys. A.* **780**, 112 (2006).
[55] M. Jarrio, J. L. Wood, and D. J. Rowe, *Nucl. Phys. A.* **528**, 409 (1991).
[56] D. J. Rowe, *AIP Conf. Proc.* **1541**, 104 (2013).
[57] D. J. Rowe, *Nucl. Phys. A.* **735**, 372 (2004).
[58] D. J. Rowe and J. L. Wood, *Fundamentals of Nuclear Models: Foundational Models.* Chapter 4, World Scientific, Singapore (2010).

Chapter 2

The Anatomy of Atomic Nuclei: Illuminating Many-body Wave Functions Through Group-theoretical Decomposition

Calvin W. Johnson
San Diego State University, 5500 Campanile Drive
San Diego, CA 92182, USA

With modern computers we can compute nuclear many-body wave functions with an astounding number of components, $\gtrsim 10^{10}$. But, aside from reproducing and/or predicting experiments, what do we learn from vectors with tens of billions of components? One way to characterize wave functions is through irreducible representations of groups. I discuss briefly the history of group-theoretical characterization of nuclear wave functions, with an emphasis of using Lanczos-type methods to efficiently dissect arbitrary wave functions into group irreps. Although the resulting decompositions are often fragmented over many irreps, one nonetheless finds powerful patterns. First, group decompositions along rotational bands show coherent commonalities, supporting the picture of a shared "intrinsic shape"; this is also called *quasi-dynamical symmetry.* Second, group decompositions for wave functions using both phenomenological and *ab initio* forces are often very similar, despite vastly different origins and dimensionalities. Both of these results suggest a group theoretical decomposition can provide a robust "anatomy" of many nuclear wave functions. This in turn supports the idea of using symmetry-based many-body frameworks for calculations.

Contents

1. The Blessing and the Curse of Large-scale Computing

The quantum theory of atomic nuclei is challenging. The force between nucleons is strong, short-ranged, and complicated [1]. For many decades nuclei were primarily modeled through phenomenological descriptions [2]. This changed starting two decades ago, thanks to the confluence of several factors: the development of forces which describe nucleon–nucleon scattering to high precision [3, 4], especially the more recent introduction of chiral effective field theory [5–8]; application of improved and more rigorous many-body methods, including Green's function Monte Carlo [9–11] and coupled cluster [12–15] calculations and the no-core shell model (NCSM) [16–18]; and improved and more rigorous effective interaction theories such as Okubo–Lee–Suzuki–Okamoto theory [19–22], the unitary correlation operator method [23], and the similarity renormalization group (SRG) [24–28]. Powering all these developments has been geometrically growing computational power. Although the problems are far from settled, nuclear theorists now routinely perform *ab initio* structure calculations.

(Herein by *ab initio* I mean interactions fitted to two- and possibly three-body data, in particular scattering phase shifts along with deuteron and triton/^{3}He binding energies [and triton half-life], and then apply to many-body systems; by *phenomenological* I mean those fitted to many–body excitation spectra and binding energies.)

For the purposes of this paper, I will focus on one particular many-body method, the configuration-interaction method, also sometimes called the interacting shell model [2, 16, 17, 29–31]. The basic idea is quite simple: in order to solve the many-body Schrödinger equation,

$$\hat{H}\Psi(r_1, r_2, \ldots, r_A) = E\Psi(r_1, r_2, \ldots, r_A), \tag{1}$$

one expands in a convenient, orthonormal basis $\{\Phi_\alpha\}$,

$$\Psi(r_1, r_2, \ldots, r_A) = \sum_\alpha c_\alpha \Phi_\alpha(r_1, r_2, \ldots, r_A), \tag{2}$$

so the Schrödinger equation becomes a matrix eigenvalue equation,

$$\sum_\beta H_{\alpha,\beta} c_\beta = E c_\alpha, \tag{3}$$

with the Hamiltonian matrix element

$$H_{\alpha,\beta} = \langle \Phi_\alpha | \hat{H} | \Phi_\beta \rangle. \tag{4}$$

This deceptively simple beginning masks a multitude of related technical issues. What is meant by a "convenient" basis? Is it better to have as compact a basis as possible to reduce the dimensionality, or to have as simple a basis as possible, to make calculation of matrix elements efficient?

For example, one often starts with a basis of Slater determinants, or, more accurately, the occupation-representation of Slater determinants, that is, antisymmetrized products of single-particle states [29,30]. Working with a rotationally invariant Hamiltonian, we choose single-particle states that are spherical tensors, that is, have good total angular momentum j and z-component m; it is then easy to construct Slater determinants with fixed total z-component M. This is called an M-scheme basis. M-scheme bases are easy to work with, but any given M-scheme basis state is an admixture of states of different total angular momentum J.

Some configuration-interaction codes instead work in the J scheme, with basis states with total angular momentum J fixed rather than M. The resulting dimensionality is typically an order of magnitude smaller. On the surface this looks to be a good thing. But each J-scheme basis state itself is a sum of many M-scheme states, and so computing the Hamiltonian matrix elements becomes significantly more time-consuming. Furthermore, in the M scheme in particular the Hamiltonian matrix is very sparse; depending on the details on the many-body calculation, only a few matrix elements out of a million will be nonzero. Because the eigenvalue problem (1) for large cases is typically solved using an Arnoldi type method such as the Lanczos algorithm [32, 33], based upon matrix-vector multiplications, the computational burden is not just the dimensions of the vectors but the number of non-zero matrix elements. J-scheme matrices are significantly denser than M-scheme ones, and in some cases there are more J-scheme non-zero matrix elements than in the M-scheme case. Furthermore, the simplicity of the M scheme allows one to avoid storage of the non-zero matrix elements but to efficiently reconstruct the Hamiltonian matrix elements as needed on the fly, primarily by loops over spectators.

It is not the purpose of this paper to argue the superiority of either the M scheme or the J scheme. Both have advantages and trade-offs, and over the decades codes have been written in both schemes. With modern parallel computing, one can distribute both the wave function vector and the non-zero matrix elements of the many-body Hamiltonian (or the arrays used to reconstruct it) over many processors. As far as I can tell, at the time of this writing the largest configuration-interaction calculations, certainly for nuclear structure physics, had dimensions of roughly 20 billion (M-scheme) basis states.

Such large calculations pose significant computational challenges. Even in single precision, a vector of 2×10^{10} components requires 80 gigabytes of storage. Assuming an M-scheme sparsity of 10^{-6}, the non-zero matrix elements would require roughly 5 petabytes (and indeed, the needle-in-the-haystack problem of identifying the non-zero matrix elements efficiently is a non-trivial computational problem), and one matrix–vector multiplication would require something like 3 trillion floating-point operations. In the J scheme the vector dimensions are an order of magnitude smaller but the matrices are denser.

The most crucial question of all is pressed upon us by Richard Hamming's dictum [34]: "The purpose of computing is insight, not numbers". What insight, then, do we gain from ever-larger matrix eigenvalue problems? The observables we can compare to experiment, after all, are few in number: binding and excitation energies and a handful of static and transition matrix elements.

One answer is to look at observables we are unlikely to measure in experiments but yet help us to conceptualize the nuclear wave function. Here we fall back on phenomenology from the early days of nuclear physics. The liquid drop model, for example, envisioned nuclides as having quadrupole oscillations and deformations leading to vibrational and rotational spectra [35]. Variants, such as the Goldhaber–Teller model for giant $E1$ resonances [36] and the scissors mode for giant M1 resonances [37,38], pictured neutrons and protons oscillating coherently against each other. But how do we connect such classical images to modern many-body theory with discrete numbers of fermions?

The answer is group representation theory. As described in section 2.4, the group SU(3) arises naturally in the description of quadrupole deformation of a nuclear droplet, both static (rotation) and dynamic (vibration). But one can also construct representations of SU(3) using number-conserving fermion operators [39, 40]. This means, in an appropriate basis with a fixed, finite number of fermions, one can construct irreducible representations (also called "irreps") of SU(3), and thus in a framework with a fixed number of particles one can arrive rigorously at rotational (and vibrational) band spectra just as in the droplet picture.

Unlike the rotation group SU(2), SU(3) is generally not an exact or *dynamical* symmetry of the nuclear Hamiltonian, and hence the generators of SU(3) do not commute with the Hamiltonian, in which case eigenstates of the Hamiltonian are not eigenstates of the Casimir invariants of SU(3). In fact, SU(3) is strongly mixed, especially by pairing and above all by the spin-orbit force [41, 42]. But, as will be discussed in the next section, the mixing is not incoherent but displays strong regularities within spectral "bands", leading to so-called *quasi-dynamical symmetry* [43–45]. (Another concept is that of *partial* dynamical symmetries [46, 47], but that is beyond our brief here.)

In this paper, I will discuss how one can dissect and illuminate nuclear wave functions using Casimir operators of different groups, that is, operators which commute with all the generators of the group and which one can use to separate components. And not only illuminations, but also construction: one hopes to choose the most important irreps to go beyond the J-scheme basis to an even more compact, symmetry-adapted basis for more efficient, most physics-based calculations.

Aside from better, not bigger, calculations, group decompositions of nuclear wave functions are often remarkably robust. The group-theoretical decomposition using decades-old phenomenological forces in tiny model spaces often agree surprisingly well with those using the freshest, independently-derived *ab initio* interactions in model spaces five or six orders of magnitude larger. I find it heartening that vastly different calculations arrive at similar results: it suggests we are doing the right thing; it suggests we really *can* have insight into the structure of nuclear wave functions and have a robust picture of them.

2. A Brief History of Group Theory in Nuclear Many-body Physics, with Illustrations

Applying group representation theory to quantum wave functions, including nuclear wave functions, has a long history. This will by no means be an exhaustive review. Useful sources containing far more technical details are the books by Talmi [48] and by Rowe and Wood [49].

I restrict myself to groups applied to the fermion shell model. Group theory is of course at the heart of the interacting boson model (IBM) [48, 50], as well as the fermion dynamical symmetry model (FDSM) arising from SO(8) [51–53], primarily designed to connect the fermion shell model to the IBM. Both the IBM and FDSM largely start from group theory to build their models. Because quasi-dynamical symmetries can mimic true dynamical symmetries, and conversely deducing symmetries directly from data can mask quasi-dynamical symmetries, I prefer to start with the fermion shell model with minimal assumptions and then arrive at the group theory.

The phenomenological interactions I use are in model spaces spanned by a single harmonic-oscillator shell: Cohen–Kurath (CK) for the $0p$ shell [54], the universal $1s0d$-shell interaction version B (USDB) [55], and a modified G-matrix interaction (GXPF1) for the $1p0f$ shell [56]. I also use an *ab initio* interaction based upon chiral effective field theory fitted to nucleon-nucleon scattering data and deuteron properties (N^3LO) [6]. The *ab initio* interaction is softened by the to an evolution parameter value of $\lambda = 2.0$ fm^{-1} and are expressed in a harmonic oscillator single-particle basis of a given frequency, here between 16 and 22 MeV depending upon the nuclide. The phenomenological interactions do not have a specified radial basis for single-particle states, though it is a commonplace to assume a harmonic-oscillator basis; for the calculations here the frequency does not matter. All *ab initio* calculations were computed in the NCSM formalism [16–18], with the model space defined by the harmonic-oscillator frequency as well as $N_{\max}$, which denotes the maximum excitation, in harmonic-oscillator quanta, allowed above the lowest shell-model configuration.

All the many-body calculations described here used the `BIGSTICK` configuration-interaction code [57]. The method of decomposition is described below in Sec. 3.

2.1. *Spin-orbit*

The nuclear Hamiltonian is invariant under rotation, but ever since the Dirac equation it's been clear that angular momentum has two pieces, orbital angular momentum $\vec{l} = \vec{r} \times \vec{p}$ and intrinsic spin s, such that, for a given particle, the angular momentum j is a tensor sum of l and s. In the electronic structure of atoms the spin component makes only a tiny contribution through magnetic interactions to the Hamiltonian, and so the orbital angular momentum and spin nearly decouple. This naturally leads to L–S or *Russell–Saunders* coupling scheme, where one carries out a tensor sum of all the l's for the particles to get the total orbital angular momentum L, and separately sum the individual spins to get the total spin S; then L and S couple to form the total angular momentum J.

In medium- to heavy-nuclei, large spin-orbit splitting, which arises naturally in relativity, energetically separates orbits with the same l but different j, leading to the j–j coupling scheme: for each particle one couples l and s to j and then sum directly to total J. In fact it was only when Maria Goeppert-Mayer [58–60] and others [61, 62] realized that strong spin-orbit splitting of levels could explain the closure at "magic numbers" as well as the magnetic dipole moments of doubly-magic-plus(minus)-one nuclides that j–j coupling took hold in nuclear physics.

For open p-shell nuclides, it was not immediately clear which coupling was superior: L–S [63, 64] or j–j [65, 66]. To give a preview of the answer, let's consider some phenomenological calculations. Table 1 shows several semi-magic nuclides. In the simplest j–j coupling picture, the ground states should be described by filled orbits. Indeed, for the neutron-rich semi-magic nuclei in the $1p0f$ and

Table 1. Analysis of ground-state (g.s.) wave functions for "simple" nuclides in the $0p$, $1s0d$, and $1p0f$ spaces.

Nuclide	space	g.s.	
^{8}He	$0p$	$53\% (0p_{3/2})^4$	$96\% L = 0$
^{12}C	$0p$	$37\% (0p_{3/2})^8$	$82\% L = 0$
^{22}O	$1s0d$	$75\%\ (0d_{5/2})^6$	$38\% L = 0$
^{24}O	$1s0d$	$91\%\ (0d_{5/2})^6 (1s_{1/2})^2$	$34\% L = 0$
^{28}Si	$1s0d$	$21\%\ (0d_{5/2})^{12}$	$36\% L = 0$
^{32}S	$1s0d$	$29\%\ (0d_{5/2})^{12} (1s_{1/2})^4$	$34\% L = 0$
^{48}Ca	$1p0f$	$90\%\ (0f_{7/2})^8$	$20\% L = 0$

$1s0d$ spaces we see this is the case. For the $0p$ nuclei, however, they are poorly described by filling the $0p_{3/2}$ orbit, but better described by an $L = 0$ state.

To illustrate, let's start in the $0p$ shell with a non-trivial nuclide, ^{11}B, using both the phenomenological CK interaction (dating from 1965!) in the $0p$ shell assuming an inert ^{4}He core, which has a basis dimension of 62 M-scheme states, and a NCSM with a modern *ab initio* interaction, computed in a harmonic-oscillator basis with oscillator frequency $\hbar\Omega = 22\,\text{MeV}$ and allowing excitations up to $N_{\text{max}} = 6$.

Figure 1 shows the decomposition of the first four states into L-components. There are a number of things to note here. First, the very good, even surprising agreement between the phenomenological calculation with the CK interaction and the *ab initio* NCSM calculation. Second, while the $1/2^-$ and $5/2^-$ states are dominated by a single L-value, both the first and second $3/2^-$ states have non-trivial — and contrasting — secondary components. Not shown is the S-decomposition, which is nearly trivial with nearly all the wave function in the $S = 3/2$ irrep. This is not

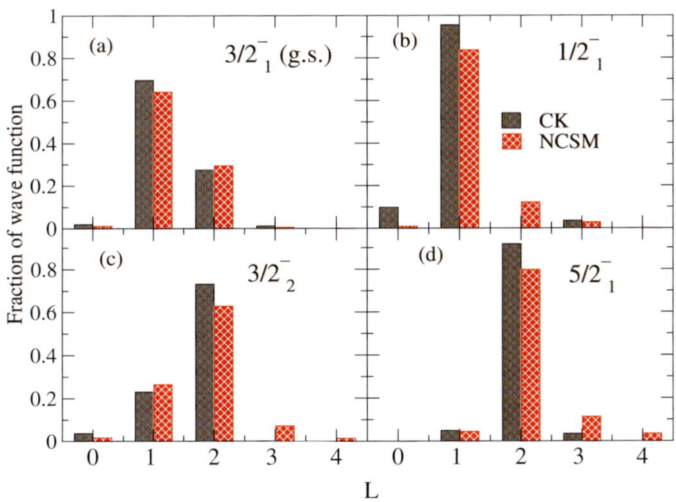

Fig. 1. L-decomposition of the $0p$-shell nuclide ^{11}B.

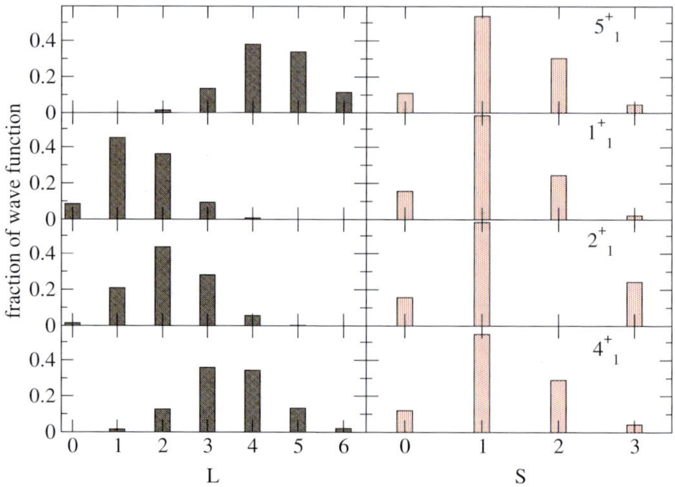

Fig. 2. L- (left) and S-decomposition (right) of the $1p0f$-shell nuclide ^{48}V, for the first four states.

surprising, given that the spin-orbit splitting is relatively weak in the $0p$ shell.

Therefore for contrast Fig. 2 shows the L- and S-decomposition for the first four states of the odd–odd nuclide ^{48}V in the $1p0f$ shell. Because an NCSM calculation is not currently practical for this nuclide, I only show results from a phenomenological calculation. While relatively more complicated than for ^{11}B, the wave functions nonetheless show regularity and simplicity: the S-decompositions, while fragmented, are strikingly similar, especially for the 4^+ (g.s.), the 1^+, and the 5^+ states. The L-decompositions also exhibit similarity across states, differing mostly in coherent shifts along the L-axis, suggesting an intrinsic shape being spun up. We'll see more of this behavior when we look at rotational bands in more depth below.

An L–S description of nuclei never really took direct hold because of the strong spin-orbit splitting, although recent work [67] showed it nonetheless illuminates $0p$-shell wave functions. But an SU(3) basis for nuclei implicitly includes L–S decompositions because SU(3) is cast in the spatial part of the wave function and separated from the spin part (cf. Chapter 8).

One can of course go further, and decompose for the proton portion of L or the neutron portion of S, but to the sake of a streamlined narrative I won't show such detailed decompositions.

2.2. *Isospin and* SU(4)

Heisenberg's introduction of the concept of isospin [68] was one of the first applications of group theory to nuclear physics. In this case isospin was a straightforward analog to spin. Isospin is not exactly conserved but is broken at only the few percent level and so can be treated as an exact symmetry. The most immediate evidence for isospin symmetry in nuclear spectra is through mirror symmetry (i.e., the same level scheme for nuclei for the same A and opposite $T_z = \frac{N-Z}{2}$, e.g., ^{13}C and ^{13}N), and isobar analog states (levels with the same T in same-mass nuclei, i.e., ^{12}C has low-lying levels with $T = 0$ but also has $T = 1$ levels mirroring those in ^{12}B and ^{12}N).

Because one can decompose the group SU(4) $\supset$ SU(2) $\times$ SU(2), Wigner suggested [69,70] looking for an SU(4) symmetry built upon $\mathrm{SU}_S(2) \times \mathrm{SU}_T(2)$, sometimes called a *supermultiplet.* The irreps of SU(4) are labeled by the quantum numbers P, P', and P'', which arise from the Young tableaux [48] and are found by the Casimir operator (for individual particle spin $\vec{s}_i$ and total spins $\vec{S}$, and similarly for isospin)

$$C_2(\mathrm{SU}(4)) = \vec{S}^2 + \vec{T}^2 + 4\sum_{ij}(\vec{s}_i \cdot \vec{s}_j)(\vec{t}_i \cdot \vec{t}_j), \tag{5}$$

which has eigenvalues

$$P(P+4) + P'(P'+2) + \left(P''\right)^2. \tag{6}$$

In the highest-weight states, $P = S$ and $P' = T$ [70].

Wigner's suggestion led to efforts to look for applications where SU(4) symmetry is approximately valid [71,72] (see also, Chapter 6). Of course, from the very start this search looks doomed, as the nuclear force in the spin–triplet, isospin-singlet channel, which gives us the bound deuteron, is very different from the spin-singlet, isospin-triplet channel. Indeed at low energies, spin-isospin composition of

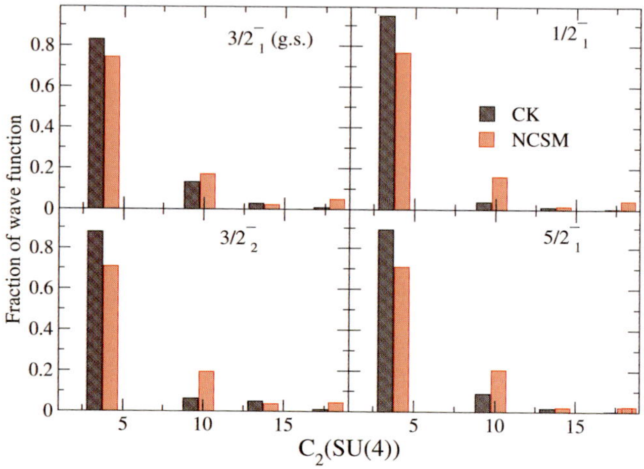

Fig. 3. SU(4)-decomposition of the $0p$-shell nuclide ^{11}B.

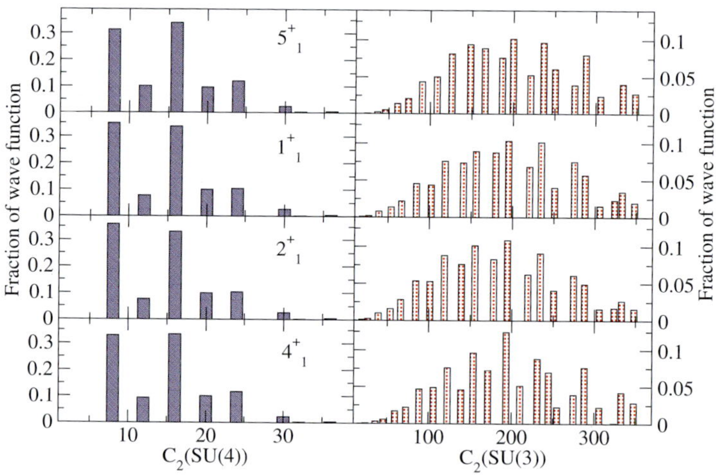

Fig. 4. SU(4)- (left) and SU(3)-decomposition (right) of the first four states of the $1p0f$-shell nuclide ^{48}V.

medium- to heavy-nuclei show significant fragmentation of the wave function over many irreps [73, 74]. We see fragmentation in Fig. 3 for ^{11}B (again with CK and NCSM calculations) and for ^{48}V in the left-hand panels of Fig. 4; both cases also exhibit evidence of coherent quasi-dynamical symmetry.

2.3. *Pairing, Seniority, and Quasi-spin*

Racah introduced the seniority scheme [65, 75, 76] to aid in the understanding of complex spectra, where the seniority quantum number ν denotes the number of "unpaired" nucleons, and where the "pairing force" came to be understood as an approximation for a zero-range δ-force [77]. The introduction of the quasi-spin scheme [78] made clear one could represent seniority through SU(2).

The problem with Racah's seniority is that the pairing operator is equally weighted over all orbits, a scheme immediately and badly broken by single-particle energies. One can then appeal to generalized seniority as a description of states [48], but one loses the quasi-spin formalism and the power of group theory (reinstated by Richardson and Gaudin's exactpairing approach [79, 80], see Chapter 11).

2.4. SU(3) *and Beyond*

Unlike atoms, atomic nuclei deform quite easily, and the liquid drop model provides a good phenomenological description of rotational and vibrational bands. If one thinks about a quadrupole-deformed body, its surface is given by

$$R(\theta,\phi) = R_0 \left(1 + \sum_m \alpha_{2,m} Y_{2,m}(\theta,\phi) \right), \tag{7}$$

where the Y_{2m} are the spherical harmonics for $l = 2$. In the Bohr–Mottelson model [35], the dynamical amplitudes $\alpha_{2,m}$ are quantized. Of the five degrees of freedom, two are the traditional deformation parameters β and γ, which are roughly deformation magnitude and triaxiality, respectively, while the remaining three are orientation through the Euler angles.

It turns out the five quadrupole operators $Q_m = \sqrt{16\pi/5}r^2$ $Y_{2m}(\theta,\phi)$, when restricted within a single harmonic-oscillator shell, and the three generators of orbital angular momentum $\vec{L} = \vec{r} \times \vec{p}$ have closed commutation relations and form the eight generators of the group SU(3). The sheer number of papers on applications of SU(3) preclude a complete review, but the basic story is straightforward. The key step was Elliott's introduction [39, 40] of a finite representation of SU(3) using one-body fermion operators for $\hat{Q}$ and $\vec{L}$. This allowed a direct connection between the well-known phenomenology of deformation and the fermion shell model.

The irreps of SU(3) are labeled by the quantum numbers λ and μ via their Young tableaux [48], and which can be interpreted in terms of the standard deformation parameters β and γ (see Fig. 1 in Ref. [81]). The second-order Casimir operator,

$$C_2(\mathrm{SU}(3)) = \frac{1}{4}\left(\hat{Q}\cdot\hat{Q} + 3\vec{L}^2\right), \tag{8}$$

has eigenvalues $\lambda^2 + \lambda\mu + \mu^2 + 3\lambda + 3\mu$, where

$$Q_m = \sqrt{\frac{4\pi}{5}}\left(\frac{r^2}{b^2}Y_{2m}(\theta_r, \phi_r) + b^2p^2Y_{2m}(\theta_p, \phi_p)\right) \tag{9}$$

is the (dimensionless) so-called Elliott quadrupole operator, whose matrix elements are non-zero only within a major harmonic-oscillator shell. One can distinguish between different combinations of λ and μ by including the third-order Casimir operator, although for simplicity here I only use the second-order Casimir operator.

For realistic interactions, SU(3) is not a dynamical symmetry because the generators of SU(3) do not commute with the Hamiltonian. Two important sources of SU(3) symmetry breaking is spin-orbit splitting [41, 42] and pairing [81]. In the limit of very strong spin-orbit splitting one can take the so-called pseudo-SU(3) limit [82, 83] in phenomenological calculations.

Decomposition of ^{11}B into SU(3) irreps is shown in Fig. 5. Because of the weak spin-orbit splitting, SU(3) is nearly a good symmetry here and all these states are dominated by the same irrep. The *ab initio* NCSM calculation is more fragmented than the phenomenological CK results, but that is largely because it includes higher harmonic-oscillator shells. By contrast, the SU(3)-decomposition of the $1p0f$ nuclide ^{48}V, as shown on the right-hand

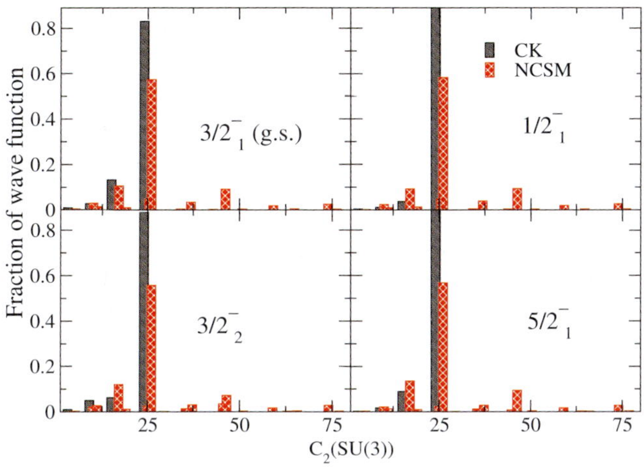

Fig. 5. SU(3)-decomposition of the $0p$-shell nuclide ^{11}B.

side of Fig. 4 is highly fragmented (as expected for a larger $1p0f$ model space), in large part due to the large spin-orbit force, although remnants of quasi-dynamical symmetry remain. Rotational $1p0f$ spectra, such as for ^{44}Ti and ^{48}Cr, are still highly fragmented but nonetheless exhibit the coherent structures [42] of quasi-dynamical symmetry [43–45].

While the Elliott SU(3) model provides a direct connection between deformation and the fermion shell model, it is limited by the fact that the operators do not connect across major harmonic-oscillator shells. A natural generalization of SU(3) is the symplectic group or Sp(3, $\mathbb{R}$) [84–86], which adds raising and lowering operators to connect different oscillator shells. Furthermore, one can naturally separate out spurious center-of-mass motion in the symplectic model, meaning one can have a translationally invariant theory.

Despite these appealing features, work using the symplectic representation has been limited, in no small part due to the difficulty of the group theory. Nonetheless, there have been several promising developments and initial applications in this direction, as described in Sec. 6.

3. Lanczos Decomposition of Wave Functions

Before giving more examples, let me explain a useful technique for carrying out these decompositions.

Given eigenstates $|\Psi\rangle$ of the Schrödinger equation (1), we want to decompose them according to irreducible representations of a group. To be specific, we divide up any space into subspaces labeled by the eigenvalues γ of a Casimir operator $\hat{\mathcal{C}}$ of a group, that is, we label states by

$$\hat{\mathcal{C}}|\Gamma\rangle = \gamma|\Gamma\rangle. \tag{10}$$

While we generally have a physical or mathematical motivation for looking at group irreps, there is also a very practical reason: in general, a Casimir operator has a highly degenerate spectrum, by which I mean, relative to the dimension of any model space, there are only a small number of distinct eigenvalues γ, with a large number of states associated with each eigenvalue. Dividing a space into subspaces labeled by γ is thus a straightforward task.

Because each irrep has many states, in principle we need some additional index i to label the states $|\Gamma; i\rangle$ of a given irrep. How to best do this can be a non-trivial task. It will turn out that this problem will not concern us.

Instead, for a given eigenstate $|\Psi\rangle$ of the Hamiltonian, one can define the fraction in a given irrep by

$$\text{frac}(\gamma) = \sum_{i \in \Gamma} |\langle \Gamma; i|\Psi\rangle|^2. \tag{11}$$

At first this looks very daunting: the number of states in any given irrep can be very large. Fortunately for us, the Lanczos algorithm [33] comes to the rescue.

For those readers not fluent with the Lanczos algorithm, used in its standards form to find extremal eigensolutions, let me describe it. Starting from an initial vector $|v_1\rangle$, often called the *pivot* and using a Hermitian (typically real, symmetric) matrix $\hat{H}$, the Lanczos algorithm carries out a sequence of matrix-vector multiplications to iteratively constructs an orthonormal set of basis vectors:

$$
\begin{aligned}
\hat{H}|v_1\rangle &= \alpha_1|v_1\rangle + \beta_1|v_2\rangle, \\
\hat{H}|v_2\rangle &= \beta_1|v_1\rangle + \alpha_2|v_2\rangle + \beta_2|v_3\rangle, \\
\hat{H}|v_3\rangle &= \qquad\qquad \beta_2|v_2\rangle + \alpha_3|v_3\rangle + \beta_3|v_4\rangle, \\
\hat{H}|v_4\rangle &= \qquad\qquad\qquad\qquad \beta_3|v_3\rangle + \alpha_4|v_4\rangle + \beta_4|v_5\rangle, \\
&\ldots
\end{aligned}
$$

The next step is to find the eigenpairs of the resulting tridiagonal matrix. The downside of the Lanczos algorithm is that the orthogonality of the Lanczos vectors $\{|v_i\rangle\}$ must be numerically enforced; hence when finding all the eigenpairs of a matrix, one typically uses the Householder algorithm, which also constructs an intermediate tridiagonal matrix through a more stable, though less straightforward, sequence of unitary transformations.

The Lanczos algorithm has its uses, however. Suppose one truncates the tridiagonal matrix at the n-th iteration, with $n \ll N$, the dimension of the full space. By the variational principle one can easily see the eigenvalues of the truncated tridiagonal are bounded by the eigenvalues of $\hat{H}$ in the full space. Furthermore, the extremal eigenvalues of the truncated tridiagonal quickly converge, as a function of n, to the extremal eigenvalues of $\hat{H}$. This is very useful in applications such as nuclear structure physics where typically one does not need all the eigenpairs but is concerned only with the lowest few states. While the exact convergence depends on the system, the ground state typically converges in 50–70 iterations and the lowest 10 states in less than 300 iterations. The Householder algorithm goes like N^3 while Lanczos goes like nN^2. Furthermore, as the Lanczos algorithm and its variants are built around matrix–vector multiplication followed by orthogonalization, it is easy to conceptualize. The Lanczos algorithm is simple, beautiful, and powerful, with an enormous literature devoted to it.

The Lanczos algorithm has applications beyond simply finding eigenpairs. To understand this, let's look in depth at how one constructs the eigen*vectors*. We want to find

$$\hat{H}|\Psi_a\rangle = E_a|\Psi_a\rangle.$$

In the Lanczos algorithm, we first construct the tridiagonal matrix $\mathbf{L}^{(n)}$ which is truncated to dimension n. It is important to note that $\mathbf{L}$ is in the space of the Lanczos vectors $\{|v_i\rangle\}$, also called the *Krylov subspace*, that is

$$L_{ij} = \langle v_i|\hat{H}|v_j\rangle. \tag{12}$$

$\mathbf{L}^{(n)}$ in turn is diagonalized

$$\mathbf{L}^{(n)}|\lambda_a\rangle = \tilde{E}_a|\lambda_a\rangle \tag{13}$$

where the extremal $\tilde{E}_a \to E_a$ as n increases. Because $\mathbf{L}$ is in the Krylov subspace, in the computer the eigenvectors $|\lambda_a\rangle = \sum_i d_{i,a}|v_i\rangle$. By knowing the Lanczos vectors in the original basis, one can then reconstruct $|\Psi_a\rangle$ in the original basis.

But suppose the pivot, $|v_1\rangle$, is a special vector (e.g., the NCSM ground state of ^{11}B of Fig. 5), such that we want to know the overlap $\langle v_1|\Psi_a\rangle$, e.g., for a $|\Psi_a\rangle$ eigenvector of a Casimir operator $\hat{\mathcal{C}}$, which for the case of the SU(3)-decomposition of Fig. 5 is given by Eq. (8). This one can just read off directly as $d_{1,a}$, that is, $\text{frac}(\gamma) = |d_{1,\Gamma}|^2$ where Γ labels the eigenvector with eigenvalue γ. With a judicious choice of Lanczos iterations, roughly the number of unique eigenvalues for the Casimir operator, each eigenvalue is associated with only one eigenvector.

This technique was originally introduced to calculate strength functions [87–90] but was later adapted for symmetry-guided decompositions [42, 67].

4. Rotational Bands

Historically, one of the strongest motivations for group-theoretical approaches to nuclear structure has been regular band structures, not only the pattern of excitation energies but also electromagnetic transitions. These patterns led first to the liquid drop picture and its quantized incarnation the Bohr–Mottelson model [35] and later to the Nilsson model [91], the SU(3) model [39, 40] and the [50] and other algebraic models [48]. As described above, detailed calculations showed that we seldom have pure dynamical symmetries, but instead quasi-dynamical symmetries are commonplace. Of all the band structures, rotational bands built upon static deformations are the most common, and so I would like to showcase further two more examples, ^{9}Be and ^{20}Ne.

^{9}Be I compute both in the $0p$ space with the phenomenological CK interaction, with dimension of 62 M-scheme states (this happens to be the same as ^{11}B as they are, in this space, particle-hole conjugates of each other), and in the NCSM with $N_{\max} = 6$ excitations allowed in a harmonic-oscillator basis with frequency $\hbar\Omega = 20\,\text{MeV}$, with a dimension of 20 million states; rotational bands in this nuclide and its sister isotope have been studied in depth in *ab initio* calculations [67, 92–94]. I show both the ground state and excited state bands. Figure 6 shows the wave functions are dominated by a single L value increasing steadily. Both bands are nearly exclusively $S = 1/2$. The SU(3)-decomposition, shown in Fig. 7, is similarly dominated by a single, consistent irrep for both bands. The natural interpretation is a fixed intrinsic shape being spun up, with different couplings between the $S = 1/2$ component and the L component. The dominance by a single L value and a single SU(3) irrep can be

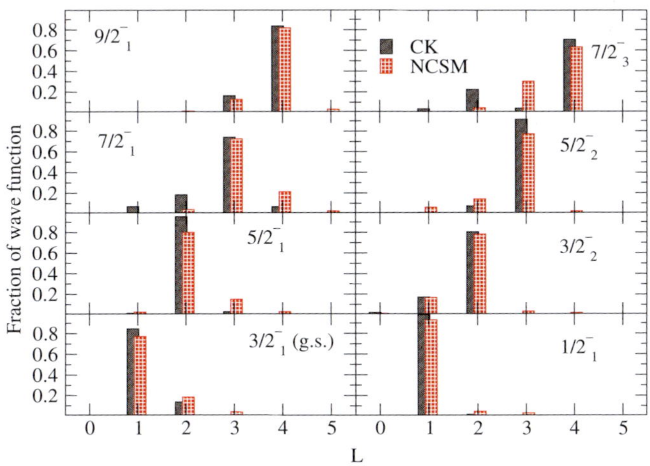

Fig. 6. L-decomposition of the ground-state (left) and first excited (right) rotational bands of ^{9}Be, using both phenomenological CK interaction (black, shaded) and *ab initio* NCSM spaces and interactions.

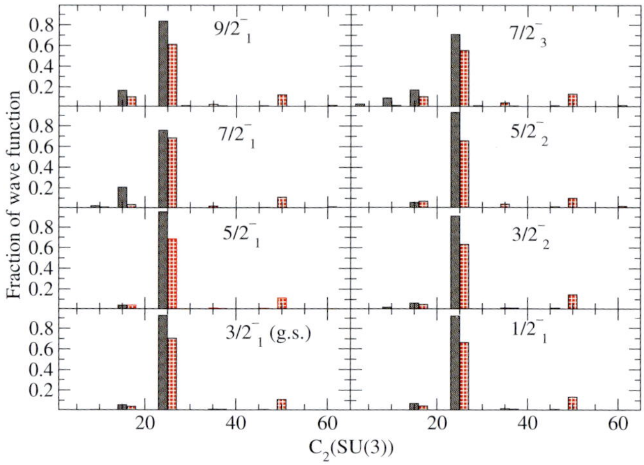

Fig. 7. SU(3)-decomposition of the ground-state (left) and first excited (right) rotational bands of ^{9}Be, using both phenomenological CK interaction (black, shaded) and *ab initio* NCSM spaces and interactions.

understood as the spin-orbit force is still too weak to dramatically fragment the distribution.

This decomposition can even be used to identify members of a band. The ground state band claims the first $7/2^-$ state as a member, while the excited state band takes not the second but the third $7/2^-$ state. The second $7/2^-$ has a quite different structure, dominated by $S = 3/2$ (not shown). This differentiation was also found elsewhere by looking at systematics of electric quadrupole and magnetic dipole moments and transitions [92, 93].

^{20}Ne I compute in the $1s0d$ space with the phenomenological USDB interaction [55] with an M-scheme dimension of 640 states, and in the NCSM with $N_{\text{max}} = 4$ excitations allowed in a harmonic-oscillator basis with frequency $\hbar\Omega = 16\,\text{MeV}$, dimension of 75 million basis states. Figure 8 shows its decomposition into SU(3) irreps. While like ^{9}Be it is also dominated by a single irrep, unlike ^{9}Be the ground-state and excited bands are dominated by very different intrinsic shapes.

In both cases, both the phenomenological and the *ab initio* calculations give similar results, despite vast differences in origin and in model space size. Indeed, for the most part such decompositions

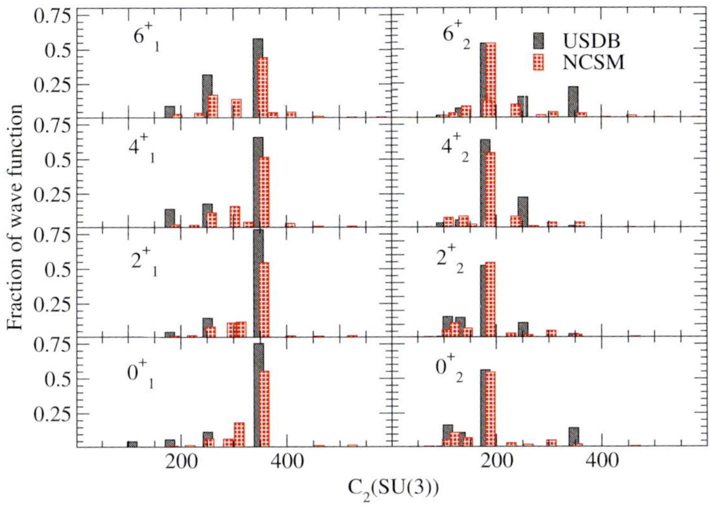

Fig. 8. SU(3)-decomposition of the ground-state(left) and first excited (right) rotational bands of ^{20}Ne, using both phenomenological USDB interaction (black, shaded) and *ab initio* NCSM spaces and interactions.

are robust against choice of model space size, basis parameters (i.e., harmonic-oscillator frequency $\hbar\Omega$), and even SRG evolution [67]. It also seems likely we have just begun to apply these tools to understanding nuclear wave functions. For example, SU(4) symmetry is of obvious importance to Gamow–Teller β-decay and decomposition into irreps could possibly help us understand, for example, unusually small matrix elements [95]. That will have to be left to future investigations.

5. Nature vs. Nurture — Do Random Interactions Naturally Exhibit Quasi-dynamical Symmetries, or Do They Have to be Coaxed?

In my discussion above I excluded the IBM and the FDSM because they implicitly assumed dynamical symmetries. Certainly we know that nuclear spectra suggest symmetries, especially SU(3). And textbooks typically lay the well-known fact that all even–even nuclides have ground states with angular momentum $J = 0$ at the feet of the pairing force.

The problem is that one can get similar results without those being good symmetries. Rotational spectra, not just the energy levels but also ratios of reduced $E2$ transition matrix elements $B(E2)$, persist even when the SU(3)-decomposition is badly fragmented, an observation which lead to the concept of quasi-dynamical symmetries. Worse, numerical experiments demonstrated that randomly chosen two-body interactions would tend to have $J = 0$ ground states, even if one sets all pairing-like matrix elements to zero [96–98].

Detailed investigations suggest random interactions echo many of the rich phenomena of real nuclei. Not only do random interactions tend to lead to $J = 0$ ground states, one sees an odd–even staggering in binding energies along an "isotopic chain", pairing gaps, and both particle-hole and pairing collectivity [97–99]. The ratios of excitation energies for the first $J = 0, 2, 4, 6, 8$ states are fairly tightly constrained to a single family spanning pairing-like, vibrational-like, and rotational-like spectra [100]. Boson models with random interactions show a tight correlation between ratios of excitation energies and ratios of $B(E2)$'s [101], a correlation not found in fermion models.

These phenomena are not the work of true dynamical symmetries; for example, even with degenerate single-particle energies the $J = 0$ ground states are not dominated by seniority zero (Zelevinsky, private communication). Very little work has been expended on looking for quasi-dynamical symmetries in random interactions; to do so one would need a robust way to compare decompositions into irreps in a fast, efficient, and above all appropriate manner. It may be worthwhile to do so, as despite numerous papers written on the topic no general understanding has yet arisen [102].

6. Symmetry-guided Bases

One motivation for decomposing wave functions into group irreps is the possibility of more compact bases for many-body calculations. By using a basis with more correlations, and especially of the *right* correlations built in, one may describe nuclei with not billions but just tens of states. Of course those correlated states are themselves complex, but if they belong to group irreps we know many of their properties in advance.

The main effort, in case you have not guessed, is in SU(3); the main barriers are, first, the high fragmentation across irreps, and, second, the difficulty of the group theory. The first has lead to the so-called pseudo-SU(3) basis [82, 83, 103], a discussion of which is beyond the scope of this paper. Because j–j coupling is much easier to implement and to harness to geometrically growing computational power, the strongest application of SU(3) has been not in phenomenology but in *ab initio* calculations. One barrier to standard NCSM calculations in the j–j scheme is that the harmonic-oscillator basis allows for exact projection of spurious center-of-mass motion but does poorly in describing the exponential tail of the wave function; brute force inclusion of many oscillator shells leads to a crippling explosion in dimensions. By contrast, the multi-shell generalization of SU(3), the symplectic group Sp(3, $\mathbb{R}$) [84, 85], naturally couples to higher shells while still projecting spurious states, leading to better radial wave functions and description of, for example, electromagnetic interactions [104] (see Chapter 4).

More recent work has demonstrated Sp(3, $\mathbb{R}$) is a strong quasi-dynamical symmetry arising naturally in *ab initio* calculations [105, 106]. Success in applying this insight to structure calculations [107–109] did not use symplectic symmetry directly, but instead used a tower of SU(3) irreps, built in different harmonic-oscillator shells (see also Chapter 8). Such work opens up the possibility of carrying out calculations not currently possible in the standard j–j coupled NCSM, such as describing the Hoyle state in ^{12}C, known to require many excitations in a harmonic-oscillator space [110]. Although this approach is very promising, only time will tell for sure if the advantages gained by group theory will outweigh the technical difficulties needed to implement. This question is very much like that one I brought up at the beginning of this paper, the simplicity of the M scheme against the compact correlations of the J scheme.

7. Summary

I have given a very brief overview of the many rich applications of group-theoretical decompositions to nuclear structure. The most important points are that while group symmetries are often badly broken, leading to fragmented distribution over many group irreducible representations, we still see strong quasi-dynamical symmetry, that is, coherent structures persistent across many states. Quasi-dynamical symmetry is most evident in rotational bands, but we can see it in non-rotational bands as well, illustrated here by ^{48}V. Assuming pure dynamical symmetries, that is dominance by a single irrep, is an oversimplification, but the ubiquity and persistence of quasi-dynamical symmetry is a strong motivation for using symmetry–adapted bases for nuclear structure. That story is not over, as we still have to find the balance between elegance and practicality.

Acknowledgments

I thank P. Navratil for use of his code to generate and SRG evolve the *ab initio* chiral interaction. I would also like to thank D. J. Millener, J. Escher, and M. A. Caprio for useful conversations in recent years, especially regarding SU(3) and SU(4), though any errors in describing group theory are mine alone. This material is based upon work supported by the U.S. Department of Energy, Office of Science, Office of Nuclear Physics, under Award Number DE-FG02-96ER40985.

References

[1] R. Machleidt, K. Holinde, and C. Elster, The bonn meson-exchange model for the nucleon–nucleon interaction, *Phys. Rep.* **149**, 1 (1987).
[2] B. A. Brown and B. H. Wildenthal, Status of the nuclear shell model, *Annu. Rev. Nucl. Part. Sci.* **38**, 29 (1988).
[3] V. Stoks, R. Klomp, C. Terheggen, and J. De Swart, Construction of high-quality NN potential models, *Phys. Rev. C.* **49**, 2950 (1994).
[4] R. B. Wiringa, V. Stoks, and R. Schiavilla, Accurate nucleon–nucleon potential with charge-independence breaking, *Phys. Rev. C.* **51**, 38 (1995).
[5] U. Van Kolck, Effective field theory of short-range forces, *Nucl. Phys. A.* **645**, 273 (1999).

[6] D. R. Entem and R. Machleidt, Accurate charge-dependent nucleon-nucleon potential at fourth order of chiral perturbation theory, *Phys. Rev. C.* **68**, 041001 (2003).
[7] E. Epelbaum, Few-nucleon forces and systems in chiral effective field theory, *Progr. Part. Nucl. Phys.* **57**, 654 (2006).
[8] R. Machleidt and D. R. Entem, Chiral effective field theory and nuclear forces, *Phys. Rep.* **503**, 1 (2011).
[9] J. Carlson, Green's function monte carlo study of light nuclei, *Phys. Rev. C.* **36**, 2026 (1987).
[10] B. Pudliner, V. Pandharipande, J. Carlson, and R. B. Wiringa, Quantum Monte Carlo calculations of $A \leq 6$ nuclei, *Phys. Rev. Lett.* **74**, 4396 (1995).
[11] S. C. Pieper, Quantum monte carlo calculations of light nuclei, *Nucl. Phys. A.* **751**, 516 (2005).
[12] F. Coester, Bound states of a many-particle system, *Nucl. Phys.* **7**, 421 (1958).
[13] J. H. Heisenberg and B. Mihaila, Ground state correlations and mean field in ^{16}O, *Phys. Rev. C.* **59**, 1440 (1999).
[14] D. J. Dean and M. Hjorth-Jensen, Coupled-cluster approach to nuclear physics, *Phys. Rev. C.* **69**, 054320 (2004).
[15] G. Hagen, T. Papenbrock, D. J. Dean, and M. Hjorth-Jensen, *Ab initio* coupled-cluster approach to nuclear structure with modern nucleon–nucleon interactions, *Phys. Rev. C.* **82**, 034330 (2010).
[16] P. Navrátil, J. Vary, and B. Barrett, Large-basis *ab initio* no-core shell model and its application to ^{12}C, *Phys. Rev. C.* **62**, 054311 (2000).
[17] B. R. Barrett, P. Navrátil, and J. P. Vary, *Ab initio* no core shell model, *Progr. Part. Nucl. Phys.* **69**, 131 (2013).
[18] I. Stetcu and J. Rotureau, Effective interactions and operators in the no-core shell model, *Progr. Part. Nucl. Phys.* **69**, 182 (2013).
[19] S. Okubo, Diagonalization of hamiltonian and tamm-dancoff equation, *Progr. Theor. Phys.* **12**, 603 (1954).
[20] K. Suzuki and S. Y. Lee, Convergent theory for effective interaction in nuclei, *Progr. Theor. Phys.* **64**, 2091 (1980).
[21] K. Suzuki, Unitary-model-operator approach to nuclear effective interaction. II. Effects of three-body cluster term, *Progr. Theor. Phys.* **68**, 1999 (1982).
[22] K. Suzuki and R. Okamoto, Degenerate perturbation theory in quantum mechanics, *Progr. Theor. Phys.* **70**, 439 (1983).
[23] H. Feldmeier, T. Neff, R. Roth, and J. Schnack, A unitary correlation operator method, *Nucl. Phys. A.* **632**, 61 (1998).
[24] S. D. Głazek and K. G. Wilson, Renormalization of Hamiltonians, *Phys. Rev. D.* **48**, 5863 (1993).
[25] F. Wegner, Flow-equations for Hamiltonians, *Ann. Physik.* **506**, 77 (1994).
[26] S. K. Bogner, R. J. Furnstahl, and R. J. Perry, Similarity renormalization group for nucleon–nucleon interactions, *Phys. Rev. C.* **75**, 061001 (2007).
[27] S. Bogner, R. Furnstahl, and A. Schwenk, From low-momentum interactions to nuclear structure, *Progr. Part. Nucl. Phys.* **65**, 94 (2010).

[28] E. D. Jurgenson, P. Navrátil, and R. J. Furnstahl, Evolution of nuclear many-body forces with the similarity renormalization group, *Phys. Rev. Lett.* **103**, 082501 (2009).

[29] P. Brussard and P. Glaudemans, *Shell-Model Applications in Nuclear Spectroscopy.* North-Holland Publishing Company, Amsterdam (1977).

[30] I. Shavitt, The history and evolution of configuration interaction, *Mol. Phys.* **94**, 3 (1998).

[31] E. Caurier, G. Martinez-Pinedo, F. Nowacki, A. Poves, and A. P. Zuker, The shell model as a unified view of nuclear structure, *Rev. Mod. Phys.* **77**, 427 (2005).

[32] B. N. Parlett, *The Symmetric Eigenvalue Problem.* Vol. 7, SIAM (1980) http://epubs.siam.org/doi/10.1137/1025067.

[33] R. R. Whitehead, A. Watt, B. J. Cole, and I. Morrison, Computational methods for shell model calculations, *Adv. Nucl. Phys.* **9**, 123 (1977).

[34] R. Hamming, *Numerical Methods for Scientists and Engineers.* Courier Corporation (2012).

[35] A. Bohr and B. R. Mottelson, *Nuclear Structure.* Vol. 2, World Scientific, Singapore (1998).

[36] M. Goldhaber and E. Teller, On nuclear dipole vibrations, *Phys. Rev.* **74**, 1046 (1948).

[37] N. L. Iudice and F. Palumbo, New isovector collective modes in deformed nuclei, *Phys. Rev. Lett.* **41**, 1532 (1978).

[38] A. Richter, Probing the nuclear magnetic dipole response with electrons, photons and hadrons, *Progr. Part. Nucl. Phys.* **34**, 261 (1995).

[39] J. P. Elliott, Collective motion in the nuclear shell model. I. Classification schemes for states of mixed configurations, *Proc. Roy. Soc. A.* **245**, 128 (1958).

[40] M. Harvey. The nuclear SU(3) model. In *Advances in Nuclear Physics.* Springer (1968), Plenum Press, New York, p. 67.

[41] P. Rochford and D. Rowe, The survival of rotor and SU(3) bands under strong spin-orbit symmetry mixing, *Phys. Lett. B.* **210**, 5 (1988).

[42] V. G. Gueorguiev, J. P. Draayer, and C. W. Johnson, SU(3) symmetry breaking in lower *fp*-shell nuclei, *Phys. Rev. C.* **63**, 014318 (2000).

[43] C. Bahri, D. J. Rowe, and W. Wijesundera, Phase transition in the pairing-plus-quadrupole model, *Phys. Rev. C.* **58**, 1539 (1998).

[44] D.J. Rowe, "Quasi-Dynamical Symmetry — a new use of symmetry in nuclear physics", in "The Nucleus: New Physics for the New Millennium", eds. F.D. Smit *et al.*, Springer US (1999).

[45] C. Bahri and D. Rowe, SU(3) quasi-dynamical symmetry as an organizational mechanism for generating nuclear rotational motions, *Nucl. Phys. A.* **662**, 125 (2000).

[46] A. Leviatan, J. N. Ginocchio, and M. W. Kirson, Proton–neutron deformations and *F*-spin symmetry in nuclei, *Phys. Rev. Lett.* **65**, 2853 (1990).

[47] J. Escher and A. Leviatan, Partial dynamical symmetry in a fermion system, *Phys. Rev. Lett.* **84**, 1866 (2000).

[48] I. Talmi, *Simple Models of Complex Nuclei.* Vol. 7, CRC Press (1993).
[49] D. J. Rowe and J. L. Wood, *Fundamentals of Nuclear Models: Foundational Models.* World Scientific (2010).
[50] F. Iachello and A. Arima, *The Interacting Boson Model.* Cambridge University Press (1987).
[51] C.-L. Wu, X.-G. Chen, J.-Q. Chen, M. W. Guidry, *et al.*, Fermion dynamical symmetries and the nuclear shell model, *Phys. Lett. B.* **168**, 313 (1986).
[52] C.-L. Wu, X.-G. Chen, J.-Q. Chen, M. W. Guidry, *et al.*, Fermion dynamical symmetry model of nuclei: Basis, Hamiltonian, and symmetries, *Phys. Rev. C.* **36**, 1157 (1987).
[53] C.-L. Wu, M. Guidry, *et al.*, *The Fermion Dynamical Symmetry Model.* Springer, New York (1994).
[54] S. Cohen and D. Kurath, Effective interactions for the 1p shell, *Nucl. Phys.* **73**, 1 (1965).
[55] B. A. Brown and W. A. Richter, New "USD" Hamiltonians for the sd shell, *Phys. Rev. C.* **74**, 034315 (2006).
[56] M. Honma, T. Otsuka, B. A. Brown, and T. Mizusaki, Effective interaction for pf-shell nuclei, *Phys. Rev. C.* **65**, 061301 (2002).
[57] C. W. Johnson, W. E. Ormand, and P. G. Krastev, Factorization in large-scale many-body calculations, *Comput. Phys. Commun.* **184**, 2761 (2013).
[58] M. G. Mayer, On closed shells in nuclei, *Phys. Rev.* **74**, 235 (1948).
[59] M. G. Mayer, On Closed Shells in Nuclei. II, *Phys. Rev.* **75**, 1969 (1949).
[60] M. G. Mayer, Nuclear configurations in the spin-orbit coupling model. I. empirical evidence, *Phys. Rev.* **78**, 16 (1950).
[61] O. Haxel, J. H. D. Jensen, and H. E. Suess, On the "magic numbers" in nuclear structure, *Phys. Rev.* **75**, 1766 (1949).
[62] E. Feenberg, K. C. Hammack, and L. W. Nordheim, Note on proposed schemes for nuclear shell models, *Phys. Rev.* **75**, 1968 (1949).
[63] E. Feenberg and E. Wigner, On the structure of the nuclei between helium and oxygen, *Phys. Rev.* **51**, 95 (1937).
[64] E. Feenberg and M. Phillips, On the structure of light nuclei, *Phys. Rev.* **51**, 597 (1937).
[65] B. H. Flowers, Studies in jj-Coupling. I. Classification of nuclear and atomic states, *Proc. Roy. Soc. A.* **212**, 248 (1952).
[66] D. Kurath, States of light nuclei from the jj coupling model, *Phys. Rev.* **88**, 804 (1952).
[67] C. W. Johnson, Spin-orbit decomposition of *ab initio* nuclear wave functions, *Phys. Rev. C.* **91**, 034313 (2015).
[68] W. Heisenberg, On the structure of atomic nuclei, *Z. Phys.* **77**, 1 (1932).
[69] E. Wigner, On the consequences of the symmetry of the nuclear hamiltonian on the spectroscopy of nuclei, *Phys. Rev.* **51**, 106 (1937).
[70] K. Hecht and S. C. Pang, On the wigner supermultiplet scheme, *J. Math. Phys.* **10**, 1571 (1969).
[71] P. Franzini and L. Radicati, On the validity of the supermultiplet model, *Phys. Lett.* **6**, 322 (1963).

[72] K. Hecht and J. Draayer, Spectral distributions and the breaking of isospin and supermultiplet symmetries in nuclei, *Nucl. Phys. A.* **223**, 285 (1974).
[73] P. Vogel and W. E. Ormand, Spin–isospin SU(4) symmetry in *sd*- and *fp*-shell nuclei, *Phys. Rev. C.* **47**, 623 (1993).
[74] N. Frazier, B. Brown, D. Millener, and V. Zelevinsky, Gamow-teller strength as a function of excitation energy, *Phys. Lett. B.* **414**, 7 (1997).
[75] G. Racah, Theory of complex spectra. II, *Phys. Rev.* **62**, 438 (1942).
[76] G. Racah, in *L. Farkas Memorial Volume* (Research Council of Israel, Jerusalem, 1952), p. 294.
[77] G. Racah and I. Talmi, The pairing property of nuclear interactions, *Physica.* **18**, 1097 (1952).
[78] A. Kerman, Pairing forces and nuclear collective motion, *Ann. Phys.* **12**, 300 (1961).
[79] R. W. Richardson, A restricted class of exact eigenstates of the pairing-force Hamiltonian, *Phys. Lett.* **3**, 277 (1963).
[80] M. Gaudin, Diagonalisation d'une classe d'hamiltoniens de spin, *J. Physique.* **37**, 1087 (1976).
[81] C. Bahri, J. Escher, and J. Draayer, Monopole-pairing and deformation in atomic nuclei, *Nucl. Phys. A.* **592**, 171 (1995).
[82] A. Arima, M. Harvey, and K. Shimizu, Pseudo *LS* coupling and pseudo SU(3) coupling schemes, *Phys. Lett. B.* **30**, 517 (1969).
[83] R. R. Raju, J. Draayer, and K. Hecht, Search for a coupling scheme in heavy deformed nuclei: The pseudo SU(3) model, *Nucl. Phys. A.* **202**, 433 (1973).
[84] G. Rosensteel and D. J. Rowe, Nuclear Sp(3, $\mathbb{R}$) Model, *Phys. Rev. Lett.* **38**, 10 (1977).
[85] G. Rosensteel and D. Rowe, On the algebraic formulation of collective models. I. The mass quadrupole collective model, *Ann. Phys.* **123**, 36 (1979).
[86] P. Park, J. Carvalho, M. Vassanji, D. Rowe, and G. Rosensteel, The shell-model theory of nuclear rotational states, *Nucl. Phys. A.* **414**, 93 (1984).
[87] E. Caurier, A. Poves, and A. Zuker, A full $0\hbar\omega$ description of the $2\nu\beta\beta$ decay of ^{48}Ca, *Phys. Lett. B.* **252**, 13 (1990).
[88] E. Caurier, A. Poves, and A. P. Zuker, Missing and quenched gamow–teller strength, *Phys. Rev. Lett.* **74**, 1517 (1995).
[89] E. Caurier, G. Martínez-Pinedo, F. Nowacki, A. Poves, J. Retamosa, and A. P. Zuker, Full $0\hbar\omega$ shell model calculation of the binding energies of the $1f_{7/2}$ nuclei, *Phys. Rev. C.* **59**, 2033 (1999).
[90] W. C. Haxton, K. M. Nollett, and K. M. Zurek, Piecewise moments method: Generalized lanczos technique for nuclear response surfaces, *Phys. Rev. C.* **72**, 065501 (2005).
[91] S. G. Nilsson, Binding states of individual nucleons in a strongly deformed field, *Mat. Fys. Medd. Dan. Vid. Selsk.* **29** (16) (1955).

[92] M. A. Caprio, P. Maris, and J. P. Vary, Emergence of rotational bands in *ab initio* no-core configuration interaction calculations of light nuclei, *Phys. Lett. B.* **719**, 179 (2013).

[93] P. Maris, M. A. Caprio, and J. P. Vary, Emergence of rotational bands in *ab initio* no-core configuration interaction calculations of the Be isotopes, *Phys. Rev. C.* **91**, 014310 (2015).

[94] R. B. Wiringa, Pair counting, pion-exchange forces and the structure of light nuclei, *Phys. Rev. C.* **73**, 034317 (2006).

[95] P. Maris, J. P. Vary, P. Navrátil, W. E. Ormand, H. Nam, and D. J. Dean, Origin of the anomalous long lifetime of ^{14}C, *Phys. Rev. Lett.* **106**, 202502 (2011).

[96] C. W. Johnson, G. F. Bertsch, and D. J. Dean, Orderly spectra from random interactions, *Phys. Rev. Lett.* **80**, 2749 (1998).

[97] V. Zelevinsky and A. Volya, Nuclear structure, random interactions and mesoscopic physics, *Phys. Rep.* **391**, 311 (2004).

[98] Y. Zhao, A. Arima, and N. Yoshinaga, Regularities of many-body systems interacting by a two-body random ensemble, *Phys. Rep.* **400**, 1 (2004).

[99] C. W. Johnson, G. F. Bertsch, D. J. Dean, and I. Talmi, Generalized seniority from random Hamiltonians, *Phys. Rev. C.* **61**, 014311 (1999).

[100] C. W. Johnson and H. A. Nam, New puzzle for many-body systems with random two-body interactions, *Phys. Rev. C.* **75**, 047305 (2007).

[101] R. Bijker and A. Frank, Band structure from random interactions, *Phys. Rev. Lett.* **84**, 420 (2000).

[102] A. Volya, Emergence of symmetry from random n-body interactions, *Phys. Rev. Lett.* **100**, 162501 (2008).

[103] O. Castaños, M. Moshinsky, and C. Quesne, Transformation to pseudo-SU(3) in heavy deformed nuclei, *Phys. Lett. B.* **277**, 238 (1992).

[104] J. Escher and J. P. Draayer, Electron scattering and Multi-$\hbar\omega$ correlations, *Phys. Rev. Lett.* **82**, 5221 (1999).

[105] T. Dytrych, K. D. Sviratcheva, C. Bahri, J. P. Draayer, and J. P. Vary, Evidence for symplectic symmetry in *Ab Initio* no-core shell model results for light nuclei, *Phys. Rev. Lett.* **98**, 162503 (2007).

[106] T. Dytrych, K. D. Sviratcheva, J. P. Draayer, C. Bahri, and J. P. Vary, *Ab initio* symplectic no-core shell model, *J. Phys.* **35**, 123101 (2008).

[107] T. Dytrych, K. D. Launey, J. P. Draayer, P. Maris, J. P. Vary, *et al.*, Collective modes in light nuclei from first principles, *Phys. Rev. Lett.* **111**, 252501 (2013).

[108] T. Dytrych, A. C. Hayes, K. D. Launey, J. P. Draayer, P. Maris, J. P. Vary, D. Langr, and T. Oberhuber, Electron-scattering form factors for ^{6}Li in the *ab initio* symmetry-guided framework, *Phys. Rev. C.* **91**, 024326 (2015).

[109] K. D. Launey, A. C. Dreyfuss, J. P. Draayer, T. Dytrych, and R. Baker, Emergence of cluster structures and collectivity within a no-core shell-model framework, *J. Phys.: Conf. Ser.* **569**, 012061 (2014).

[110] T. Neff, Clusters and halos in light nuclei, *J. Phys.: Conf. Ser.* **403**, 012028 (2012).

PART 2

Emergent Simplicity within Intricate Multi-particle Dynamics: Nuclear Structure and Reactions

Chapter 3

The Emergence and Use of Symmetry in the Many-nucleon Model of Atomic Nuclei

David J. Rowe
Department of Physics, University of Toronto
Toronto ON M5S 1A7, Canada

This Chapter explores the properties of nuclei that are indicators of the emergence of simple dynamics associated with the symmetries that are available to a nucleus as a many-nucleon system. In particular, it focuses on the dynamical symmetries associated with nuclear collective motions. The term "symmetry" is interpreted in a broad sense to mean any set of transformations of the states of a nucleus that leave some important attributes of the nucleus invariant. Symmetries that provide good quantum numbers, or even approximately good quantum numbers, are clearly useful. However, symmetry relations can be used more widely such as, for example, to decompose the states of a nucleus into a sequence of invariant subsets ordered such that their contributions to the observed states of nuclei are of decreasing importance. A standard example of this is given by the spherical harmonic-oscillator shell model which proves to be particularly relevant for many states of near closed shell and singly closed shell nuclei. An alternative, based on the dynamical harmonic-oscillator symmetry which proves to be much more appropriate for the rotational states of deformed nuclei, is reviewed in this paper. A particularly significant result is the emergence of low-energy states of open shell nuclei with the dynamical symmetries of a rotor whose properties depend very little on the details of the nuclear interactions other than that they are basically attractive, of relatively short range, and favor spatially symmetric states.

Contents

1. Introduction

The properties of nuclei are very much emergent phenomena. Even the existence of the nucleus as an extremely dense central core of an atom was an unexpected discovery [1–3]. An early perception of the nucleus, after the discovery of the neutron [4], was that, because it was made up of strongly interacting nucleons in an exceedingly small volume, it was inconceivable that it could be described in terms of many-nucleon quantum mechanics with anything like the simplicity of the atomic shell model. It was believed that nuclear dynamics must surely be chaotic. Thus, liquid-drop and compound nuclear models were pursued [5]. However, within a relatively short period of time, the observation of patterns in the properties of nuclei, such as the spins of odd-mass nuclei and single-particle separation energies, led to their interpretation in terms of a nuclear shell model [6, 7]. Patterns in the quadrupole moments of odd-mass nuclei [8, 9] and their interpretation [10] similarly led to a simple collective model [11] and then to a unified model [12, 13] in which nucleons move in a deformable shell-model potential with vibrational and rotational degrees of freedom. Thus, the first major breakthroughs in the construction of successful models of nuclear structure was due to the observation of patterns in nuclear data and the emergence of simplicity in their interpretation.

Patterns are signatures of underlying symmetries and their recognition can be of enormous help in understanding nuclear properties. Several symmetries of the nucleus (some approximate) as a system of non-relativistic interacting nucleons are in common use in nuclear physics and it is worthwhile to observe how indispensable they really are. The symmetries that give rise to the total angular momentum, parity, and isospin quantum numbers of nuclear states immediately come to mind. The commutation relations that underlie the basic algebraic structures of sets of symmetry-related operators in quantum mechanics are also unavoidable. For example, the basic commutation relations between the position and momentum coordinates, $\{x_i, p_i; i = 1, 2, 3\}$, which are interpreted as operators $\{\hat{x}_i, \hat{p}_i; i = 1, 2, 3\}$ in quantum mechanics with commutation relations

$[\hat{x}_i, \hat{p}_j] = \mathrm{i}\hbar\delta_{i,j}$, are expressions of the Heisenberg algebra on which many-particle quantum mechanics is founded. A remarkable property of this relationship between nucleon coordinates is that, for a given value of $\hbar$, the Heisenberg algebra for a nucleon has a single unique irreducible unitary representation [14]. This unique representation is realized, for example, in terms of the states of the nucleon in a spherical harmonic-oscillator potential, which, as a consequence, also has a single irreducible unitary representation. If it were not for this profound, and extremely rare property of the Heisenberg algebra, the quantum mechanics of elementary particles, as nucleons are assumed to be in nuclear physics, would be infinitely more complex.

Another essential symmetry of many-nucleon quantum mechanics, that is customarily taken for granted, is the exchange symmetry of identical nucleons. Classical intuition would lead one to expect that the exchange of two identical particles, in states with wave functions that are identical in all respects, should leave the combined wave functions of the particles unchanged. However, group theory reveals the possibility that a pair of identical wave functions for two particles can change sign under such an exchange and experimental observations show that indeed this happens when the particles are spin-$\frac{1}{2}$ fermions. The implication of this symmetry, which underlies the Pauli exclusion principle, in combination with the uniqueness of a single-nucleon Hilbert space, is that the many-nucleon Hilbert space is likewise unique and spanned by totally antisymmetric Slater determinants of single-nucleon states. The objective of this paper is to explore the implications of these and other symmetries that emerge in experimental and theoretical studies of nuclei.

There are many advantages to the explicit use of symmetries in nuclear physics. At an observational level, the properties of nuclei are most readily amenable to interpretation when they exhibit recognizable patterns. If the patterns and associated properties are consistent with the representations of a symmetry associated with a phenomenological model, it is particularly advantageous. Theoretical studies can then focus on a microscopic interpretation of the model. It is important, however, to be aware that a given symmetry may have many interpretations and that the success of a model does not necessarily justify any particular interpretation.

In this paper, the nucleus is considered as a system of many nucleons regarded as elementary particles with effective interactions that are presumed to be derivable from high-energy particle theory, nucleon scattering data, and the properties of few-nucleon bound states. Symmetry-based models that have interpretations as sub-models of many-nucleon quantum mechanics can be used in multiple ways. They can be used in algebraic model interpretations of nuclear properties. They can also be used as coupling schemes for microscopic many-nucleon calculations and to interpret the physical significance of results obtained in arbitrary bases.

2. Symmetry Concepts

The following concepts are in common use, *albeit* with variations on their precise definitions [15–17].

A *symmetry of an object or set of objects* is basically a set of transformations of the object, or set of objects, that leave it invariant. Such sets of transformations have group theoretical properties that can be exploited with advantage in many problems in physics.

A *symmetry group of a model Hamiltonian* is a set of transformations of the states of the model that leave the Hamiltonian invariant. Such symmetry groups and their subgroups are important because they provide conserved quantum numbers for the model's states.

A *dynamical group for a model* is a group of transformations of the Hilbert space $\mathcal{H}$ of the model such that $\mathcal{H}$ carries an irreducible unitary representation of this group. Such a dynamical group is then rich enough to relate all states of a model's representation. This is appropriate because if a Hilbert space has states that are not related in some way by the operators of the model, the unrelated states are more usefully regarded as belonging to different model representations.

A *spectrum generating algebra* for a model Hamiltonian is a set of elements of a Lie algebra having the essential property that the

Hamiltonian is expressible in terms of its elements. The irreducible unitary representations of a spectrum generating algebra provide the information needed to determine the eigenstates of the Hamiltonian. A given Hamiltonian may have several spectrum generating algebras. A large spectrum generating algebra would be the Lie algebra of a dynamical group for the model. However, a Hamiltonian of the model will generally be expressible in terms of the elements of a subalgebra. A smaller spectrum generating algebra is generally more useful for determining the eigenstates of the Hamiltonian. However, a larger spectrum generating algebra facilitates the calculation of transition matrix elements between energy eigenstates of a larger number of operators.

In the characterization of a model, it is useful to focus on its *observables*. An observable is some attribute of a state of the system that could conceivably be measured. In quantum mechanics, it is represented by a Hermitian linear operator on the model Hilbert space. Thus, an observable that is represented by a Hermitian linear operator $\hat{X}$, will have a precise real value if and only if the system is in an eigenstate of $\hat{X}$. In general, if the system is in a quantum-mechanical state $|\psi\rangle$, it will have a distribution of values about a mean given by the so-called expectation value $\langle\psi|\hat{X}|\psi\rangle$, which will also be real. For present purposes, we avoid the philosophical subtleties of measurement theory by regarding any set of Hermitian operators on the Hilbert space of a model as observables of the model if they have well-defined commutation relations and well-defined expectation values in the states of the model. However, in speaking of the observables of model, it will generally be useful to focus on a basic subset from which all others can be generated. For example, a basic set of observables of the many-nucleon model of a nucleus is given by the Hermitian position, momentum, and spin operators of its nucleons.

A *Lie algebra of observables* in quantum mechanics is a set of observables having the property that the commutator of any two elements of the set is a linear combination of the elements of the set. Note, however, that the commutator of two Hermitian operators is necessarily skew-Hermitian and needs to be multiplied by $\mathrm{i} = \sqrt{-1}$

to become Hermitian. For example, the commutation relations of the position and momentum coordinates $\{x_j, p_j, j = 1, 2, 3\}$ of a single particle are given by the Heisenberg equations

$$[\hat{x}_j, \hat{p}_k] = \mathrm{i}\hbar\delta_{j,k}\hat{I}, \tag{1}$$

where $\hat{I}$ is the identity operator. Thus, the Heisenberg algebra of observables for the particle comprises the set $\{\hat{x}_j, \hat{p}_j, \hat{I}\}$. A useful property is that from a Lie algebra of basic observables, whose unitary representations define the possible Hilbert spaces of the model, one can construct infinitely many other observables for the model as Hermitian combinations of the basis observables, e.g., $\hat{X}\hat{Y} + \hat{Y}\hat{X}$. It should be noted, however, that because the many-nucleon states of a nucleus are totally antisymmetric, any observable for a particular nucleus must be totally symmetric in order that its action leaves this antisymmetry invariant. This symmetry constraint on the observables of nuclear physics is automatically respected when they are expressed in the language of fermion second quantization.

The above concepts are employed explicitly in the so-called algebraic models. However, essentially all models in use in nuclear physics have underlying algebraic structures and, to a significant extent, make use of them. For example, the standard many-nucleon model of a nucleus has a dynamical group which is the infinite-dimensional unitary group of one-body unitary transformations and which has a single irreducible representation (or "irrep") on the whole Hilbert space of a nucleus; its Hilbert space is spanned by totally antisymmetric Slater determinants. The Lie algebra of this group, is a Lie algebra of one-body operators which is a spectrum-generating algebra for any microscopic nuclear Hamiltonian. Thus, non-relativistic nuclear many-nucleon theory is founded on these fundamental underlying structures.

3. The Origin and Nature of Emergent Symmetries

Symmetries of condensed matter systems are generally most evident at temperatures approaching zero kelvin. Similarly, the symmetries of nuclei are observed primarily in their low-energy states. Some symmetries arise because the interactions between nucleons are of short range and strongest in spatially symmetric states. As a consequence, states of lowest energy are predominantly of maximal space symmetry and have the corresponding supermultiplet symmetry required to form totally antisymmetric states. In fact, this symmetry has a profound influence on the character of the low-energy states of nuclei as discussed long ago by Mottelson [18] in terms of the competition between pairing and aligned coupling schemes in the spherical shell model.

Two nucleons of the same angular momentum j interact most strongly either when their j's are in opposite directions and they form an antisymmetric, spatially spherical $J = 0$ state, or when their j's are parallel and they form a maximally deformed, but exchange symmetric, $J=2j$ state. However, two nucleons of the same kind, i.e., both neutrons or both protons, must be in antisymmetric states. As a result, it turns out that for the spherical shell-model states of an even-mass singly closed shell nucleus, the lowest-energy states are dominated by those in which the valence shell nucleons, which are all protons or all neutrons, are coupled pairwise to spherical $J=0$ states. This gives rise to the so-called jj pair-coupling scheme, introduced by Racah [19] in atomic physics and adapted for use in the nuclear shell model by Flowers and Edmonds [20–23], in which states are classified by the representations of a compact symplectic group $\mathrm{Sp}(2j+1)$. In contrast, for doubly open shell nuclei, the aligned coupling of the nucleons to form maximally deformed combinations tends to be favored and the many-nucleon states that emerge are more likely to be those of LS coupling in which the many-nucleon states are classified by the representations of Elliott's SU(3) group. The competition between aligned and pair coupling was first considered by Belyaev [24].

The spherical shell model of Mayer–Jensen [6, 7] was based on the expectation that nucleons would generally pair off as far as possible to give states of zero angular momentum. However, the observation of large quadrupole moments of many odd-mass nuclei [8, 9] implied that such nuclei were non-spherical [10] and led to the collective model [11, 12]. More recent observations [25, 26] reveal that, whereas the ground states of closed shell and singly closed shell nuclei are predominantly spherical in accord with the Mayer–Jensen model, the states of doubly open shell nuclei and, moreover, many excited states of closed and singly closed shell nuclei are strongly deformed and exhibit rotational bands (see also Chapter 1). Thus, it would appear that the spherical shell model and the collective models have different, although sometimes overlapping, domains of relevance.

It is instructive to note that the centre-of-mass states of a nucleus, like those of any object, reflect the translational and Galilean invariance of its dynamics. Centre-of-mass motions of a nucleus are decoupled from its intrinsic dynamics and are not generally of interest. However, it is worth considering the implications of what to expect when the nucleus is intrinsically non-spherical. Because of the rotational invariance of the nuclear Hamiltonian, the set of rotated states of a deformed nuclear state must all have the same energy. However, there is no analog of Galilean invariance for rotational dynamics. Consequently, one does not expect the intrinsic state of a rotating deformed nucleus to be independent of its rotational angular momentum. Nevertheless, for intrinsic states of large deformation, and relatively small values of the angular momentum, the angular velocity of the rotational dynamics could be small enough that the centrifugal and Coriolis forces are minor perturbations and the symmetries of a rigid rotor are able to survive.

The weak coupling of low angular-momentum rotations to other dynamics of a nucleus is a prototype of the more general so-called *adiabatic decoupling* of fast and slow degrees of freedom considered in many areas of physics and as exploited in a quantum mechanical context by Born and Oppenheimer [27]. A study of this decoupling was considered from the perspective of group theory [28] and it was observed that the perturbation of the slow dynamics by its coupling to fast degrees of freedom, if not too strong, would commonly result in the dynamical symmetry associated with the slow degrees of freedom being replaced by a *quasi-dynamical symmetry*, i.e., by states of mixed symmetry that are almost indistinguishable from those of the unperturbed dynamics. Examples of this were reviewed in

Ref. [29]. Thus, in studying the emergence of dynamical symmetry in nuclear physics, it is also of interest to consider the emergence of quasi-dynamical symmetry, which is orders of magnitude more robust in the presence of symmetry-breaking interactions and for practical purposes almost as useful.

4. The Collective Models as Algebraic Many-nucleon Models

A standard paradigm of research in nuclear physics starts with observations of patterns in the properties of nuclei and their interpretation in terms of models. In order that the several models used in this way can contribute to an overall understanding of nuclear structure, they should be related and separately understood as submodels or approximations to submodels of a single more realistic model of the nucleus which, for present purposes, we take to be many-nucleon quantum mechanics. In this way, the limitations of a model are better understood and, more significantly, they can be explored in further situations. As often happens, understanding and exploring a model's limitations can be more insightful than demonstrating its successes especially when a successful model has more than one physical interpretation.

The Mayer–Jensen-based shell model was formulated as a submodel of many-nucleon quantum mechanics. Its successes were nevertheless of initial concern [30, 31] because it was difficult to understand how wave functions for nucleons with hard repulsive cores could have convergent expansions in an independent-particle basis. The resolution of these concerns led to major advances in many-body quantum mechanics and the theory of effective interactions. The Bohr collective model was formulated [11] in parallel with the shell model as a deformable quantum-fluid model with shape vibrational and rotational degrees of freedom having irrotational-flow moments of inertia. It was soon extended to the Bohr-Mottelson unified model [12], in which nucleons were added to a polarizable Bohr–model core, and further extended by Nilsson [13] who introduced a shell-model-like Hamiltonian with an axially-symmetric harmonic-oscillator potential for the wave functions of the extra-core particles in the deformed field of the polarized core.

The Bohr–Mottelson–Nilsson unified model [32–34] and its many developments [35–40], have been extraordinarily successful in the interpretation of collective structure in nuclei. Moreover, they continue to provide the basic concepts and language in terms of which nuclear collective phenomena are described. Thus, a great deal of effort has been devoted by numerous researchers from a variety of geometrical, mean-field, and algebraic perspectives, to give this phenomenologically conceived model a microscopic foundation in many-nucleon quantum mechanics. From these many developments, reviewed in Ref. [41], there emerged an algebraic symmetry-based many-nucleon realization of the collective models.

This section gives a succinct but precise expression of these microscopic versions of the collective models from an algebraic perspective. The approach is to identify the observables of the model, by which we mean the variables of the model that are interpreted in quantum mechanics as Hermitian operators, express them in terms of nucleon position and momentum coordinates, and thereby derive their commutation relations and the Lie algebras to which they belong. In applications of these models, one will of course need to go further and, for example, introduce Hamiltonians and other observables, usually expressed as polynomials in a model's basic algebra of observables.

4.1. *The Elliott* SU(3) *Model*

Elliott's SU(3) model [42, 43] was a first attempt to understand the emergence of rotational states within the framework of the spherical shell model. In retrospect, it can be understood as the projected image of a rigid-rotor model (cf. following section) onto a finite shell-model subspace of states of minimum spherical harmonic-oscillator energy. Thus, it has many properties that resemble those of a truncated rigid rotor. Moreover, because of its rotor-like properties, the SU(3) model can be regarded as an effective shell model of nuclear rotations, with effective quadrupole moments and an effective rotational Hamiltonian as defined by Lee–Suzuki methods [44, 45]. The SU(3) model was particularly significant in revealing the mechanism by which states with deformed equilibrium states would emerge in the shell model for nucleons with attractive interactions that favored maximal space symmetry. However, in spite of its importance, it did not provide the many-nucleon description of nuclear rotations that was being sought. The problem was that it failed to give any information on the dynamics of nuclear rotational flows or their moments of inertia. This was because the expectation values of the many-nucleon kinetic energy, which are trivially determined in Elliott's model, take a single constant value for all states of an SU(3) irrep. A remarkable result is that a multi-shell version of the SU(3) model subsequently re-emerged as an essential component of the many-nucleon theory of nuclear collective dynamics that later developed and for which the symmetry methods developed for the SU(3) model proved to be invaluable.

4.2. *The Rigid-rotor Model*

The rigid-rotor model was formulated as an algebraic model by Ui [46] with a Lie algebra of observables that contains a set of commuting quadrupole operators, expressed as components of an $L = 2$ spherical tensor, $\{\hat{Q}_{2\nu}, \nu = 0, \pm 1, \pm 2\}$, and angular momentum operators $\{\hat{L}_0, \hat{L}_{\pm 1} = \mp \hat{L}_\pm/\sqrt{2}\}$ with commutation relations

$$\begin{aligned} &[\hat{L}_0, \hat{L}_\pm] = \pm \hat{L}_\pm, \quad [\hat{L}_+, \hat{L}_-] = 2\hat{L}_0, \\ &[\hat{L}_\pm, \hat{Q}_{2\nu}] = \sqrt{(2 \mp \nu)(2 \pm \nu + 1)}\,\hat{Q}_{2,\nu\pm 1}, \\ &[\hat{L}_0, \hat{Q}_{2\nu}] = \nu \hat{Q}_{2\nu}, \quad [\hat{Q}_{2\mu}, \hat{Q}_{2\nu}] = 0. \end{aligned} \tag{2}$$

The quadrupole observables are elements of an Abelian Lie algebra and the angular momenta span an SO(3) Lie algebra. Together they form a so-called ROT(3) Lie algebra whose elements have expressions

$$\hat{L}_0 = \hat{L}_{23}, \quad \hat{L}_{\pm 1} = \mp\frac{1}{\sqrt{2}}(\hat{L}_{31} \pm \mathrm{i}\hat{L}_{12}), \tag{3}$$

$$\begin{aligned} \hat{Q}_{2,0} &= \frac{1}{\sqrt{2}}(2\hat{Q}_{11} - \hat{Q}_{22} - \hat{Q}_{33}), \\ \hat{Q}_{2,\pm 1} &= \sqrt{3}(\hat{Q}_{12} \pm \mathrm{i}\hat{Q}_{13}), \\ \hat{Q}_{2,\pm 2} &= \frac{\sqrt{3}}{2}(\hat{Q}_{22} - \hat{Q}_{33} \pm 2\mathrm{i}\hat{Q}_{23}), \end{aligned} \tag{4}$$

in terms of the microscopic position and momentum observables $\{\hat{x}_{ni}, \hat{p}_{ni}, i, j = 1, 2, 3\}$

$$\hat{Q}_{ij} = \sum_{n=1}^{A} \hat{x}_{ni}\hat{x}_{nj}, \quad \hbar\hat{L}_{ij} = \sum_{n=1}^{A}(\hat{x}_{ni}\hat{p}_{nj} - \hat{x}_{nj}\hat{p}_{ni}). \tag{5}$$

From a modern perspective, the construction of the unitary representations of this Lie algebra is straightforward. This is pedagogically useful because it provides simple examples of an induced representation [47] in which a trivial one-dimensional representation of an Abelian Lie algebra of commuting quadrupole operators defines the intrinsic state for a representation of a full rotor-model algebra. Thus, it serves as a prototype for the systematic construction of more general rotor models with dynamical intrinsic degrees of freedom such as those of the unified model. However, as a many-nucleon model without vibrational degrees of freedom, the rigid-rotor model is unrealistic. This is because its intrinsic states are eigenstates of the nuclear quadrupole moments which, with no vibrational fluctuations, can only be realized on a many-nucleon Hilbert space in a limit in which their intrinsic wave functions approach delta functions in their quadrupole moments. Consequently, this model is not as useful as might have been hoped. In particular, the kinetic energies of its states are infinite. However, from geometric considerations, one can infer that its moments of inertia should be those of a rigid rotor [48, 49]. Thus, whereas the rotor limit of the Bohr model has irrotational-flow moments of inertia which are too small, Ui's rotor model has rigid-body moments of inertia which are too large. It appears that Ui's rigid-rotor model is too rigid; vibrational shape fluctuations, even if small, are essential.

4.3. *The Tomonaga Approach*

The search for a microscopic version of the collective model, initiated by Tomonaga [50] in two dimensions and extended by Miyazima and Tamura [51] to three, successfully identified a microscopic harmonic-oscillator algebra of shape observables and canonical momenta. Their approach has much in common with the random-phase approximation of Bohm and Pines [52], in which the elementary excitation operators are treated in a macroscopic limit as boson operators that are derivable by equations-of-motion methods [53]. With the benefit of hindsight, the Tomonaga approach can be completed as follows.

A first essential step is the replacement of the surface monopole quadrupole shape coordinates of the Bohr model, as in Ui's model, with many-nucleon Cartesian quadrupole moments, $\{Q_{ij} = \sum_{n=1}^{A} x_{ni}x_{nj}\}$. Time derivatives and corresponding momenta are then given by

$$\begin{aligned} \dot{Q}_{ij} &= \frac{dQ_{ij}}{dt} = \sum_n (\dot{x}_{ni}x_{nj} + x_{ni}\dot{x}_{nj}), \\ P_{ij} &= M\dot{Q}_{ij} = \sum_n (p_{ni}x_{nj} + x_{ni}p_{nj}), \end{aligned} \tag{6}$$

where M is the nucleon mass. These moments and momenta are now quantized as appropriate for nucleons by replacing the x_{ni} and p_{ni} coordinates by operators $\hat{x}_{ni}$ and $\hat{p}_{ni}$ with commutation relations $[\hat{x}_{ni}, \hat{p}_{mj}] = \mathrm{i}\hbar\delta_{i,j}\delta_{m,n}$, to obtain quantum observables

$$\hat{Q}_{ij} = \sum_n \hat{x}_{ni}\hat{x}_{nj}, \quad \hat{P}_{ij} = \sum_n (\hat{p}_{ni}\hat{x}_{nj} + \hat{x}_{ni}\hat{p}_{nj}), \tag{7}$$

that satisfy the commutation relations

$$[\hat{Q}_{ij}, \hat{P}_{kl}] = \mathrm{i}\hbar(\delta_{il}\hat{Q}_{jk} + \delta_{ik}\hat{Q}_{jl} + \delta_{jl}\hat{Q}_{ik} + \delta_{jk}\hat{Q}_{il}). \tag{8}$$

Following a procedure similar to that of İnönü and Wigner [54], the monopole and quadrupole moments and momenta of a nucleus are now expressed in units of a small parameter ϵ which will be assigned a value such that $4\epsilon^2 = 1/\langle\hat{Q}_0\rangle$, where $\langle\hat{Q}_0\rangle$ is the mean value of the close-to-constant monopole moment of the nucleus in its low-energy states. In these units, the monopole/quadrupole observables are given by

$$\hat{q}_0 = \epsilon\hat{Q}_0 = \epsilon(\hat{Q}_{11} + \hat{Q}_{22} + \hat{Q}_{33}),$$

$$\hat{p}^0 = \epsilon\hat{P}^0 = \epsilon(\hat{P}_{11} + \hat{P}_{22} + \hat{P}_{33}), \tag{9}$$

$$\hat{q}_{2\nu} = \epsilon\hat{Q}_{2\nu}, \quad \hat{p}^{2\nu} = \epsilon(-1)^\nu\hat{P}_{2,-\nu}. \tag{10}$$

With the commutation relations of Eq. (8), it is then determined that

$$[\hat{q}_0, \hat{p}^0] = \mathrm{i}\hbar\,\hat{I} + O(\epsilon^2), \quad [\hat{q}_{2\mu}, \hat{p}^{2\nu}] = \mathrm{i}\hbar\,\delta^\nu_\mu + O(\epsilon), \tag{11}$$

and that the commutators $[\hat{q}_0, \hat{p}^{2\nu}]$, $[\hat{q}_{2\nu}, \hat{p}^0]$, $[\hat{q}_{2\mu}, \hat{q}_{2\nu}]$, and $[\hat{p}^{2\mu}, \hat{p}^{2\nu}]$ are of order $O(\epsilon)$. These are microscopic versions of the Heisenberg commutation relations assumed in the Bohr model.

Two significant observations follow for the Bohr model derived in this way. One is that its quadrupole moments are the infinitesimal generators of irrotational flows, consistent with Bohr's formulation of the model as a quantum fluid model. The second is that the simplest shell-model representation of the model in its spherical vibrational limit is given for a closed harmonic-oscillator shell-model nucleus. In such a representation, the one-phonon monopole excitation is at a high excitation energy as expected for almost incompressible nuclear matter and the quadrupole excitations are high-energy giant-quadrupole resonance excitations. It is also apparent that, without additional degrees of freedom as introduced in the unified model, the microscopic Bohr model cannot provide a description of low-energy rotational states.

4.4. *A Microscopic Bohr Model with Vorticity*

Observe that the commutation relations for the observables $\{\hat{Q}_{ij}, \hat{P}_{ij}, i, j = 1, 2, 3\}$ of the above microscopic version of the

Bohr model, only close to form the Bohr model algebra, given by Eq. (11), in a constant volume limit of the nucleus. This is no doubt a reasonable approximation in practice. However, it is not necessary because the commutation relations of the three antisymmetric angular-momentum operators, defined in Eq. (5), together with the six symmetric $\{\hat{P}_{ij}\}$ operators close on the Lie algebra of a nine-dimensional general linear algebra. Moreover, the operators of the set $\{\hat{Q}_{ij}, \hat{P}_{ij}, \hat{L}_{ij}, i, j = 1, 2, 3\}$ are the infinitesimal generators of a so-called GCM(3) Lie group, where GCM(3) is a mnemonic for the dynamical group of a general collective motion model in three-dimensional space. This model is a minor extension of a CM(3) model introduced by Weaver *et al.* [55] to include the giant monopole degrees of freedom and which, as a result, fits more naturally within the chain of models under discussion.

Construction of the irreducible representations of the GCM(3) model [56, 57] shows that the model augments the irrotational-flow representations of the Bohr model by including an intrinsic angular momentum $\mathscr{V}$ corresponding to quantized vorticity which can take any non-negative integer value. The $\mathscr{V} = 0$ representation has states that are in one-to-one correspondence with those of the Bohr model. However, they are obtained in the CM(3) and GCM(3) representations without the need for an incompressibility approximation. The new representations with $\mathscr{V} \neq 0$ have spectra of angular momentum states that are the same as that of the unique irrep of an augmented Bohr model with an added intrinsic $\mathrm{SU}(2)_S$ spin $S = \mathscr{V}$.

A kinetic energy operator for the GCM(3) model is derived [49] in parallel with the way it was derived for the Bohr model as proportional to the Laplacian on the space generated by the momentum operators, now given by the operators $\{\hat{L}_{ij}, \hat{P}_{ij}\}$. For the special case of zero vorticity, this returns irrotational-flow moments of inertial for the model's rotations as for the Bohr model. Expressions were also obtained for representations with non-zero vorticity [49, 58–60].

4.5. *The Symplectic Model*

In spite of the successes of the CM(3) and GCM(3) models in giving a microscopic extension of the Bohr model that includes vorticity degrees of freedom, it did not solve the problem of determining the many-nucleon states corresponding to collective model descriptions of nuclear properties. One obstacle was that only for a zero-vorticity representation were basis states easily expressed in conventional shell-model terms. Another was a difficulty in seeing how the addition of vorticity to states that were realized as giant-resonance vibrational states, as in the irrotational-flow Bohr model, could result in the emergence of low-energy rotational bands of states. A third problem was the finding that vorticity is not conserved by the many-nucleon kinetic energy [49].

Looking back, with the evolutionary steps laid out as above in the development of a microscopic version of the collective model, the next step appears to be obvious. Having identified algebraic models with subsets of the $\{\hat{Q}_{ij}, \hat{L}_{ij}, \hat{P}_{ij}\}$ observables, it is natural to extend them to a model that includes all these observables plus the many-nucleon kinetic energy $\sum_{n=1}^{A} \hat{\mathbf{p}}_n^2/2M$, where M is the mass of a nucleon. The Lie algebra that emerges then contains all symmetric bilinear combinations of the nucleon position and momentum coordinates. It is also the smallest Lie algebra that contains the nuclear quadrupole moments and the nuclear kinetic energy, both of which are essential components of a complete microscopic model of nuclear collective states. It is the Lie algebra of the symplectic group $\mathrm{Sp}(3,\mathbb{R})$, which is the group of all real linear transformations of a nucleon's six position and momentum coordinates that preserve the Heisenberg commutation relations,

$$[\hat{x}_{ni}, \hat{p}_{nj}] = \mathrm{i}\hbar\delta_{i,j}, \quad i,j = 1,2,3. \tag{12}$$

The symplectic model was then defined [61,62] as an extension of the GCM(3) model with an $\mathrm{Sp}(3,\mathbb{R})$ dynamical group and a set of basic observables given by the infinitesimal generators $\{\hat{Q}_{ij}, \hat{L}_{ij}, \hat{P}_{ij}, \hat{K}_{ij}\}$

of this Lie group, where

$$\hat{K}_{ij} = \sum_{n=1}^{A} \hat{p}_{ni}\hat{p}_{nj}. \tag{13}$$

The $\mathrm{Sp}(3,\mathbb{R})$ Lie algebra containing these observables is a classical semi-simple Lie algebra with well-known representations on many-nucleon Hilbert spaces. The corresponding Lie group contains the chains of dynamical subgroups

$$\mathrm{Sp}(3,\mathbb{R}) \supset \mathrm{GCM}(3) \supset \mathrm{ROT}(3) \supset \mathrm{SO}(3) \tag{14}$$

$$\mathrm{Sp}(3,\mathbb{R}) \supset \mathrm{U}(3) \supset \mathrm{SU}(3) \supset \mathrm{SO}(3) \tag{15}$$

for each of the models considered above. The first of these chains relates specifically to the collective models and the second gives its expression as a multi-shell decomposition in terms of $\mathrm{U}(3) \supset \mathrm{SU}(3)$ subspaces. Thus, although the symplectic model was derived as a logical progression in the development of microscopically based algebraic models expressed in terms of physical observables, the end result was a model with a dynamical group, $\mathrm{Sp}(3,\mathbb{R})$, which is also the dynamical group of the spherical harmonic oscillator on which the shell model is based. As a consequence, it also contains as a subgroup, the symmetry group U(3) of the spherical harmonic oscillator.

An alternative route to the symplectic model follows the developments by Filippov, Vanagas and their many colleagues [63–65] of a translationally invariant shell model for which $\mathrm{O}(A-1)$, where A is the nuclear mass number, is a symmetry group. The symplectic model is obtained in the framework of the translationally invariant shell model from the observation that $\mathrm{Sp}(3,\mathbb{R})$ is a complementary dual of the $\mathrm{O}(A-1)$ group as defined by Moshinsky and Quesne [66] (see also Ref. [67] for a review of this and other duality relations of importance in nuclear physics).

4.6. *The Symplectic Shell Model*

In addition to serving as the Lie algebra for a dynamical group, whose irreps define a submodel of many-nucleon quantum mechanics, a Lie algebra of microscopic observables also defines a coupling scheme for the many-nucleon Hilbert space and a basis for more complete mixed-representation calculations. In combination with the spin and isospin groups, the U(3) symmetry group of a spherical harmonic oscillator already defines a coupling scheme for the spherical harmonic-oscillator shell model. Moreover, the U(3)-invariant subspaces of the many-nucleon Hilbert space is partially ordered by their spherical harmonic-oscillator energies. The $\mathrm{Sp}(3,\mathbb{R})$ symplectic group, which is a dynamical group for both the spherical and anisotropic harmonic oscillators, as well as a microscopic collective model, goes further and defines an alternative coupling scheme of the many-nucleon Hilbert space by expressing it as a sum of $\mathrm{Sp}(3,\mathbb{R})$-invariant subspaces. The many-nucleon Hilbert space is thereby realized as a sum of irreducible collective model subspaces, which can similarly be ordered with respect to suitably-defined intrinsic energies as discussed in Sec. 7.

5. Calculations in Single and Mixed Symplectic Model Spaces

The many-nucleon Hilbert space is a tensor product of a center-of-mass subspace and a complementary center-of-mass-invariant subspace. Only the center-of-mass-invariant component is of interest. However, suppression of the spurious center-of-mass degrees of freedom is simple in the symplectic-model coupling scheme because the elements of the $\mathrm{Sp}(3,\mathbb{R})$ algebra are quadratic sums of single-nucleon operators. Thus, removal of the center-of-mass contribution to these operators amounts to summing over one less linear combination of the nucleon coordinates and is equivalent to reducing the effective number of nucleons in a nucleus from A to $A-1$, where A is the nuclear mass number. This minor change in the effective mass number of a nucleus effects only the zero-point energy of the nucleus. In particular, it does not change the commutation relations of the $\mathrm{Sp}(3,\mathbb{R})$ Lie algebra in any way. Neither does it change the nucleon permutation symmetry of the model's wave functions because the center-of-mass ground-state wave function is fully symmetric and has vanishing angular momentum and intrinsic spin. The center-of-mass-invariant component of the many-nucleon Hilbert space is also a product of neutron and proton subspaces, each of which has states of a range of neutron and proton spins of corresponding $\mathrm{SU}(2)_{S_n}$ and $\mathrm{SU}(2)_{S_p}$ irreps.

The symplectic model can be deployed at many levels. With the inclusion of neutron and proton spins, it can be used as a phenomenological algebraic model, like any other collective model, in the interpretation of nuclear data. However, by construction, all its observables are microscopically defined and all its representations are defined on subspaces of the many-nucleon Hilbert space. Thus, it can also be used as a coupling scheme to define a symplectic shell model as an ordered sequence of irreducible symplectic model subspaces for the many-nucleon Hilbert space. It can also be applied with different kinds of model interactions. The possibilities include: algebraic interactions that are expressed in terms of the infinitesimal generators of the model's dynamical group; schematic interactions such as pairing-plus-quadrupole operators, that are caricatures of interactions designed to emphasize particular, often competing, coupling schemes; and interactions that are the restrictions of more general shell-model interactions to the model space.

Observe also that the symplectic model and the symplectic shell model are fundamentally no-core models [68]. The many-nucleon space of a calculation, regardless of what basis is used, is necessarily a subspace of the many-nucleon Hilbert space. However, no fully occupied subspace of the many-nucleon space is excluded in the model and treated as inert.

5.1. *The Structure of the Symplectic-model Algebra*

Raising and lowering operators of a spherical harmonic oscillator, $c^\dagger_{ni}$ and c_{ni}, which satisfy boson commutation relations, are defined for a nucleon in terms of its position and momentum coordinates by

$$\hat{x}_{ni} = \frac{1}{\sqrt{2}a}(c^\dagger_{ni} + c_{ni}), \quad \hat{p}_{ni} = i\hbar\frac{a}{\sqrt{2}}(c^\dagger_{ni} - c_{ni}), \tag{16}$$

where $a = \sqrt{M\omega/\hbar}$ is the harmonic-oscillator unit of inverse length. Basic elements of the $\mathrm{Sp}(3,\mathbb{R})$ algebra then have expansions in terms of these operators given by

$$\hat{Q}_{ij} = \frac{1}{2a^2}\left(2\hat{\mathcal{Q}}_{ij} + \hat{\mathcal{A}}_{ij} + \hat{\mathcal{B}}_{ij}\right), \quad \hat{P}_{ij} = \mathrm{i}\hbar(\hat{\mathcal{A}}_{ij} - \hat{\mathcal{B}}_{ij}),$$
$$\hat{K}_{ij} = \frac{1}{2}a^2\hbar^2(2\hat{\mathcal{Q}}_{ij} - \hat{\mathcal{A}}_{ij} - \hat{\mathcal{B}}_{ij}), \quad \hat{L}_{ij} = -\mathrm{i}(\hat{\mathcal{C}}_{ij} - \hat{\mathcal{C}}_{ji}). \tag{17}$$

in which

$$\hat{\mathcal{A}}_{ij} = \hat{\mathcal{A}}_{ji} = \sum_n c^\dagger_{ni}c^\dagger_{nj}, \quad \hat{\mathcal{B}}_{ij} = \hat{\mathcal{B}}_{ji} = \sum_n c_{ni}c_{nj},$$
$$\hat{\mathcal{C}}_{ij} = \sum_n \left(c^\dagger_{ni}c_{nj} + \frac{1}{2}\delta_{i,j}\right), \quad \hat{\mathcal{Q}}_{ij} = \frac{1}{2}\left(\hat{\mathcal{C}}_{ij} + \hat{\mathcal{C}}_{ji}\right), \tag{18}$$

with n summed over the effective number of $A-1$ nucleons, thereby excluding the linear combinations that would involve the nuclear center-of-mass degrees of freedom. It is seen that the $\mathrm{Sp}(3,\mathbb{R})$ Lie algebra is a linear combination of the elements $\{\hat{L}_{ij}, \hat{\mathcal{Q}}_{ij}\}$ of a U(3) subalgebra and the giant monopole/quadrupole raising and lowering operators $\{\hat{\mathcal{A}}_{ij}\}$ and $\{\hat{\mathcal{B}}_{ij}\}$. The elements of the U(3) subalgebra commute with the spherical harmonic-oscillator Hamiltonian

$$\hat{H}_0 = \hbar\omega \sum_{ni}\left(c^\dagger_{ni}c_{ni} + \frac{1}{2}\right) \tag{19}$$

and the giant-resonance raising and lowering operators satisfy the commutation relations

$$[\hat{H}_0, \hat{\mathcal{A}}_{ij}] = 2\hbar\omega\hat{\mathcal{A}}_{ij}, \quad [\hat{H}_0, \hat{\mathcal{B}}_{ij}] = -2\hbar\omega\hat{\mathcal{B}}_{ij}. \tag{20}$$

It follows that an irreducible $\mathrm{Sp}(3,\mathbb{R})$ representation is uniquely defined by a so-called *lowest-grade* U(3) *irrep*, the states of which are annihilated by the giant-resonance lowering operators $\{\hat{\mathcal{B}}_{ij}\}$ [see Fig. 1 of chapter 4, where the lowest-grade U(3) irrep is illustrated by the lowest line and is often referred to as a symplectic "bandhead"]. The *highest-weight state* $|\sigma\rangle$ of the lowest-grade U(3) irrep[a] is then

[a]This is analogous to SU(2), where for an SU(2) irrep labeled by angular momentum J, the highest-weight state is $|J, M_J = J\rangle$, which is annihilated by the raising J_+ operator.

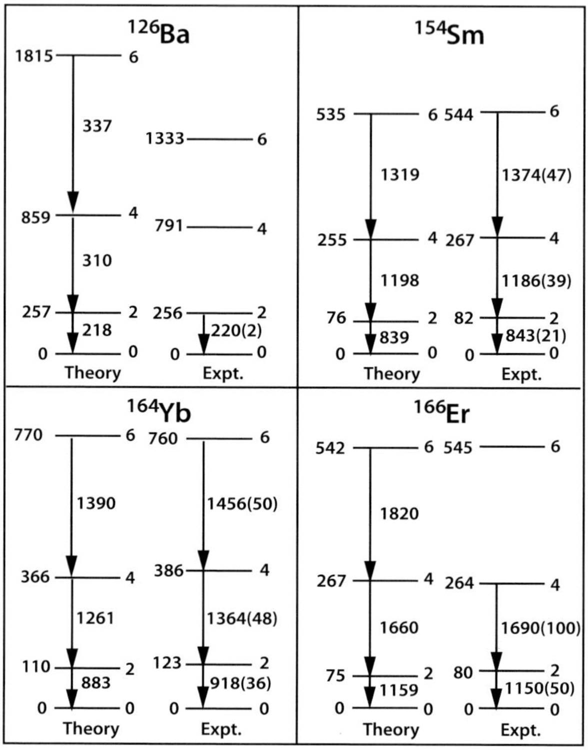

Fig. 1. The spectrum of states for some rotational nuclei in the symplectic model calculations of Park *et al.* [69]. The figure is from Ref. [41].

a state of weight $\sigma = (\sigma_1, \sigma_2, \sigma_3)$ defined by the identities

$$\hat{\mathcal{C}}_{ij}|\sigma\rangle = 0, \quad i < j, \quad \hat{\mathcal{C}}_{ii}|\sigma\rangle = \sigma_i|\sigma\rangle, \quad i = 1, 2, 3, \tag{21}$$

$$\hat{\mathcal{B}}_{ij}|\sigma\rangle = 0, \quad i, j = 1, 2, 3. \tag{22}$$

Thus, the state $|\sigma\rangle$ is said to be simultaneously the highest-weight state for an U(3) irrep denoted by $\{\sigma_1, \sigma_2, \sigma_3\}$ and the lowest-weight state for an $\mathrm{Sp}(3, \mathbb{R})$ irrep denoted by $\langle\sigma_1, \sigma_2, \sigma_3\rangle$, or equivalently by $\langle N_0(\lambda_0\, \mu_0)\rangle$, where $\lambda_0 = \sigma_1 - \sigma_2$ and $\mu_0 = \sigma_2 - \sigma_3$ are the familiar quantum numbers of the Elliott model (Sec. 4.1), and $N_0 = \sigma_1 + \sigma_2 + \sigma_3$. It also follows that the shell-model subspace for an $\mathrm{Sp}(3, \mathbb{R})$ irrep is constructed by adding to its lowest-grade

U(3) states the infinite set of multiple giant-resonance excitations generated by the $\{\hat{\mathcal{A}}_{ij}\}$ raising operators (cf. Fig. 1 of chapter 4).

5.2. Calculations with Algebraic and Schematic Interactions

Early symplectic model and submodel calculations were reviewed in Ref. [41]. They were restricted to representations with $S_n = S_p = 0$ and most of them were for light nuclei, with the exception of those of Park *et al.* [69], which were for the ground-state rotational bands of ^{126}Ba, ^{154}Sm, ^{164}Yb, and ^{166}Er. They were also restricted to single symplectic model representations, with the exception of those of Draayer *et al.* [70], which included multiple representations coupled by pairing and spin-orbit interactions.

The calculations of Park *et al.* [69] showed that rotational spectra and $E2$ transition rates could be obtained for heavy deformed nuclei with a microscopic Hamiltonian

$$\hat{H} = \hat{H}_0 + \alpha\hat{Q}_2 \cdot \hat{Q}_2 + \beta\left[\hat{Q}_2 \times \hat{Q}_2 \times \hat{Q}_2\right]_0, \tag{23}$$

where $\hat{H}_0$ is an independent-particle spherical harmonic-oscillator Hamiltonian and $\hat{Q}_2$ is the nuclear quadrupole tensor. The potential of this Hamiltonian contains two parameters: their ratio determines the deformation at which the potential energy is a minimum and their magnitude determine the energy level spacings as shown in Fig. 1. Their close fits to the properties of rare-earth nuclei reflect the fact that, like the experimental spectra, the results of the symplectic model with this interaction are those of a rotor. Thus, they identified the kinds of shell-model configurations needed to describe the rotational states of well-deformed nuclei. For example, Fig. 2 shows the decomposition of the ground state calculated for ^{166}Er as a sum of states labeled by their spherical harmonic-oscillator energies. The maximally deformed states of ^{166}Er in the valence-shell space of the spherical shell model belong to an SU(3) irrep $(\lambda\,\mu) = (28\,16)$ of spherical harmonic-oscillator energy $801\hbar\omega$. However, the measured quadruple moments and $E2$ transitions between the low-energy states of ^{166}Er are much too large to belong to the symplectic model irrep

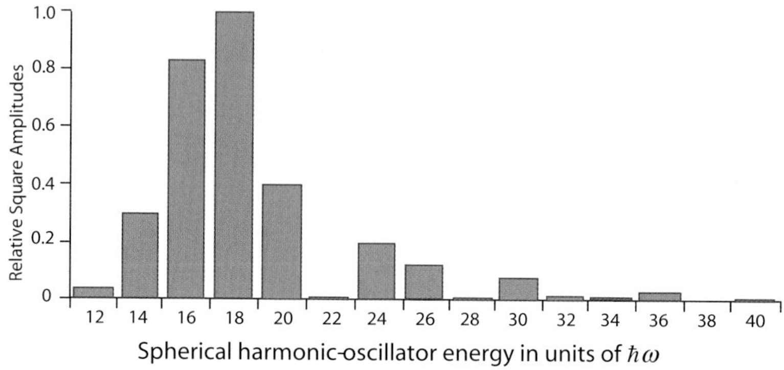

Fig. 2. Squared amplitudes of the spherical harmonic-oscillator components of the ^{166}Er ground-state wave function in the symplectic model calculations of Park *et al.* [69]. The amplitudes are expressed as functions of their spherical harmonic-oscillator energies relative to those of the valence-shell states for this nucleus (set to $0\hbar\omega$).

based on these lowest-grade states. A more appropriate irrep was determined to be one with lowest-grade states that belong to an SU(3) irrep (108 0). The lowest spherical harmonic-oscillator energy at which such an SU(3) irrep can be found is at $813\hbar\omega$. Thus, the lowest-grade U(3) states of Park's calculation were at a spherical harmonic-oscillator energy of $12\hbar\omega$ above those of the standard shell-model states.The other contributing states, shown in Fig. 2, were of spherical harmonic-oscillator energies ranging from $12\hbar\omega$ to $\sim 36\hbar\omega$, and peaking at $\sim 18\hbar\omega$, above those of the spherical shell model.

The close relationship of the microscopically based symplectic model to a phenomenological rotor model, was demonstrated even more clearly in the calculations of Bahri *et al.* [71] for a Hamiltonian based on the Davidson potential

$$\hat{H} = \hat{H}_0 + \chi\left(\hat{Q}_2 \cdot \hat{Q}_2 + \frac{\epsilon}{\hat{Q}_2 \cdot \hat{Q}_2}\right). \tag{24}$$

Its application to the ground-state rotational band of ^{166}Er included all states of angular momentum $L \leq 18$. The notable characteristic of Bahri's results, shown in comparison with corresponding rigid-rotor-model results in Fig. 3, is that they agree to a much higher degree of accuracy with those of a rigid rotor than with the

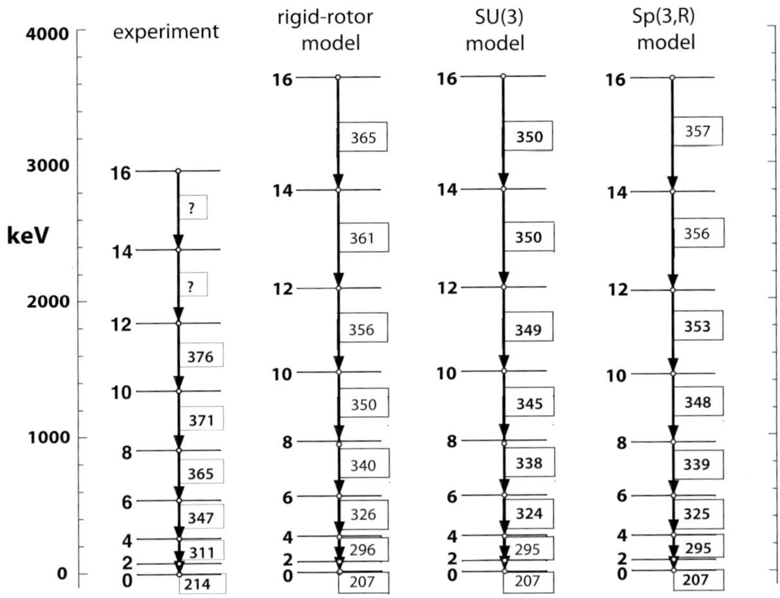

Fig. 3. Experimental energy levels and $E2$ transition rates for ^{166}Er and their interpretation in terms of a rigid-rotor model, an effective SU(3) model with $(\lambda\,\mu) = (78\,0)$, and the symplectic model with $\langle N_0(\lambda_0\,\mu_0)\rangle = \langle 826.5(78\,0)\rangle$. The figure is adapted from Ref. [71].

experimentally observed states, which show signs of centrifugal stretching with increasing angular momentum. This implies that the Davidson potential has a too sharply-defined minimum, which suppresses its vibrational fluctuations and centrifugal stretching effects. Note that in neither the rigid-rotor nor the symplectic model calculations was an effective charge needed to adjust the predicted $B(E2)$ transition rates to agree with the measured values.The values $\lambda_0 = 78$ and $\mu_0 = 0$ of the $\mathrm{Sp}(3,\mathbb{R})$ irrep were chosen to obtain an axially symmetric rotor band of the observed deformation without an effective charge. Similarly, as in the calculations of Park *et al.*, the two parameters of the potential-energy component of the Hamiltonian were adjusted to get the deformation of the nucleus, as determined by its $E2$ transition rates, and the observed moments of inertia.

Figure 3 also shows a fit to the rigid-rotor spectrum given by the SU(3) model defined by the lowest-grade SU(3) component of the symplectic model, *albeit* with use of an effective charge of 2.04 times the mean nucleon charge to obtain the observed $E2$ transition rates. The remarkably close agreement between the rigid-rotor and SU(3) results (with an effective charge) and the results of the more complete symplectic model (without an effective charge) is an illustration of the quasi-dynamical symmetries of these models [28, 29] in describing the realistic rotations of a nucleus with small vibrations about its equilibrium deformation. It is easy to see how negligible vibrational fluctuations result in a suppression of centrifugal effects. This explains the good fit obtained with the rotor model. However, the reason for the success of the SU(3) model as an effective rotor model, for large values of its quantum numbers, is much less obvious. It is particularly remarkable, when it is seen how little the lowest-grade SU(3) component of the symplectic model states, shown in Fig. 4 as the 0(78 0) component, contributes to the symplectic model states. The figure shows the characteristic property of a quasi-dynamical symmetry in which its states of different angular momentum have almost equal components in a large distribution of SU(3) irreps and are such that they are almost indistinguishable from those given by the average over the contributing irreps. Such results express the Elliott model [42, 43] as an effective model of nuclear rotations with SU(3) as a quasi-dynamical symmetry; see Refs. [29, 72] for other examples of quasi-dynamical symmetry.

The recent interest in shape coexistence in nuclei [25, 26, 73] focuses on the observation that nuclear states of very different deformation frequently occur side by side in a single nucleus. One of the first examples is the fall of states of $4\hbar\omega$ and higher spherical harmonic-oscillator energy to become rotational states based on the first-excited state of ^{16}O at 6.04 MeV; see Refs. [74–78] and references therein. This phenomenon, was shown to emerge in a calculation by Dreyfuss *et al.* [79] of the sequences of rotational states built on the lowest three 0^+ states of ^{12}C with a Hamiltonian

$$\hat{H} = \hat{H}_0 + \frac{1}{2}\chi[e^{-\gamma(\hat{Q}_2\cdot\hat{Q}_2 - \langle\hat{Q}_2\cdot\hat{Q}_2\rangle_n)} - 1]/\gamma - \kappa\sum_i l_i\cdot s_i, \qquad (25)$$

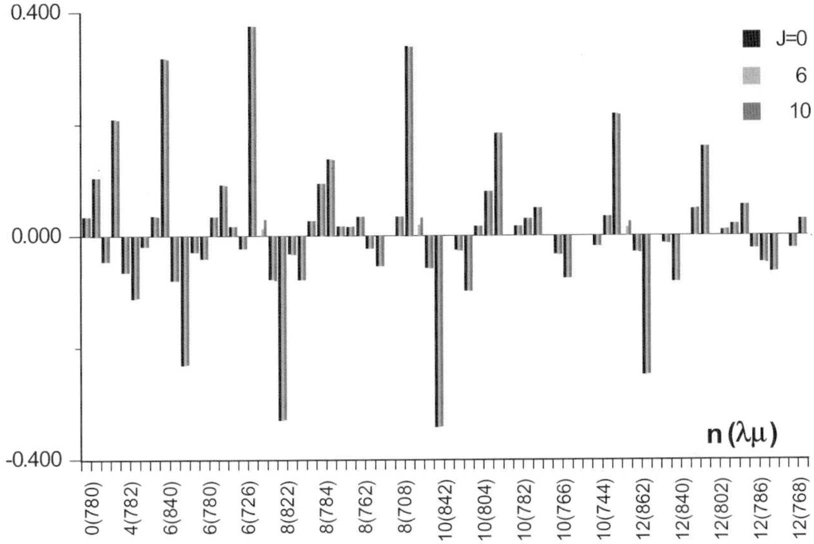

Fig. 4. Amplitudes for the wave functions of the lowest $J = 0$, 6, and 10, $\mathrm{Sp}(3, \mathbb{R})$ model states of ^{166}Er in an U(3) basis. The calculation includes all basis states of the irreducible $\mathrm{Sp}(3, \mathbb{R})$ representation $\langle 826.5(78\,0)\rangle$ of spherical harmonic-oscillator energy $\leq 12\hbar\omega$ above that of its lowest-grade states. The results are insensitive to the precise value of N_0. However, the lowest-grade SU(3) irrep $(78\,0)$, which is required from mean-field consistency conditions to give the observed deformation in the SU(3) model with an effective charge of $2.0e$, first occurs in a harmonic-oscillator shell at $\sim 10\hbar\omega$ above the lowest available spherical harmonic-oscillator shells. The figure is adapted from Ref. [71].

where $\langle \hat{Q}_2 \cdot \hat{Q}_2 \rangle_n$ is the average expectation value of the operator $\hat{Q}_2 \cdot \hat{Q}_2$ within the space of spherical harmonic-oscillator states of excitation energy $n\hbar\omega$.[b] Their ground-state band was calculated in the space of two mixed symplectic irreps coupled by a spin-orbit interaction: one based on a lowest-grade $0\hbar\omega$ SU(3) irrep $(0\,4)$ of intrinsic spin $S = 0$ and a second based on a lowest-grade $0\hbar\omega$ SU(3) irrep $(1\,2)$ of intrinsic spin $S = 1$. Rotational states were also calculated for two other symplectic model irreps: one built on a $2\hbar\omega$ lowest-grade SU(3) irrep $(6\,2)$ of spin $S = 0$, and one built on a

[b]The rationale for this Hamiltonian and its parameters is given in a publication by Tobin *et al.* [80], who extended the calculations of Dreyfuss *et al.* to $2s1d$-shell nuclei.

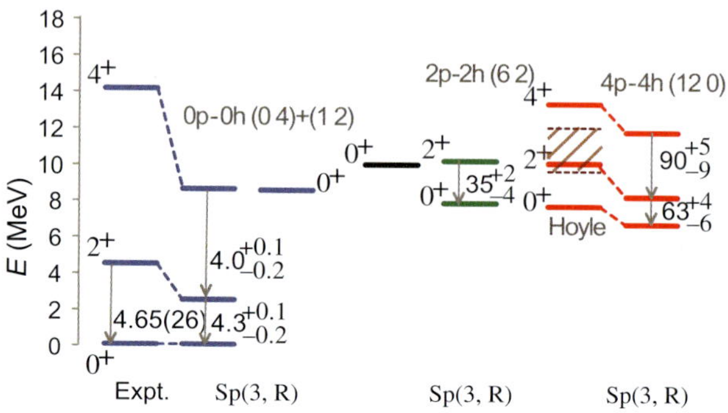

Fig. 5. The spectrum of ^{12}C states obtained in the calculations of Dreyfuss *et al.* [79]. Particularly notable is the appearance in the low-energy domain of the highly deformed Hoyle state and rotational states built on it which are of minimum spherical harmonic-oscillator energy $4\hbar\omega$. The figure is reproduced with permission, from Ref. [79].

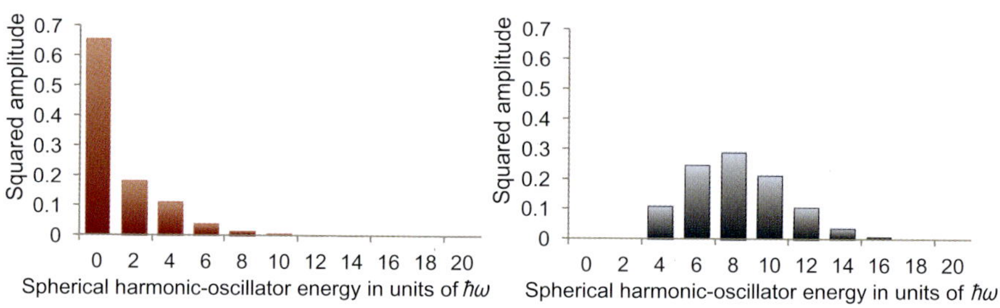

(a) Ground state (b) Hoyle state

Fig. 6. Squared amplitudes of the wave functions of the ground and excited 0^+ (Hoyle) state of ^{12}C as functions of their spherical harmonic-oscillator excitation energies calculated by Dreyfuss *et al.* [79]. The figure was provided by K. Launey.

$4\hbar\omega$ lowest-grade SU(3) irrep (12 0) of spin $S = 0$. The 0^+ state of the latter irrep is identified with the so-called Hoyle state, which is known to be the ground state of a strongly-deformed rotational band and is of major importance in the theory of nucleosynthesis [81]. The spectrum of states obtained are shown in Fig. 5.

The decomposition of the calculated Hoyle state in terms of spherical harmonic-oscillator states, shown in Fig. 6(b) is similar to that for the rare-earth rotational state in Figs. 2 and 4. Both figures show the lowest-grade U(3) states of the symplectic irrep to be small components of the total. This is in contrast to that shown for the ^{12}C ground state in Fig. 6(a). Another significant observation is that the Hoyle state is observed and calculated to be at an energy $\sim 8\,$MeV, which is very much less than the $4\hbar\omega$ spherical harmonic-oscillator energies of the (12 0) configuration on which it is based.

5.3. *Calculations with Realistic Interactions in an U(3)-coupled Spherical Shell-model Basis*

The standard shell-model approach to the many-nucleon description of low-energy states of a nucleus starts with a decomposition of the infinite-dimensional Hilbert space into an ordered sequence of subspaces. This makes it possible to diagonalize a nuclear Hamiltonian in a space spanned by a selection of some number of the leading subspaces in the decomposition.

In the Mayer–Jensen-based shell model, such a decomposition is obtained by ordering in increasing energy the eigenstates of an independent-particle model Hamiltonian with an adjusted spherical harmonic-oscillator potential and a spin-orbit interaction. The parameters of the model are chosen such that the sequential filling of the independent-particle states of the model gives nuclear ground states of the measured radii and reproduces the magic numbers of so-called closed shell nuclei. It is then presumed that the low-energy eigenstates of a realistic nuclear Hamiltonian have sufficiently large components in a finite shell-model space, consisting of some number of leading subspaces of the shell-model decomposition, that they can be described to an acceptable level of approximation with an effective Hamiltonian restricted to such a space.

The present-day availability of powerful computer resources and the facility to carry out shell-model calculations in huge spaces in uncoupled so-called M-scheme bases [82, 83], make it possible to examine the extent to which it is realistic to treat nucleons as elementary particles with effective interactions derived from high-energy physics and from few-nucleon scattering and bound-state data [84–87]. The no-core shell-model (NCSM) calculations [88–90] that followed are outstanding developments in substantiating the many-nucleon foundations of nuclear physics. However, they do not replace the need for calculations aimed at exposing the physical properties of nuclear states. As pointed out by Whitehead [82], a limitation of the M scheme is that "by replacing virtually everything by elementary computer operations the calculations simply produce numbers at the end and leave little room for gaining physical insight along the way". Thus, a need to understand the physical content of the results that emerge from such shell-model calculations remains. This motivated the construction of symmetry-based schemes for calculating nuclear properties and thereby interpreting the results of the NCSM calculations. Other motivations were to extend the range of *ab initio* theory to heavier nuclei and to examine in large multi-shell studies the validity of simple model interpretations of nuclear phenomena. These challenges were taken up in the symmetry-adapted NCSM (SA-NCSM) calculations of the LSU team [91–95] for light nuclei, e.g., ^{6}Li, ^{8}Be, ^{12}C and ^{16}O, in an $\mathrm{U}(3) \times \mathrm{U}(2)_{S_n} \times \mathrm{U}(2)_{S_p}$ coupling scheme, where $\mathrm{U}(2)_{S_n}$ and $\mathrm{U}(2)_{S_p}$ are the neutron and proton intrinsic spin groups. The most significant result to emerge from the calculations is the simplicity of their low-energy states when expressed in this coupling scheme in contrast to their unrecognizably complicated expression in an M scheme. Specifically, the many $\mathrm{U}(3) \times \mathrm{U}(2)_{S_n} \times \mathrm{U}(2)_{S_p}$ representations that contribute to the eigenstates were found to be consistent with them being predominantly those of a single $\mathrm{Sp}(3,\mathbb{R})$ irrep coupled to intrinsic spin states of complementary symmetry. A sample result of such a large multi-shell calculation [95] is shown for the ground state of ^{8}Be with the so-called N^{3}LO interaction of Entem and Machleidt [84] in Fig. 7 (see also Chapter 8).

5.4. *Application of the Symplectic Model to Heavier Nuclei*

The emergence of states at low excitation energies that are spread over standard shell-model basis states of high spherical harmonic-oscillator energies is of particular relevance for understanding the structure of rotational states in nuclei. The

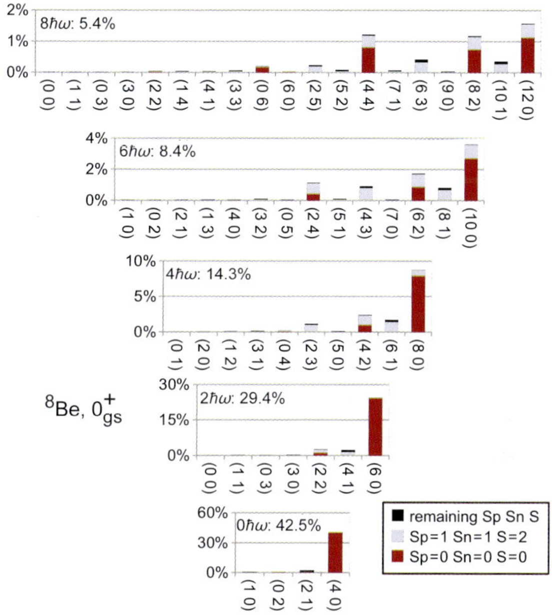

Fig. 7. Probability distributions for the states of proton and neutron spins (S_pS_nS) of $(\lambda\,\mu)$ SU(3) irreps of spherical harmonic energy ranging from the minimum value $N_0\hbar\omega$ available for ^{8}Be up to $(N_0+8)\hbar\omega$ for the so-called chiral N^{3}LO interaction [84]. The total probability for each $n\hbar\omega$ subspace is given in the upper left-hand corner of each histogram. The figure shows the dominance of a few $\mathrm{U}(3)\times\mathrm{U}(2)_{S_n}\times\mathrm{U}(2)_{S_p}$ representations at each level. Moreover the U(3) irreps that contribute are consistent with being almost entirely those of a single $\mathrm{Sp}(3,\mathbb{R})$ irrep. The figure was adapted from Ref. [95] by Launey.

appearance of states with rotational properties and deformations much larger than those of neighboring states that exist within the leading shell-model spaces is known as shape coexistence and is observed to be widespread [26, 73]. From the perspective of shell-model calculations, the appearance of such states in multi-shell calculations brings to light a concern of fundamental importance.

Given sufficient computer resources, it is in principle possible, to diagonalize a Hamiltonian repeatedly in a sequence of calculations in which the size of the active space is increased a step at a time until some lowest-energy states of the calculations and their properties cease to change significantly with increasing size of the space [90].

It would then be reasonable to suppose, as far as these lowest-energy states were concerned, that the calculations had converged to some level of accuracy and that the states were, in fact, eigenstates of the Hamiltonian to a corresponding level of accuracy. However, one would not be justified in assuming that the low-energy states obtained in this way were in fact, the lowest-energy states of the system for the given Hamiltonian. This is understood in the many-body theory of phase transitions. If one does calculations in a basis of states defined by a model of a system with a particular dynamical symmetry, it is unlikely that one will obtain states of the system with a very different symmetry. They are more likely to be spread thinly over a wide range of high-energy states in the chosen basis. Thus, the observation of such low-energy states of a system with properties very different from those expected from a standard model perspective are appropriately described as *emergent phenomena*.

It is now clear that examples of such emergent phenomena are given by observations of shape coexistence and strongly-deformed rotational states in many relatively light nuclei [26, 73]. In fact, although rotational states of modest deformation do appear in the spherical shell model and are understood in terms of a multi-shell U(3) coupling scheme, these states often have negligible overlaps with the far more deformed states that can fall into the lower energy region but have their origins in quite high-lying spherical shell-model spaces. As discussed in Sec. 5.2, it appears [69, 71, 96, 97] that the spherical shell-model subspaces that contribute to the low-lying rotational states of strongly-deformed rare-earth nuclei are distributed over independent-particle spherical harmonic-oscillator spaces of energies ranging from $\sim 10\hbar\omega$ to $\sim 30\hbar\omega$ or more above the energies of the leading shell-model subspaces. Thus, for such strongly deformed rotational states, the standard spherical shell-model approach to their description is wildly inadequate.

6. The Symplectic Model as a Unified Model

From a mean-field perspective, the symplectic model emerges as a microscopic unified model, with a structure similar to that of the Bohr–Mottelson–Nilsson model [12, 13, 98]. In its mean-field treatment, the symplectic model is mapped to a classical representation [99, 100] for which a unitary representation is its quantization. This perspective exposes the physical content of the symplectic model and provides practical methods for its applications. It is particularly useful because nuclear properties have a long history of being interpreted in unified-model terms.

The mean-field perspective makes use of a fundamental relationship between Hartree–Fock (HF) theory and many-nucleon quantum mechanics. HF theory is defined on a continuous smooth manifold of Slater determinants whereas many-nucleon quantum mechanics is defined on the Hilbert space spanned by these determinants. Both spaces can be generated from a single so-called particle-hole vacuum state, which can be considered as a lowest-weight state for an irrep of the infinite-dimensional Lie algebra of 1-body Hermitian operators. The HF manifold is generated by the Lie group of 1-body unitary transformations of the lowest-weight state, whereas the Hilbert space of the nucleus is an infinite-dimensional vector space spanned by the set of all Slater determinants generated by the repeated action of the one-body raising operators on the particle-hole vacuum state.

In mathematical terminology, the HF manifold of Slater determinants is said to be a coadjoint orbit (see also Chapter 10). It has the property that it spans the Hilbert space of many-nucleon quantum mechanics, in parallel with the way the points on a sphere span a three-dimensional vector space. However, it is also a curved phase space with classical properties. Thus, time-dependent Hartree–Fock (TDHF) dynamics obey classical Lagrangian equations and Hamiltonian equations of motion, whereas quantum dynamics obeys the Schrödinger equation and Heisenberg equations of motion. Thus, many-nucleon quantum mechanics is a quantization of the classical HF dynamics.

The dynamics of HF theory as constrained quantum mechanics, was derived in geometric terms in Ref. [99] and other references therein. Simple introductions are given in Ref. [33] and in Chapter 8 of Ref. [101]. The application of TDHF theory to the low-lying rotor-vibrator states of deformed nuclei has recently been given by Matsuyanagi *et al.* [102].

6.1. *The Mean-field Manifold of Lowest-weight States for a Symplectic-model Representation*

A lowest-weight state $|\sigma\rangle$ for a symplectic model irrep, defined by Eqs. (21) and (22), is an eigenstate of an independent-particle Hamiltonian and has the necessary properties of a vacuum state for the construction of both an irreducible unitary representation of $\mathrm{Sp}(3,\mathbb{R})$ and for a mean-field representation of the symplectic-model dynamics. However, a given symplectic model irrep $\langle\sigma\rangle$ has many lowest-weight states. Thus, we now label a particular choice of lowest weight with an added index $|\sigma,\omega\rangle$, to signify that it is defined for a spherical harmonic-oscillator of frequency ω. A manifold of mean-field states for a symplectic model irrep is then a set of lowest-weight states of a symplectic model representation defined as follows.

The standard lowest-weight state $|\sigma,\omega\rangle$ for an irreducible $\mathrm{Sp}(3,\mathbb{R})$ representation $\langle\sigma\rangle \equiv \langle\sigma_1,\sigma_2,\sigma_3\rangle$ is defined in a spherical harmonic-oscillator basis as the highest-weight state of a set of lowest-grade U(3) states of the representation. It is an eigenstate $|\sigma,\omega\rangle$ of the independent-particle Hamiltonian

$$\hat{h}(\omega) = \sum_n \left[\frac{\hat{\mathbf{p}}_n^2}{2M} + \frac{1}{2}M\omega^2\left(\hat{x}_{n1}^2 + \hat{x}_{n2}^2 + \hat{x}_{n3}^2\right)\right] \tag{26}$$

of energy

$$E_\sigma(\omega) = (\sigma_1+\sigma_2+\sigma_3)\hbar\omega, \tag{27}$$

where $\hat{x}_{ni}$ is the multiplicative operator defined such that $\hat{x}_{ni}\psi(x) = x_{ni}\psi(x)$. Other lowest-weight states $\{|\sigma,\omega_d\rangle\}$ for an $\mathrm{Sp}(3,\mathbb{R})$ irrep $\langle\sigma\rangle$ are given by the eigenstates of the generally anisotropic harmonic-oscillator Hamiltonians

$$\hat{h}(\omega_d) = \sum_n \left[\frac{\hat{\mathbf{p}}_n^2}{2M} + \frac{1}{2}M\left(\omega_1^2\hat{x}_{n1}^2 + \omega_2^2\hat{x}_{n2}^2 + \omega_3^2\hat{x}_{n3}^2\right)\right] \tag{28}$$

with energy eigenvalues

$$E_\sigma(\omega_d) = \sigma_1\hbar\omega_1 + \sigma_2\hbar\omega_2 + \sigma_3\hbar\omega_3, \tag{29}$$

where ω_d now denotes the triple of frequencies $\{\omega_1, \omega_2, \omega_3\}$.

Such an eigenstate $|\sigma, \omega_d\rangle$ continues to satisfy equations [cf. Eqs. (21) and (22)]

$$\hat{\mathcal{C}}_{ij}(d)|\sigma,\omega_d\rangle = 0, \quad i<j, \quad \hat{\mathcal{C}}_{ii}(d)|\sigma,\omega_d\rangle = \sigma_i|\sigma,\omega_d\rangle, \quad i = 1,2,3, \tag{30}$$

$$\hat{\mathcal{B}}_{ij}(d)|\sigma,\omega_d\rangle = 0, \quad i,j = 1,2,3, \tag{31}$$

with $\hat{\mathcal{C}}_{ij}(d)$ and $\hat{\mathcal{B}}_{ij}(d)$ defined as in Eq. (18) but now in terms of the raising and lowering operators, $\{c^\dagger_{ni}, c_{ni}\}$, of the generally anisotropic harmonic oscillator related to the nucleon position and momentum coordinates by

$$\hat{x}_{ni} = \sqrt{\frac{\hbar}{2M\omega_i}}(c^\dagger_{ni} + c_{ni}), \quad \hat{p}_{ni} = \mathrm{i}\hbar\sqrt{\frac{M\omega_i}{2\hbar}}(c^\dagger_{ni} - c_{ni}). \tag{32}$$

Thus, it is a deformed lowest-weight state for the same $\mathrm{Sp}(3,\mathbb{R})$ irrep and a highest-weight state for a (now deformed) $\mathrm{U}(3)_d \subset \mathrm{Sp}(3,\mathbb{R})$ subgroup.

6.2. *A Microscopic Unified Model*

An intrinsic state for a microscopic version of the Bohr–Mottelson unified model can now be defined, in a variation of HF theory, as the minimal-energy lowest-weight state of a symplectic model irrep. Its rotational states are then expressed as linear combinations of the states given by rotations of its intrinsic state, and its vibrational states are given by generalized TDHF methods. This is outlined in Ref. [103] and will be presented in more depth in a forthcoming book [104]. For present purposes, only the rotational states of the microscopic unified model are considered.

It is known [105–111] and shown explicitly in Ref. [112] that, in a suitably-defined basis for an SU(3) irrep, the matrix elements of the quadrupole operators rapidly approach those of a rigid-rotor model as the dimension of the irrep becomes large. Moreover, in this limit, they relate linearly to the quadrupole moments of the highest-weight state for this SU(3) irrep in the same way that those of a rotor model relate to its intrinsic state. Thus, because the minimum-energy lowest-weight state of a symplectic model irrep is the intrinsic state of a more realistic rotor model, it is expected that angular momentum states projected from it will resemble those of the corresponding more-deformed rotor even more closely.

For a Hamiltonian $\hat{H}$ with reasonably short-range nuclear interactions, mean-field self-consistency implies that, for the minimum-energy solution, $|\sigma, \omega_d\rangle$, ellipsoids of constant potential of the anisotropic harmonic-oscillator Hamiltonian $\hat{h}(\omega_d)$ of Eq. (28) will be of the same shape, to a high level of accuracy, as the ellipsoid

$$\frac{x_1^2}{\langle\sigma, \omega_d|\hat{x}_1^2|\sigma, \omega_d\rangle} + \frac{x_2^2}{\langle\sigma, \omega_d|\hat{x}_2^2|\sigma, \omega_d\rangle} + \frac{x_3^2}{\langle\sigma, \omega_d|\hat{x}_3^2|\sigma, \omega_d\rangle} = 1, \tag{33}$$

defined by the mean-square values of the $\{x_i\}$ coordinates. Imposing this condition corresponds [103], not surprisingly, to an equipartition of the harmonic-oscillator energy relative to the three axes of the ellipsoids and is such that

$$\sigma_1\omega_1 = \sigma_2\omega_2 = \sigma_3\omega_3 = k = (\sigma_1\sigma_2\sigma_3)^{\frac{1}{3}}\omega, \tag{34}$$

for a value of

$$\omega^3 = \omega_1\omega_3\omega_3 \tag{35}$$

that determines the volume of the nucleus. Thus, for a given value of ω the shape-consistent mean-field state for the $\mathrm{Sp}(3,\mathbb{R})$ representation $\{\sigma_1, \sigma_2, \sigma_3\}$ is given by the eigenstate $|\sigma, \omega_d\rangle$ of the Hamiltonian $\hat{h}(\omega_d)$ of Eq. (28) with $\omega_i = k/\sigma_i$ and eigenvalue $3\hbar k$.

From this result, one also obtains [103, 112, 113] a simple prediction of the renormalization of quadrupole moments and $E2$ transitions probabilities resulting from the coupling of the U(3) states to the giant-resonance excitations. Quadrupole moments of the lowest-grade U(3) highest-weight state $|\sigma, \omega\rangle$, in a spherical harmonic-oscillator basis, are given in terms of $1/a^2$, where as defined above $a = \sqrt{M\omega/\hbar}$ is the inverse oscillator unit of length, by

$$\langle \hat{Q}_{2,0} \rangle_\omega = a^2 \langle (2x_1^2 - x_2^2 - x_3^2) \rangle_\omega = 2\sigma_1 - \sigma_2 - \sigma_3 = 2\lambda_0 + \mu_0, \tag{36}$$

$$\sqrt{\frac{2}{3}} \langle \hat{Q}_{2,\pm 2} \rangle_\omega = a^2 \sqrt{\frac{2}{3}} \langle (x_2^2 - x_3^2) \rangle_\omega = \sigma_2 - \sigma_3 = \mu_0, \tag{37}$$

where $\lambda_0 = \sigma_1 - \sigma_2$ and $\mu_0 = \sigma_2 - \sigma_3$ are the usual labels for an SU(3) representation. However, for a triaxial harmonic oscillator with frequencies $\omega_d = \{\omega_1, \omega_2, \omega_3\}$, there are three inverse units of length a_1, a_2, and a_3 with $a^2/a_i^2 = \omega/\omega_i$. It follows that, for the shape-consistent state $|\sigma, \omega_d\rangle$, the quadrupole moments become

$$\langle \hat{Q}_{2,0} \rangle_{\omega_d} = \frac{2\sigma_1\omega}{\omega_1} - \frac{\sigma_2\omega}{\omega_2} - \frac{\sigma_3\omega}{\omega_3} = \frac{2\sigma_1^2 - \sigma_2^2 - \sigma_3^2}{(\sigma_1\sigma_2\sigma_3)^{\frac{1}{3}}}, \tag{38}$$

$$\sqrt{\frac{2}{3}} \langle \hat{Q}_{2,\pm 2} \rangle_{\omega_d} = \frac{\sigma_2\omega}{\omega_2} - \frac{\sigma_3\omega}{\omega_3} = \frac{\sigma_2^2 - \sigma_3^2}{(\sigma_1\sigma_2\sigma_3)^{\frac{1}{3}}}. \tag{39}$$

An expansion in powers of λ_0/N_0 and μ_0/N_0, with $N_0 = \sigma_1 + \sigma_2 + \sigma_3$, then gives

$$\langle \hat{Q}_{2,0} \rangle_{\omega_d} = 2(2\lambda_0 + \mu_0) + (2\lambda_0^2 + 2\lambda_0\mu_0 - \mu_0^2)/N_0 + \cdots, \tag{40}$$

$$\sqrt{\frac{2}{3}} \langle \hat{Q}_{2,\pm 2} \rangle_{\omega_d} = 2\mu_0 - \mu_0(2\lambda_0 - \mu_0)/N_0 + \cdots, \tag{41}$$

which, to leading order in λ_0/N_0 and μ_0/N_0, imply that the effective charge of quadrupole moments and $E2$ transitions in the SU(3) model is predicted to be

$$e_{\text{eff}} = \big(2 + \text{higher order terms}\big) \times \frac{Ze}{A}, \tag{42}$$

where e the charge of a proton.

These results show that the quadrupole moments and matrix elements of the quadrupole operators for the states obtained by strong coupling of the lowest-grade U(3) states of a symplectic model representation to its giant monopole/quadrupole resonance excitations will, to a good approximation, be twice the values of those for the uncoupled lowest-grade states.

More precise results can be obtained with realistic Hamiltonians and their expectation values in the quantum-mechanical states of good angular momentum projected from minimum-energy lowest-weight states or in the more-easily determined shape-consistent lowest-weight states. Such angular-momentum projected states could be derived, for example, by the methods of Lee and Cusson [114]. However, simplifications arise, as illustrated in Ref. [115], when the states are projected from a lowest-weight state that is an eigenstate of a harmonic-oscillator Hamiltonian and expressible as a linear combination of U(3) states.

The above considerations have ignored the $\mathrm{SU}(2)_S$ spin degrees of freedom and are appropriate for the dominant lowest-energy representations of even nuclei which are expected to be of maximal space symmetry and have intrinsic spin $S_n = S_p = 0$. The $S \neq 0$ states nevertheless contribute in an essential way to the structures of nuclei, and studies within the framework of the unified model, notably by Stephens and colleagues [38, 116, 117], are indicative of ways to include non-zero spins.

7. Ordering the Subspaces of the Symplectic Shell Model

The first step in a symplectic shell-model calculation is to identify which symplectic-model representations should be included. Ideally, the representations available to a given nucleus should be listed in order with the most relevant heading the list. In principle, such an ordering could be obtained by means of calculations done within the spaces of single $\mathrm{Sp}(3,\mathbb{R}) \times \mathrm{SU}(2)_{S_n} \times \mathrm{SU}(2)_{S_p}$ representations. If there were relatively little mixing of symplectic irreps, it might also be possible to identify the relevant irreps by a combination of experimental observations and theoretical considerations.

Indications of what to expect follow from the observation that a lowest-weight state for an $\mathrm{Sp}(3,\mathbb{R})$ irrep $\langle\sigma\rangle$ has spherical harmonic-oscillator energy $N_0\hbar\omega$, where $N_0 = \sigma_1 + \sigma_2 + \sigma_3$, and that, within a space of states of spherical harmonic-oscillator energy $N_0\hbar\omega$, calculations invariably determine that the most deformed SU(3) representations lie lowest in energy. This suggests that for each value of N_0, the available $\mathrm{Sp}(3,\mathbb{R}) \times \mathrm{SU}(2)_{S_n} \times \mathrm{SU}(2)_{S_p}$ irreps should be ordered with those $(\lambda_0\,\mu_0)$ that yield the largest values of the second-order SU(3) Casimir invariant coming first; recall that this Casimir invariant for an SU(3) irrep $(\lambda_0\,\mu_0)$, with $\lambda_0 = \sigma_1 - \sigma_2$ and $\mu_0 = \sigma_2 - \sigma_3$, has value given to within a factor by

$$C_{\lambda_0\mu_0} = \lambda_0^2 + \lambda_0\mu_0 + \mu_0^2 + 3\lambda_0 + 3\mu_0. \tag{43}$$

Listing the most deformed SU(3) irreps available among the states of each $N_0\hbar\omega$ is already useful because it reveals the minimum values of N_0 for which states of the experimentally observed deformation (with an effective charge of twice the mean-nucleon charge) can be found. As the calculations reviewed in Sec. 5.2 indicate, these values of N_0 are substantially larger for well-deformed rotational nuclei than the smallest values they could have.

A simple ordering of symplectic shell-model subspaces is given [103, 113] by the eigenvalues $E_\sigma(\omega_d)$, given in Eq. (29) and denoted as E_d hereafter, of the shape-consistent lowest-weight states as eigenstates of the anisotropic independent-particle harmonic-oscillator Hamiltonians of Eq. (28). As mentioned above, these eigenvalues can also be expressed in terms of ω as $E_d = 3(\sigma_1\sigma_2\sigma_3)^{\frac{1}{3}}\hbar\omega$. Examples of such orderings are given in Table 1.

A notable result is that the highly deformed representations with large values of the Casimir invariant $C_{\lambda_0\mu_0}$ have E_d energies much below the $N_0\hbar\omega$ values they would have if the harmonic-oscillator mean field were spherical. The first table shows that for ^{12}C the lowest three Sp(3, $\mathbb{R}$) representations obtained in this way are consistent with the low-energy states observed in this nucleus and are those considered in an Sp(3, $\mathbb{R}$) model framework by Dreyfuss *et al.* [79] (Fig. 5). In particular, the highly deformed second-excited state, known as the Hoyle state, which belongs to the $\langle N_0(\lambda_0\,\mu_0)\rangle = \langle 30(12\,0)\rangle$ irrep, falls well below the unperturbed excitation energy of $4\hbar\omega$ that it would have (with approximately half the deformation) in the spherical harmonic-oscillator shell model. A similar result was found for ^{16}O, for which the lowest three representations shown in Table 1 are consistent with those observed and as understood in an

Table 1. Minimum E_d energies in order of increasing values in units of $\hbar\omega$ for irreducible $\mathrm{Sp}(3,\mathbb{R})$ representations of ^{12}C, ^{16}O, and ^{168}Er. For each minimum, the corresponding values are given for $N_0 = \sigma_1 + \sigma_2 + \sigma_3$, $\lambda_0 = \sigma_1 - \sigma_2$, $\mu_0 = \sigma_2 - \sigma_3$, and the Casimir invariant $C_{\lambda_0\mu_0}$. (The results shown in these tables are from Ref. [103].)

^{12}C					^{16}O					^{168}Er				
N_0	λ_0	μ_0	$C_{\lambda_0\mu_0}$	E_d	N_0	λ_0	μ_0	$C_{\lambda_0\mu_0}$	E_d	N_0	λ_0	μ_0	$C_{\lambda_0\mu_0}$	E_d
26	0	4	28	25.30	36	0	0	0	36.00	814	30	8	1318	812.55
30	12	0	180	25.96	40	8	4	148	37.29	826	96	20	11884	812.90
28	6	2	76	26.28	38	4	2	46	37.29	824	82	26	9856	812.99
32	14	0	238	26.89	48	24	0	648	38.10	816	40	16	2664	813.01
34	16	0	304	27.76	44	16	2	346	38.30	828	104	20	13668	813.01
36	18	0	378	28.57	42	10	4	198	38.49	822	70	28	7938	813.04
					46	18	2	424	39.25	818	52	20	4360	813.09
										820	60	26	6094	813.10
										830	114	16	15466	813.19

$\mathrm{Sp}(3,\mathbb{R})$ model framework [78]. The highly deformed first excited state of ^{16}O has long been of interest [74] and has been interpreted as predominantly an SU(3) (8 4) 4-particle–4-hole excitation [75,118, 119]. The second excited (positive parity) state was also interpreted as an SU(3) (4 2) 2-particle–2-hole excitation [120].

For a range of Erbium and Tungsten isotopes, Jarrio *et al.* [96] estimated the most appropriate $\mathrm{Sp}(3,\mathbb{R})$ irreps needed to describe the ground-state rotational bands of these nuclei from a consideration of their quadrupole moments and $E2$ transition rates. They determined the mean values of λ_0 and μ_0 for the low-energy states of ^{168}Er, for example, to be in the neighborhood of $\langle\lambda_0\rangle \approx 80$ and $\langle\mu_0\rangle \approx 12$. They also determined the mean values of these quantum numbers for Nilsson model states at the experimentally determined deformation and obtained the values $\lambda_0 \approx 95$ and $\mu_0 \approx 12$. Their results can be compared with those of Table 1.

Given that the energies in Table 1 are approximations to the mean-field energies of the shape-consistent lowest-weight states, they are remarkably successful in explaining the appearance of highly deformed rotational states in the low-energy spectra of nuclei. Especially notable is the finding that, whereas it is necessary to go to states of $\sim 12\hbar\omega$ excitation energy in a spherical harmonic-oscillator basis to obtain the lowest-weight states for an $\mathrm{Sp}(3,\mathbb{R})$ irrep with the large experimentally observed deformation, such states become the next to lowest-energy states of an anisotropic harmonic-oscillator basis. Moreover, if the shape-consistent lowest-weight state for this irrep were expanded in a spherical harmonic-oscillator basis [115], it is expected that it would extend over a large range of such basis states as determined in the calculations of Refs. [69,71] and discussed in Sec. 5.2.

Note that the results of Table 1 are not intended to be taken literally as a realistic ordering of the symplectic model subspace of the many-nucleon Hilbert space. They nevertheless show that the quadrupole component of the mean field can dramatically lower the energies of deformed states and radically change the spherical shell-model structures of strongly deformed nuclei. It is also worth noting that the separation in energy of symplectic-model irreps of different deformation will inevitably result in a substantial suppression of their mixing due to symmetry-breaking interactions.

8. Summary and Implications for Further Studies

The NCSM calculations [88–90] for light nuclei with realistic interactions in huge shell-model spaces were major developments in substantiating the many-nucleon foundations of nuclear physics. The parallel calculations in U(3)-coupled bases [91–95] are equally important in identifying the emergence of symmetry in the states calculated and in preparing the way for calculation in heavier nuclei. It is nevertheless evident that conventional shell-model calculations in symplectic bases are too computationally intensive for the routine analysis and interpretation of nuclear data. Thus, the present paper has explored practical methods for putting the symmetries and coupling schemes that have been identified to use in understanding the dynamics of rotational states of nuclei.

The perspective developed in this chapter builds on contributions from four microscopically based theories of quadrupole collective dynamics in deformed nuclei: (i) the Bohr–Mottelson–Nilsson unified model [12, 13]; (ii) HF and TDHF theory (or the equivalent random-phase approximation) as reviewed, for example, in Refs. [33, 53, 102]; (iii) the spherical shell model in an $\mathrm{U}(3) \times \mathrm{SU}(2)_{S_n} \times \mathrm{SU}(2)_{S_p}$-coupled basis [95]; and (iv) the symplectic model [41, 61, 62]. It is hoped that what is presented provides a framework for further developments.

The symplectic shell-model approach to the description of nuclear rotational states gives a decomposition of the many-nucleon Hilbert space into decoupled collective subspaces for irreducible unitary representations of, for example, the direct product groups $\mathrm{Sp}(3,\mathbb{R}) \times \mathrm{SU}(2)_{S_n} \times \mathrm{SU}(2)_{S_p}$ or, if preferred, $\mathrm{Sp}(3,\mathbb{R}) \times \mathrm{U}(4)$, where U(4) is the Wigner supermultiplet group. These decompositions have the following unique and valuable property.

Property: There can be no non-zero matrix elements of the isoscalar monopole/quadrupole operators or the nuclear kinetic energy, both of which are elements of the $\mathrm{Sp}(3,\mathbb{R})$ Lie algebra, between collective subspaces containing single $\mathrm{Sp}(3,\mathbb{R})$ irreps.

Thus, the low-energy states of these subspaces are the closest possible approaches to pure collective states and their low-energy rotational states are the closest approaches to rigid-body rotational states allowed by many-nucleon quantum mechanics. The latter is implied by the observation that the vibrational excitations of their intrinsic states occur at close to giant-resonance energies. Thus, without mixing of symplectic irreps there can be no low-energy beta and gamma vibrational states as in the Bohr–Mottelson unified model. Conversely, the observation of $E2$ transitions between a pair of physical states implies either that they belong to a common symplectic irrep or that they have components of an irrep in common. Thus, the challenge is presented of identifying the dominant symplectic model irreps to which low-lying states of heavy deformed nuclei belong and of assessing the mixing of irreps that occur.

It has been shown that, for a given nuclear Hamiltonian, the collective subspaces can be ordered by the expectation values of the energies of their minimal-energy intrinsic (lowest-weight) states. For a suitable Hamiltonian, the orderings should explain the emergence of rotational states in the low-energy spectra of nuclei with much larger deformation than the standard spherical shell model would lead one to expect but which are nevertheless observed [26, 73]. It would be insightful to explore the orderings of collective subspaces given by the expectation values in shape-consistent states of schematic Hamiltonians consisting, for example, of a kinetic energy plus $Q \cdot Q$, pairing and spin-orbit interactions with adjustable parameters. The mixing of irreps and possible fits to experimental data might also be explored in this way.

To understand the physics of actual nuclear rotational states, it is important to find out if the mixing of symplectic irreps is a relatively minor perturbation or if there are systematic processes that change the perspective of nuclear rotations and monopole/quadrupole vibrations in a fundamental way. For example, the Bohr–Mottelson unified model assumes that the intrinsic states of rotational nuclei are predominantly axially symmetric. Such axially symmetric rotational states could become more prevalent among the low-energy states of the symplectic shell model if its axially-symmetric irreps were lowered preferentially over its triaxial irreps, e.g., as a result of pairing interactions. Such interactions could also lower the energies of low-K states of triaxial irreps. They could also mix symplectic model irreps and lead to a closer approach to the rotational states of a superfluid and a reduction of the moments of inertia.

Whatever the outcome, such investigations will inevitably lead to a better understanding of the dynamics of nuclear rotations. It is expected that the mean-field approach to the symplectic model will also be useful for assessing the validity of other mean-field-based models such as cranking models and the Kumar–Baranger model [121–123].

Acknowledgements

Discussions with J. L. Wood and the help of K. Launey in the preparation of the manuscript and some of its figures are much appreciated.

References

[1] H. Geiger and E. Marsden, On a diffuse reflection of the α-particles, *Proc. Roy. Soc. A.* **82**, 495 (1909).

[2] H. Geiger, The scattering of the α-particles by matter, *Proc. Roy. Soc. A.* **83**, 492 (1910).

[3] E. Rutherford, The scattering of α and β particles by matter and the structure of the atom, *Phil. Mag.* **21**, 669 (1911).

[4] J. Chadwick, Possible existence of a neutron, *Nature.* **192**, 312 (1932).

[5] N. Bohr and F. Kalckar, On the transmutation of atomic nuclei by impact of material particles: I. General theoretical remarks, *Kgl. Danske Vid. Selsk. Skr.* **14** (10) (1937).

[6] M. G. Mayer, On closed shells in nuclei. II, *Phys. Rev.* **75**, 1969 (1949).

[7] O. Haxel, J. H. D. Jensen, and H. E. Suess, On the "Magic Numbers" in nuclear structure, *Phys. Rev.* **75**, 1766 (1949).

[8] T. Schmidt, *Naturwiss.* **28**, 565 (1940).

[9] C. H. Townes, H. M. Foley, and W. Low, Nuclear quadrupole moments and nuclear shell structure, *Phys. Rev.* **76**, 1415 (1949).

[10] J. Rainwater, Nuclear energy level argument for a spheroidal nuclear model, *Phys. Rev.* **79**, 432 (1950).

[11] A. Bohr, The coupling of nuclear surface oscillations to the motion of individual nucleons, *Mat. Fys. Medd. Dan. Vid. Selsk.* **26** (14) (1952).

[12] A. Bohr and B. R. Mottelson, Collective and individual-particle aspects of nuclear structure, *Mat. Fys. Medd. Dan. Vid. Selsk.* **27** (16) (1953).

[13] S. G. Nilsson, Binding states of individual nucleons in a strongly deformed field, *Mat. Fys. Medd. Dan. Vid. Selsk.* **29** (16) (1955).

[14] J. von Neumann, Uber Einen Satz Herrn M. H. Stone, *Ann. Math.* **33**, 567 (1932).

[15] F. Iachello and A. Arima, *The Interacting Boson Model.* Cambridge University Press, Cambridge (1987).

[16] A. Bohm, Y. Ne'eman, and A. O. Barut, (eds)., *Dynamical Groups and Spectrum Generating Algebras.* Vol. 1 & 2, World Scientific, Singapore (1988).

[17] C. E. Wulfman, *Dynamical symmetry.* World Scientific, Singapore (2011).

[18] B. R. Mottelson. In G. Racah, (ed)., *Nuclear Spectroscopy, Rendiconti Della Scuola Internazionale Di Fisica "Enrico Fermi", Corso XV*, p. 44, Zanichelli, Bologna (1962).

[19] G. Racah, Theory of complex spectra, IV, *Phys. Rev.* **76**, 1352 (1949).

[20] B. H. Flowers, Studies in jj-coupling. I. Classification of nuclear and atomic state., *Proc. Roy. Soc. A.* **212**, 248 (1952).

[21] A. R. Edmonds and B. H. Flowers, Studies in jj-coupling. II. Fractional parentage coefficients and the central force energy matrix for equivalent particles, *Proc. Roy. Soc. A.* **214**, 515 (1952).

[22] A. R. Edmonds and B. H. Flowers, Studies in jj-coupling. III. Nuclear energy level, *Proc. Roy. Soc. A.* **215**, 120 (1952).
[23] K. Helmers, Symplectic invariants and Flowers' classification of shell model states, *Nucl. Phys.* **23**, 594 (1961).
[24] S. T. Belyaev, Effect of pairing correlations on nuclear properties, *Mat. Fys. Medd. Dan. Vid. Selsk.* **31** (11) (1959).
[25] J. L. Wood, K. Heyde, W. Nazarewicz, M. Huyse, and P. van Duppen, Coexistence in even-mass nuclei, *Phys. Rep.* **215**, 101 (1992).
[26] K. Heyde and J. L. Wood, Shape coexistence in atomic nuclei, *Rev. Mod. Phys.* **83**, 1467 (2011).
[27] M. Born and J. R. Oppenheimer, Zur Quantentheorie der Molekeln, *Ann. Physik.* **389**, 457 (1927). [On the Quantum Theory of Molecules, translation by S. M. Binder].
[28] D. J. Rowe, P. Rochford, and J. Repka, Dynamic structure and embedded representation in physics: The group theory of the adiabatic approximation, *J. Math. Phys.* **29**, 572 (1988).
[29] D. J. Rowe. Embedded representations and quasi-dynamical symmetry. In J. Escher, O. Castaños, J. Hirsch, S. Pittel, and G. Stoitcheva, (eds.), *Computational and Group-Theoretical Methods in Nuclear Physics*. p. 165. World Scientific, Singapore (2004), arXiv:1106.1607 [nucl-th].
[30] K. A. Brueckner, Two-body Forces and nuclear saturation. III. Details of the Structure of the Nucleus, *Phys. Rev.* **97**, 1353 (1955).
[31] H. A. Bethe and J. Goldstone, Effect of a repulsive core in the theory of complex nuclei, *Proc. Roy. Soc. A.* **238**, 551 (1957).
[32] G. Alaga, K. Alder, A. Bohr, and B. R. Mottelson, Intensity rules for beta and gamma transitions to nuclear rotational states, *Mat. Fys. Medd. Dan. Vid. Selsk.* **29** (9) (1955).
[33] D. J. Rowe, *Nuclear Collective Motion: Models and Theory*. Methuen, London (1970). Reprinted by World Scientific, Singapore (2010).
[34] A. Bohr and B. R. Mottelson, *Nuclear Structure*. Vol. 2, Benjamin, Reading, Mass. (1975). Reprinted by World Scientific, Singapore (1998).
[35] A. S. Davydov and G. F. Filippov, Rotational states in even atomic nuclei, *Nucl. Phys.* **8**, 237 (1958).
[36] G. Gneuss and W. Greiner, Collective potential energy surfaces and nuclear structure, *Nucl. Phys. A.* **171**, 449 (1971).
[37] P. O. Hess, M. Seiwert, J. Maruhn, and W. Greiner, General collective model and its application to ${}^{238}_{92}$U, *Z. Phys. A.* **296**, 147 (1980).
[38] F. S. Stephens, Coriolis effects and rotation alignment in nuclei, *Rev. Mod. Phys.* **47**, 43 (1975).
[39] D. J. Rowe, T. A. Welsh, and M. A. Caprio, Bohr model as an algebraic collective model, *Phys. Rev. C.* **79**, 054304 (2009).
[40] T. A. Welsh and D. J. Rowe, A computer code for calculations in the algebraic collective model of the atomic nucleus, *Comput. Phys. Commun.* **200**, 220 (2016).

[41] D. J. Rowe, Microscopic theory of the nuclear collective model, *Rep. Prog. Phys.* **48**, 1419 (1985).

[42] J. P. Elliott, Collective motion in the nuclear shell model. I: Classification schemes for states of mixed configurations, *Proc. Roy. Soc. A.* **245**, 128 (1958).

[43] J. P. Elliott, Collective motion in the nuclear shell model. II: The introduction of intrinsic wave-functions, *Proc. Roy. Soc. A.* **245**, 562 (1958).

[44] S. Y. Lee and K. Suzuki, The effective interaction of two nucleons in the $s-d$ shell, *Phys. Lett. B.* **91**, 173 (1980).

[45] K. Suzuki and S. Y. Lee, Convergent theory for effective interaction in nuclei, *Prog. Theor. Phys.* **64**, 2091 (1980).

[46] H. Ui, Quantum mechanical rigid rotator with an arbitrary deformation. I, *Prog. Theor. Phys.* **44**, 153 (1970).

[47] G. Mackey, *Induced Representations of Groups and Quantum Mechanics.* Benjamin, New York (1968).

[48] D. J. Rowe, How do deformed nuclei rotate? *Nucl. Phys. A.* **152**, 273 (1970).

[49] D. J. Rowe and G. Rosensteel, Geometric derivation of the kinetic energy in collective models, *J. Math. Phys.* **20**, 465 (1979).

[50] S. Tomonaga, Elementary theory of quantum-mechanical collective motion of particles, *Prog. Theor. Phys.* **13**, 467 (1955).

[51] T. Miyazima and T. Tamura, A collective description of the surface oscillation of atomic nuclei, *Prog. Theor. Phys.* **15**, 255 (1956).

[52] D. Bohm and D. Pines, A collective description of electron interactions: III. Coulomb interactions in a degenerate electron gas, *Phys. Rev.* **92**, 609 (1953).

[53] D. J. Rowe, The equations-of-motion method and the extended shell model, *Rev. Mod. Phys.* **40**, 153 (1968).

[54] E. İnönü and E. P. Wigner, On the contraction of groups and their representations, *Proc. Nat. Acad. Sci. USA.* **39**, 510 (1953).

[55] L. Weaver, L. C. Biedenharn, and R. Y. Cusson, Rotational bands in nuclei as induced group representation, *Ann. Phys.* **77**, 250 (1973).

[56] G. Rosensteel and D. J. Rowe, The algebraic CM(3) model, *Ann. Phys.* **96**, 1 (1976).

[57] L. Weaver, R. Y. Cusson, and L. C. Biedenharn, Nuclear rotational-vibrational collective motion with nonvanishing vortex spin, *Ann. Phys.* **102**, 493 (1976).

[58] W. Zickendraht, Collective and single-particle coordinates in nuclear physics, *J. Math. Phys.* **12**, 1663 (1971).

[59] A. Y. Dzyublik, V. I. Ovcharenko, A. I. Steshenko, and G. V. Filippov, *Yad. Fiz.* **15**, 869 (1972). [*Sov. J. Nucl. Phys.* **15** 487, (1972)].

[60] P. Gulshani and D. J. Rowe, Collective motions in nuclei and the spectrum generating algebras $T_5\times$ SO(3), GL(3,R), and CM(3), *Can. J. Phys.* **54**, 970 (1976).

[61] G. Rosensteel and D. J. Rowe, Nuclear Sp(3,R) model, *Phys. Rev. Lett.* **38**, 10 (1977).

[62] G. Rosensteel and D. J. Rowe, On the algebraic formulation of collective models III: The symplectic shell model of collective motion, *Ann. Phys.* **126**, 343 (1980).

[63] G. F. Filippov, V. I. Ovcharenko, and A. I. Steshenko. Generalized Eulerian angles and the collective motion of many-particle system. In F. Calogero and C. Ciofi degli Atti, (eds.), In *Proc. Int. Sym. Present Status and Novel Developments in the Nuclear Many-Body Problem*, Roma (1973), p. 627.

[64] V. V. Vanagas, Microscopic theory of the nucleus in the framework of restricted dynamics, *Fiz. Elem. Chastits At. Yad.* **11**, 454 (1980). [*Sov. J. Part. Nucl.* **11**, 172 (1980)].

[65] V. S. Vasilevsky, Y. F. Smirnov, and G. F. Filippov, Generating function for the full basis of the irreducible representation of the group Sp(6,R), *Yad. Fiz.* **32**, 987 (1980). [*Sov. J. Nucl. Phys.* **32**, 510 (1980)].

[66] M. Moshinsky and C. Quesne, Linear canonical transformations and their unitary representations., *J. Math. Phys.* **12**, 1722 (1971).

[67] D. J. Rowe, M. J. Carvalho, and J. Repka, Dual pairing of symmetry groups and dynamical groups in physics, *Rev. Mod. Phys.* **84**, 711 (2012).

[68] J. P. Draayer, T. Dytrych, K. D. Launey, and D. Langr, Symmetry-adapted no-core shell model applications for light nuclei with QCD-inspired interactions, *Prog. Part. Nucl. Phys.* **67**, 516 (2012).

[69] P. Park, J. Carvalho, M. Vassanji, D. J. Rowe, and G. Rosensteel, The shell-model theory of nuclear rotational states., *Nucl. Phys. A.* **414**, 93 (1984).

[70] J. P. Draayer, K. J. Weeks, and G. Rosensteel, Symplectic shell-model calculations for ^{20}Ne with horizontal configuration mixing, *Nucl. Phys. A.* **413**, 215 (1984).

[71] C. Bahri and D. J. Rowe, SU(3) quasi-dynamical symmetry as an organizational mechanism for generating nuclear rotational states, *Nucl. Phys. A.* **662**, 125 (2000).

[72] P. O. Hess, A. Algora, M. Hunyadi, and J. Cseh, Configuration-mixed effective SU(3) symmetries, *Eur. Phys. J. A.* **15**, 449 (2002).

[73] J. L. Wood and K. Heyde, (eds.), *A Focus on shape coexistence in nuclei, J. Phys. G.* **43**, (2016).

[74] H. Morinaga, Interpretation of some of the excited states of $4n$ self-conjuate nuclei, *Phys. Rev.* **101**, 254 (1956).

[75] G. E. Brown and A. M. Green, Even parity states of ^{16}O and ^{17}O, *Nucl. Phys.* **75**, 401 (1966).

[76] P. J. Ellis and T. Engeland, The weak coupling model applied to the nuclei with $A = 16 - 19$: I. Energy levels, *Nucl. Phys. A.* **144**, 161 (1970).

[77] D. J. Millener. SU(3) in shell model calculations. In J. P. Draayer and J. Janecke, (eds.), *Group Theory and Special Symmetries in Nuclear Physics*, World Scientific, Singapore (1992).

[78] D. J. Rowe, G. Thiamova, and J. L. Wood, Implications of deformation and shape coexistence for the nuclear shell model, *Phys. Rev. Lett.* **97**, 202501 (2006).

[79] A. Dreyfuss, K. D. Launey, T. Dytrych, J. P. Draayer, and C. Bahri, Hoyle state and rotational features in Carbon-12 within a no-core shell-model framework, *Phys. Lett. B.* **727**, 511 (2013).
[80] G. K. Tobin, M. C. Ferriss, K. D. Launey, T. Dytrych, J. P. Draayer, A. C. Dreyfuss, and C. Bahri, Symplectic no-core shell-model approach to intermediate-mass nuclei, *Phys. Rev. C.* **89**, 034312 (2014).
[81] F. Hoyle, On nuclear reactions occurring in very hot stars. 1. the synthesis of elements from carbon to nickel., *Astrophys. J. Suppl. Ser.* **1**, 121 (1954).
[82] R. R. Whitehead, A numerical approach to nuclear shell-model calculations, *Nucl. Phys. A.* **182**, 290 (1972).
[83] R. R. Whitehead, A. Watt, B. J. Cole, and I. Morrison. Computational methods for shell-model calculations. In M. Baranger and E. Vogt, (eds.), *Advances in Nuclear Physics*, Vol. 9, Chapter 2, p. 123. Plenum, New York (1977).
[84] D. R. Entem and R. Machleidt, Accurate charge-dependent nucleon-nucleon potential at fourth order of chiral perturbation theory, *Phys. Rev. C.* **68**, 041001(R) (2003).
[85] R. Machleidt and D. R. Entem, Chiral effective field theory and nucler forces, *Phys. Rep.* **503**, 1 (2011).
[86] R. Machleidt, Nuclear forces, *AIP Conf. Proc.* **1541**, 61 (2013).
[87] A. M. Shirokov, J. P. Vary, A. I. Mazur, and T. A. Weber, Realistic nuclear Hamiltonian: *Ab exitu* approach, *Phys. Lett. B.* **644**, 33 (2007).
[88] P. Navrátil, J. P. Vary, and B. R. Barrett, Large-basis *ab initio* no-core shell model and its application to ^{12}C, *Phys. Rev. C.* **62**, 054311 (2000).
[89] P. Navratil, V. G. Gueorguiev, J. P. Vary, W. E. Ormand, and A. Nogga, Structure of $A = 10 - 13$ nuclei with two- plus three-nucleon interactions from chiral effective field theory, *Phys. Rev. Lett.* **99**, 042501 (2007).
[90] P. Maris, J. P. Vary, and A. M. Shirakov, *Ab initio* no-core full configuration calculations of light nuclei, *Phys. Rev. C.* **79**, 014308 (2009).
[91] T. Dytrych, K. D. Sviratcheva, C. Bahri, J. P. Draayer, and J. P. Vary, Evidence for symplectic symmetry in *ab initio* no-core shell model results for light nuclei, *Phys. Rev. Lett.* **98**, 162503 (2007).
[92] T. Dytrych, K. D. Sviratcheva, C. Bahri, J. P. Draayer, and J. P. Vary, Dominant role of symplectic symmetry in *ab initio* no-core shell model results for light nuclei, *Phys. Rev. C.* **76**, 014315 (2007).
[93] T. Dytrych, K. D. Sviratcheva, C. Bahri, J. P. Draayer, and J. P. Vary, Highly deformed modes in the *ab initio* symplectic no-core model, *J. Phys. G: Nucl. Part. Phys.* **35**, 095101 (2008).
[94] T. Dytrych, K. D. Sviratcheva, J. P. Draayer, C. Bahri, and J. P. Vary, *Ab initio* symplectic no-core shell model, *J. Phys. G.* **35**, 123101 (2008).
[95] T. Dytrych, K. D. Launey, J. P. Draayer, P. Maris, J. P. Vary, E. Saule, U. Catalyurek, M. Sosonkina, D. Langr, and M. A. Caprio, Collective modes in light nuclei from first principles, *Phys. Rev. Lett.* **111**, 252501 (2013).
[96] M. Jarrio, J. L. Wood, and D. J. Rowe, The SU(3) structure of rotational states in heavy deformed nuclei, *Nucl. Phys. A.* **528**, 409 (1991).

[97] J. Carvalho and D. J. Rowe, The shell-model distribution of nuclear shapes, *Nucl. Phys. A.* **548**, 1 (1992).
[98] B. R. Mottelson and S. G. Nilsson, *Mat. Fys. Skr. Dan. Vid. Selsk.* **1** (8) (1959).
[99] D. J. Rowe, Constrained quantum mechanics and a coordinate independent theory of the collective path, *Nucl. Phys. A.* **391**, 307 (1982).
[100] G. Rosensteel, Symplectic manifolds, coadjoint orbits, and mean field theory, *Int. J. Theor. Phys.* **25**, 553 (1986).
[101] D. J. Rowe and J. L. Wood, *Fundamentals of Nuclear Models: Foundational Models.* World Scientific, Singapore (2010).
[102] K. Matsuyanagi, M. Matsuo, T. Nakatsukasa, N. Hinohara, and K. Sato, Open problems in the microscopic theory of large-amplitude collective motion, *J. Phys. G.* **37**, 064018 (2010).
[103] D. J. Rowe, The emergence of deformation and rotational states in the many-nucleon quantum theory of nuclei, *J. Phys. G.* **43**, 024011 (2016).
[104] D. J. Rowe and J. L. Wood, *Fundamentals of Nuclear Models: Unified Models.* World Scientific, Singapore (2016).
[105] M. J. Carvalho. *Algebraic study of rotational features in nuclei.* PhD thesis, University of Toronto, Toronto (1984).
[106] R. Le Blanc, J. Carvalho, M. G. Vassanji, and D. J. Rowe, An effective shell-model theory of collective states, *Nucl. Phys. A.* **452**, 263 (1986).
[107] Y. Leschber and J. P. Draayer, Macroscopic limit of the microscopic SU(3) $\supset$ SO(3) integrity basis interaction, *Phys. Rev. C.* **33**, 749 (1986).
[108] Y. Leschber, J. P. Draayer, and G. Rosensteel, Connection between macroscopic and microscopic hamiltonians for nuclear rotational motion, *J. Phys. G: Nucl. Part. Phys.* **12**, L179 (1986).
[109] O. Castaños, J. P. Draayer, and Y. Leschber, Shape variables and the shell model, *Z. Phys. A.* **329**, 33 (1988).
[110] D. J. Rowe, M. G. Vassanji, and M. J. Carvalho, The coupled-rotor-vibrator model, *Nucl. Phys. A.* **504**, 76 (1989).
[111] H. A. Naqvi and J. P. Draayer, Shell-model operator for K-band splitting, *Nucl. Phys. A.* **516**, 351 (1990).
[112] D. J. Rowe, A. E. McCoy, and M. A. Caprio, Evolution of the many-nucleon theory of nuclear collective structure and its macroscopic limits: an algebraic perspective, *Phys. Scr.* **91**, 033003 (2016).
[113] D. J. Rowe, The fundamental role of symmetry in nuclear models, *AIP Conf. Proc.* **1541**, 104 (2013). arXiv:1304.6115 [nucl-th].
[114] H. C. Lee and R. Y. Cusson, Practical angular momentum projection method in multi-shell Hartree-Fock calculations; ^{22}Na as an example, *Phys. Lett. B.* **38**, 453 (1972).
[115] D. J. Rowe, S. Bartlett, and C. Bahri, Angular-momentum projection of rotational model wave angular-momentum projection of rotational model wave functions, *Phys. Lett. B.* **472**, 227 (2000).

[116] F. S. Stephens and R. Simon, Coriolis effects in the yrast states, *Nucl. Phys. A.* **183**, 257 (1972).
[117] F. S. Stephens, R. M. Diamond, and S. G. Nilsson, A coupling scheme relevant to high angular momenta and intermediate nuclear deformations, *Phys. Lett. B.* **44**, 429 (1973).
[118] W. H. Bassichis and G. Ripka, A Hartree–Fock calculation of excited states of ^{16}O, *Phys. Lett.* **15**, 320 (1965).
[119] Y. Suzuki, Structure study of $T = 0$ states in ^{16}O by $^{12}\text{C} + \alpha$ cluster-coupling model. II, *Prog. Theor. Phys.* **56**, 111 (1976).
[120] E. K. Warburton, B. A. Brown, and D. J. Millener, Large-basis shell-model treatment of $A = 16$, *Phys. Lett. B.* **293**, 7 (1992).
[121] M. Baranger and K. Kumar, Nuclear deformations in the pairing-plus-quadrupole model (I), *Nucl. Phys. A.* **62**, 113 (1965).
[122] M. Baranger and K. Kumar, Nuclear deformations in the pairing-plus-quadrupole model (II). Discussion of validity of the model, *Nucl. Phys. A.* **110**, 490 (1968).
[123] K. Kumar and M. Baranger, Nuclear deformations in the pairing-plus-quadrupole model (V). Energy levels and electromagnetic moments of the W, Os and Pt nuclei, *Nucl. Phys. A.* **122**, 273 (1968).

Chapter 4

Toward Large-scale Multi-shell Calculations with Symmetry-adapted Bases: Lessons from the Symplectic Shell Model

Jutta E. Escher
Lawrence Livermore National Laboratory
7000 East Avenue, Livermore, CA 94550
escher1@llnl.gov

Basic tenets and selected applications of the symplectic shell model are reviewed. The emphasis is on providing a brief introduction to the structure of the model space and to the tools and techniques needed to carry out model calculations. The case is made that it is necessary and possible to move beyond previously used approximations which employ schematic Hamiltonians and highly truncated model spaces. The rich structure underlying the full symplectic Sp(3, $\mathbb{R}$) model has the potential to shed light on the nature and origin of collective motion in nuclei, and to provide physically relevant truncation schemes that enable large-scale shell-model calculations.

Contents

1. Introduction

An important goal of low-energy nuclear science is to achieve reliable descriptions of the structure of and reactions with nuclei across the isotopic chart [1]. Nuclear theories range from microscopic *ab initio* approaches to macroscopic phenomenological descriptions. A variety of approaches is required in order to connect to experimental observations, identify systematic trends, and guide and test theories across the nuclear chart. Complementary approaches find application in different, but overlapping areas of the nuclear chart and need to be enhanced and extended in order to connect to each other, cover the isotopes of interest, and provide the structure input required by nuclear reaction theories.

The present paper focuses on a model that connects two complementary approaches to nuclear structure: the $\mathrm{Sp}(3,\mathbb{R})$ (or "symplectic") model provides a link between macroscopic collective descriptions of the nuclear many-body problem and microscopic shell-model approaches [2–11]. Its underlying algebraic structure facilitates the interpretation of microscopic results in terms of collective nuclear motion, and points toward physically relevant truncation schemes. Its realization in terms of fermionic operators makes it possible to explore collective motion from a microscopic perspective. It has the potential to offer valuable insights into the underlying structure of nuclear systems, to provide physically relevant truncation schemes for microscopic calculations that are limited by the the size of the model spaces needed, and to shed light on the emergence of collective phenomena.

Growing computational resources and improved numerical algorithms, together with a move from realistic phenomenological and meson–theoretical nucleon–nucleon potential models to nuclear interactions derived from chiral perturbation theory, the inclusion of three-nucleon forces, and the inclusion of the continuum have contributed greatly to the success of no-core shell-model methods in describing light nuclei [12–15]. Despite this progress, these approaches suffer from limitations associated with the tremendous increase in the size of the model space that is needed to describe nuclei beyond the p shell. To cover heavier systems, it will be necessary to search for physically relevant truncation schemes.

The Sp(3, $\mathbb{R}$) model provides a fermionic realization of the multi-shell shell-model in a symmetry-adapted basis that allows for model space truncations in both the number of major shells involved ("horizontal" truncation) and with respect to the underlying symmetry (deformation-related or "vertical" truncation). The horizontal truncation scheme is common to multi-shell shell-model calculations. The vertical structure, which originates from the underlying Sp(3, $\mathbb{R}$) structure and is built into the basis-state construction, is unique to the symplectic model. As it is closely connected to collective motion in nuclei, it provides an alternative, physics-driven, strategy for truncating the model spaces needed for medium-mass and heavy nuclei. After the introduction of the model by Rosensteel and Rowe around 1980 [2, 3, 16], much effort was devoted to the development of the mathematical structure of the model. While the earlier applications mostly relied on schematic Hamiltonians constructed from symmetry-preserving interactions, the algebraic and computational tools are now available to move toward more sophisticated calculations.

It is the purpose of this Chapter to draw lessons for future large-scale multi-shell calculations from work done in the symplectic shell model. The model will be referred to as symplectic or Sp(3, $\mathbb{R}$) model throughout the paper. It is sometimes denoted by Sp(6, $\mathbb{R}$). Recently, it has also been called the no-core symplectic shell model (NCSpM) which emphasizes that fact that its full implementation corresponds to a no-core shell model.

In Sec. 2, the construction of the model space is outlined and procedures for obtaining matrix elements of general one-body and two-body operators are described. In Sec. 3, the basic ideas of the approach are illustrated with a calculation in a highly-truncated model space. The importance of symmetry-breaking interactions and how to incorporate them in practical calculations are discussed in Sec. 4, and some possible future developments are outlined. Section 5 contains concluding remarks.

2. The Symplectic Shell Model

The symplectic shell model is an algebraic, fermionic, nuclear shell-model scheme that is based on the Sp(3, $\mathbb{R}$) dynamical group of the spherical harmonic oscillator (see also Chapter 3). Its model space construction involves multiple $2\hbar\omega$ one-particle-one-hole excitations of the monopole and quadrupole type (where $\hbar\omega$ is the harmonic-oscillator frequency). The associated group structure naturally accommodates all essential observables for a description of nuclear monopole and quadrupole collective vibrations as well as for rigid and irrotational flow rotations. It provides a link between macroscopic collective descriptions of the nuclear many-body system and microscopic approaches, since its observables are expressible in microscopic shell-model terms and arbitrary one- and two-body operators can be evaluated in its framework. The model provides a specific multi-shell realization of the nuclear shell-model that makes it possible to classify and truncate the enormous shell-model space while retaining the multi-$\hbar\omega$ quadrupole correlations that lead to deformed nuclei [2, 5, 9, 16].

2.1. *Generators of* Sp(3, $\mathbb{R}$)

The symmetry algebra of the symplectic scheme is spanned by one-body operators which are bilinear products in the (relative) position ($x_{si}, i = 1, 2, 3, s = 1, \ldots, A-1$) and momentum ($p_{si}$) observables:

$$\begin{aligned} Q_{ij} &= \sum_s x_{si} x_{sj}, \\ K_{ij} &= \sum_s p_{si} p_{sj}, \\ L_{ij} &= \sum_s (x_{si} p_{sj} - x_{sj} p_{si}), \\ S_{ij} &= \sum_s (x_{si} p_{sj} + p_{si} x_{sj}), \end{aligned} \tag{1}$$

where $A-1$ is the number of Jacobi "particles" remaining after removal of the center-of-mass contribution. Together the operators generate the twenty one-dimensional symplectic algebra Sp(3, $\mathbb{R}$), that is, the Lie algebra of linear transformations which preserve a skew-symmetric bilinear form on a six-dimensional real vector space. It is the smallest Lie algebra that contains both the quadrupole moments and the many-nucleon kinetic energy, and it has several physically relevant subalgebras. These include GCM(3) and its subalgebra $[\mathbb{R}^5]$SO(3), associated with the general collective motion model and its rotational limit, respectively, the algebra GL(3, $\mathbb{R}$) of the general linear motion group, as well as SU(3) and its subalgebra SO(3), associated with the Elliott model and the rotation group, respectively.

For many purposes, it is advantageous to express the symplectic generators in terms of harmonic-oscillator boson creation $b^\dagger_{si} = (x_{si} - \mathrm{i}p_{si})/\sqrt{2}$ and annihilation $b_{si} = (x_{si} + \mathrm{i}p_{si})/\sqrt{2}$ operators [10, 17, 18]:

$$\hat{H}_0 = \sqrt{3} \sum_s \{b_s^{\dagger(10)} \times \tilde{b}_s^{(01)}\}_{00}^{(00)} + \frac{3}{2}(A-1),$$

$$\hat{C}_{lm}^{(11)} = \sqrt{2} \sum_s \{b_s^{\dagger(10)} \times \tilde{b}_s^{(01)}\}_{lm}^{(11)} \quad (l = 1, 2),$$

$$\hat{A}^{(20)}_{lm} = \frac{1}{\sqrt{2}} \sum_s \{b^{\dagger(10)}_s \times b^{\dagger(10)}_s\}^{(20)}_{lm} \quad (l = 0, 2), \tag{2}$$

$$\hat{B}^{(02)}_{lm} = \frac{1}{\sqrt{2}} \sum_s \{\tilde{b}^{(01)}_s \times \tilde{b}^{(01)}_s\}^{(02)}_{lm} \quad (l = 0, 2).$$

Here, the notation $T^{(\lambda\mu)}_{lm}$ indicates that the operator T possesses good SU(3) [superscript $(\lambda\mu)$] and SO(3) [subscript lm] tensorial properties. The spherical components of the oscillator bosons, $b^{\dagger(10)}_{s,1,\pm1} = \mp\frac{1}{\sqrt{2}}(b^\dagger_{s1} \pm \mathrm{i} b^\dagger_{s2})$, $b^{\dagger(10)}_{s,1,0} = b^\dagger_{s3}$, and $\tilde{b}^{(01)}_{s,1,\pm1} = \mp\frac{1}{\sqrt{2}}(b_{s1} \pm \mathrm{i} b_{s2})$, $\tilde{b}^{(01)}_{s,1,0} = b_{s3}$, have been used. Since $b^\dagger_s b^\dagger_s$ adds two quanta to particle s, thereby moving it up across two major oscillator shells, $\hat{A}^{(20)}_{lm}$ creates a $2\hbar\omega$ excitation in the system. Analogously, $\hat{B}^{(02)}_{lm}$, which is related to $\hat{A}^{(20)}_{lm}$ by Hermitian conjugation, $\hat{B}^{(02)}_{lm} = (-1)^{l-m}(\hat{A}^{(20)}_{l-m})^\dagger$, annihilates a $2\hbar\omega$ excitation. The $\hat{C}^{(11)}_{lm}$ act only *within* a major harmonic-oscillator shell. They generate the group SU(3) of the well-known Elliott model [19–21]:

$$\begin{aligned}
\sqrt{3}\hat{C}^{(11)}_{2m} &= Q^E_{2m} \equiv \sqrt{\frac{4\pi}{5}} \sum_s (r^2_s Y_{2m}(\hat{r}_s) + p^2_s Y_{2m}(\hat{p}_s)) \\
&\qquad (m = 0, \pm1, \pm2), \\
\hat{C}^{(11)}_{1m} &= \hat{L}_m \quad (m = 0, \pm1),
\end{aligned} \tag{3}$$

where Q^E_{2m} denotes the symmetrized quadrupole operator of Elliott, which does not couple different major shells, and $\hat{L}_m$ is the orbital angular momentum operator. The harmonic-oscillator Hamiltonian, $\hat{H}_0 = \sum_{i=1}^3 \hat{C}_{ii}$, is an SU(3) scalar and generates U(1) in U(3) = SU(3) × U(1).

Alternatively, one can realize the symplectic generators in terms of fermionic creation and annihilation operators [10]:

$$\hat{C}^{(11)}_{lm} = \sum_\eta \sqrt{\frac{\eta(\eta+1)(\eta+2)(\eta+3)}{6}} \times \{a^\dagger_\eta \times \tilde{a}_\eta\}^{(11)S=0}_{lm\Sigma=0} + \hat{\mathcal{O}}^{cm}_{\hat{C}}(A),$$

$$\hat{A}^{(20)}_{lm} = \sum_{\eta} \sqrt{\frac{(\eta+1)(\eta+2)(\eta+3)(\eta+4)}{12}} \times \{a^{\dagger}_{\eta+2} \times \tilde{a}_{\eta}\}^{(20)S=0}_{lm\Sigma=0} + \hat{\mathcal{O}}^{cm}_{\hat{A}}(A), \tag{4}$$

$$\hat{B}^{(02)}_{lm} = \sum_{\eta} \sqrt{\frac{(\eta+1)(\eta+2)(\eta+3)(\eta+4)}{12}} \times \{a^{\dagger}_{\eta} \times \tilde{a}_{\eta+2}\}^{(02)S=0}_{lm\Sigma=0} + \hat{\mathcal{O}}^{cm}_{\hat{B}}(A),$$

where $a^{\dagger}_{\eta lm\frac{1}{2}\sigma}$ $[\tilde{a}_{\eta lm\frac{1}{2}\sigma} = (-1)^{\eta+l+m+\frac{1}{2}+\sigma} a_{\eta l-m\frac{1}{2}-\sigma}]$ is a single-particle creation [annihilation] operator, which produces [destroys] a fermion with angular momentum l with projection m and spin 1/2 with projection σ in the ηth major oscillator shell. The sums run over all shells, and the coupling to total spin $S = 0$ with projection $\Sigma = 0$ reflects the fact that the generators act on spatial degrees of freedom only. The operators $\hat{\mathcal{O}}^{cm}(A)$ remove the spurious center-of-mass content from the generators; explicit expressions are given in Ref. [10].

The fermionic realization of Sp(3, $\mathbb{R}$) underscores the connection between the collective aspects of the model, embodied in its generators, and the microscopic picture of the nucleus in terms of individual nucleons. It also provides for a mechanism to evaluate matrix elements of arbitrary one-body and two-body operators within the model, as is discussed in Sec.2.3.

2.2. *Symplectic Model Space*

A basis for the symplectic model is generated by applying symmetrically coupled products of the $2\hbar\omega$ raising operator $\hat{A}^{(20)}$ with itself to a starting many-particle shell-model configuration. For example, for the usual $0\hbar\omega$ states, each starting $0\hbar\omega$ configuration is characterized by the distribution of oscillator quanta into the three cartesian directions, $\{\sigma_1, \sigma_2, \sigma_3\}$, where $\sigma_1 \geq \sigma_2 \geq \sigma_3$. Here, σ_i denotes the eigenvalue of the U(3) weight operator $C_{ii} = \sum_s (b^{\dagger}_{si} b_{si} + 1/2)$, which essentially counts the number of oscillator bosons in the ith direction of the system. Since $s = 1, 2, \ldots, A-1$, it follows that the σ_i are half-integer numbers for even-A and integers for odd-A nuclei. Equivalently, one may employ quantum numbers $N_\sigma(\lambda_\sigma\,\mu_\sigma)$, where $\lambda_\sigma = \sigma_1 - \sigma_2$, $\mu_\sigma = \sigma_2 - \sigma_3$ are the Elliott SU(3) labels, and $N_\sigma = \sigma_1 + \sigma_2 + \sigma_3$ is the eigenvalue of the harmonic-oscillator Hamiltonian, $\hat{H}_0 = \hat{C}_{11} + \hat{C}_{22} + \hat{C}_{33}$, which takes the minimum value consistent with the Pauli exclusion principle. Each such set of U(3) quantum numbers (which labels a symplectic bandhead) uniquely determines an irreducible representation (irrep) of the symplectic algebra, since it characterizes an Sp(3, $\mathbb{R}$) lowest-weight state (see also Chapter 3).

Repeated application of the symplectic generator $\hat{A}^{(20)}$ to this bandhead allows one to build a basis for the Sp(3, $\mathbb{R}$) irrep under consideration. The basis state construction is schematically illustrated in Fig. 1 for a typical Elliott starting state with $(\lambda_\sigma\,\mu_\sigma) = (\lambda\,0)$. A similar figure for $(\lambda_\sigma\,\mu_\sigma) = (0\,\mu)$ is given in Ref. [22]. Specifically, the product of $N/2$, $N = 0, 2, 4, \ldots$, raising operators $\hat{A}^{(20)}$ is multiplicity-free and generates $N\hbar\omega$ excitations for each starting configuration $N_\sigma(\lambda_\sigma\,\mu_\sigma)$. Each such product operator $\mathcal{P}^{N(\lambda_n\,\mu_n)}$ can be labeled according to its U(3) content, $\{n_1, n_2, n_3\}$ or $N(\lambda_n\,\mu_n)$, where $(\lambda_n \mu_n)$ ranges over the set

$$\Omega = \{(n_1 - n_2, n_2 - n_3) | n_1 \geq n_2 \geq n_3 \geq 0; N = n_1 + n_2 + n_3\}, \tag{5}$$

where n_1, n_2, n_3 are even integers. The raising polynomial $\mathcal{P}^{N(\lambda_n\,\mu_n)}$ is then coupled with $|N_\sigma(\lambda_\sigma\,\mu_\sigma)\rangle$ to good SU(3) symmetry $\rho(\lambda_\omega\,\mu_\omega)$, with ρ denoting the multiplicity in the coupling $(\lambda_n\,\mu_n) \otimes (\lambda_\sigma\,\mu_\sigma)$. The quanta distribution in the associated lowest-weight state is given by $\{\omega_1, \omega_2, \omega_3\}$, with $N_\omega \equiv N_\sigma + N = \omega_1 + \omega_2 + \omega_3$, $\omega_1 \geq \omega_2 \geq \omega_3$, and $\lambda_\omega = \omega_1 - \omega_2$, $\mu_\omega = \omega_2 - \omega_3$. The states of the Sp(3, $\mathbb{R}$) $\supset$ SU(3) basis are thus labeled by three types of U(3) quantum numbers: $\Gamma_\sigma \equiv \{\sigma_1, \sigma_2, \sigma_3\} = N_\sigma(\lambda_\sigma\,\mu_\sigma)$, the symplectic bandhead or Sp(3, $\mathbb{R}$) lowest-weight U(3) symmetry, which specifies the Sp(3, $\mathbb{R}$) irreducible representation; $\Gamma_n \equiv \{n_1, n_2, n_3\} = N(\lambda_n \mu_n)$, the U(3) symmetry of the raising polynomial; and $\Gamma_\omega \equiv \{\omega_1, \omega_2, \omega_3\} = N_\omega(\lambda_\omega\,\mu_\omega)$, the U(3) symmetry of the coupled product. A given symplectic

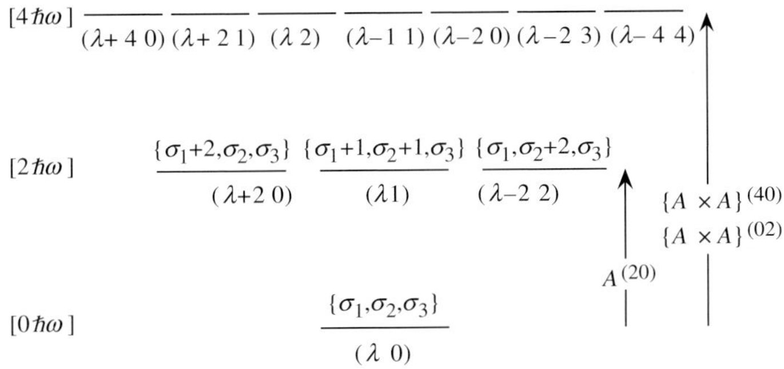

Fig. 1. Basis construction in the symplectic model. SU(3)-coupled products of the raising operator $\hat{A}^{(20)}$ with itself act on an Elliott starting state (symplectic bandhead) with $(\lambda_\sigma\, \mu_\sigma) = (\lambda 0)$ $(\{\sigma_1, \sigma_2, \sigma_3 = \sigma_2\})$ to generate symplectic $2\hbar\omega$, $4\hbar\omega$, ... excitations. Also shown are the SU(3) labels $(\lambda\, \mu)$ and quanta distributions $\{\omega_1, \omega_2, \omega_3\}$ for some excited states.

representation space $N_\sigma(\lambda_\sigma\, \mu_\sigma)$ is infinite dimensional, since $N/2$, the number of oscillator excitations, can take any non-negative integer value. In practical applications, one must therefore either truncate the symplectic Hilbert space, or restrict oneself to interactions and observables for which the matrix elements depend solely on the symplectic irrep and can be calculated analytically.

To complete the basis state labeling, additional quantum numbers α are required. This can be accomplished by reducing $\mathrm{Sp}(3, \mathbb{R}) \supset \mathrm{SU}(3)$ states with respect to a suitable subgroup. States with good angular momentum values can be obtained from the $\mathrm{SU}(3) \supset \mathrm{U}(1) \times \mathrm{SU}(2)$ (canonical) basis by projection [19, 20, 23]. The associated quantum numbers are $\alpha = \kappa L M$, where κ is a multiplicity index, which enumerates multiple occurrences of a particular L value in the SU(3) irrep $(\lambda\, \mu)$ from 1 to $\kappa_L^{\max}(\lambda\, \mu)$,

$$\kappa_L^{\max}(\lambda\, \mu) = [(\lambda + \mu + 2 - L)/2] - [(\lambda + 1 - L)/2] \\ - [(\mu + 1 - L)/2], \tag{6}$$

where $[\ldots]$ is the greatest non-negative integer function. The $\kappa_L^{\max}(\lambda\, \mu)$ occurrences of L can be distinguished in a variety of ways. The physically most significant scheme is that of Elliott [19, 20], in which case the projection of L along the body-fixed 3-axis, denoted

K, is used to sort the L-values into the familiar K-bands of the rotational model, but the resulting basis states are not orthonormal with respect to K. Vergados [24] gives a prescription to construct orthogonal basis states in a systematic manner for all $(\lambda\,\mu)$, such that the physical interpretation of K as a band label can be approximately maintained. Here, we use the running index $\kappa = 1, 2, \ldots, \kappa_L^{\max}$ to distinguish multiple occurrences of L in a given SU(3) irrep $(\lambda\,\mu)$. When applying the formalism to realistic nuclei, one typically assigns rotational band labels according to the calculated $B(E2)$ rates.

The dynamical symmetry chain and the associated quantum labels of the scheme described above are given by [10]:

$$\begin{array}{cccccc} \mathrm{Sp}(3,\mathbb{R}) & \supset & \mathrm{U}(3) & \supset \mathrm{SO}(3) & \supset \mathrm{SO}(2) \\ N_\sigma(\lambda_\sigma\mu_\sigma) & N(\lambda_n\,\mu_n)\rho & N_\omega(\lambda_\omega\,\mu_\omega)\;\kappa & L & M \end{array} \tag{7}$$

The quadratic Casimir operators of SU(3) and Sp(3, $\mathbb{R}$),

$$\hat{C}_{\mathrm{SU}(3)} = \frac{1}{2}[C_2^{(11)} \cdot C_2^{(11)} + C_1^{(11)} \cdot C_1^{(11)}], \tag{8}$$

$$\begin{aligned} \hat{C}_{\mathrm{Sp}(3,\mathbb{R})} &= -2\hat{A}_0^{(20)}\hat{B}_0^{(02)} - 2\hat{A}_2^{(20)} \cdot \hat{B}_2^{(02)} + \hat{C}_{\mathrm{SU}(3)} \\ &\quad + \frac{1}{3}\hat{H}_0^{\,2} - 4\hat{H}_0 \end{aligned} \tag{9}$$

have the following eigenvalues in the dynamical symmetry basis:

$$\langle\hat{C}_{\mathrm{SU}(3)}\rangle[(\lambda_\omega\,\mu_\omega)] = \frac{2}{3}(\lambda_\omega^2 + \mu_\omega^2 + \lambda_\omega\mu_\omega + 3\lambda_\omega + 3\mu_\omega), \tag{10}$$

$$\begin{aligned} \langle\hat{C}_{\mathrm{Sp}(3,\mathbb{R})}\rangle[N_\sigma(\lambda_\sigma\,\mu_\sigma)] &= \frac{2}{3}(\lambda_\sigma^2 + \mu_\sigma^2 + \lambda_\sigma\mu_\sigma + 3\lambda_\sigma + 3\mu_\sigma) \\ &\quad + \frac{N_\sigma^2}{3} - 4N_\sigma. \end{aligned} \tag{11}$$

The collection of all $0\hbar\omega$ configurations provides a complete Hilbert space for the Elliott SU(3) submodel of the Sp(3, $\mathbb{R}$) model and is referred to as the $0\hbar\omega$ horizontal shell-model space (or valence shell). The set of states built on a given U(3) irrep $N_\sigma(\lambda_\sigma\mu_\sigma)$ is called the vertical extension of that irrep. Each vertical extension can be partitioned into horizontal slices with the states within the

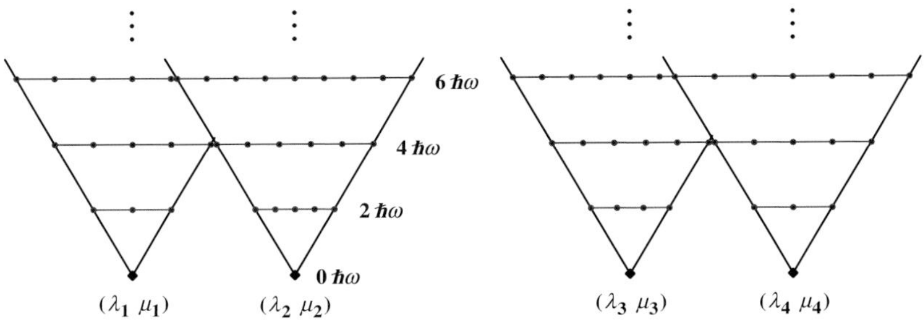

Fig. 2. Symplectic shell-model space. The schematic plot illustrates a model space with multiple symplectic representations. Each "cone" corresponds to an Sp(3, $\mathbb{R}$) irrep and is uniquely characterized by U(3) quantum numbers $N_\sigma(\lambda_\sigma\,\mu_\sigma)$, where $(\lambda_\sigma\,\mu_\sigma)$ denotes the Elliott SU(3) quantum labels for the associated $0\hbar\omega$ shell-model configuration (cones that start at $2\hbar\omega, 4\hbar\omega, 6\hbar\omega, \ldots$ are part of the symplectic model space, but are not shown in the figure). For a given starting representation $(\lambda_\sigma\,\mu_\sigma)(\sigma = 1, 2, 3, 4$ here), one obtains multiple SU(3) configurations, $(\lambda_\omega\,\mu_\omega)$, at each $N\hbar\omega$ level of excitation ($N > 0$), indicated here by small filled circles.

$N/2$th slice representable as a homogeneous polynomial of degree $N/2$ in the $\hat{A}^{(20)}$ tensors acting on the parent $0\hbar\omega$ configuration (see Fig. 2). Similarly, symplectic irreps can be built on parent $2\hbar\omega, 4\hbar\omega, 6\hbar\omega, \ldots$ configurations, and hence span the entire Hilbert space. Interactions can thus be classified according to their effect on this structure; pairing, for example, causes horizontal mixing, both within each "cone" (symplectic irrep) and between different cones, while the quadrupole–quadrupole interaction induces horizontal and vertical mixing, but does not connect different cones.

2.3. *Hamiltonians and Observables*

The construction scheme outlined in the previous section provides a specific, symmetry-adapted basis for the full multi-shell shell-model space. Like a traditional no-core shell-model space, it can be truncated at a specific $N_\omega\hbar\omega$ (or $N_{\max}$) level ("horizontal" truncation). Unlike its traditional counterpart, however, the Sp(3, $\mathbb{R}$) space is also endowed with a vertical organizational scheme that is based on its symplectic structure and that allows one to consider alternative truncation strategies. This is of particular value if nuclear systems have indeed Sp(3, $\mathbb{R}$) as an approximate symmetry. In that case the organizing structure for the Hilbert space provides a (physics-driven) truncation scheme which reduces the number of the necessary configurations in each $\hbar\omega$ subspace, while allowing one to include excitations to energy levels at excitations that are higher than would otherwise be reachable. An application to the triaxially deformed nucleus ^{24}Mg, which makes use of this strategy, will illustrate the vertical truncation scheme in the next section.

A second advantage of employing symplectic basis states lies in the possibility to systematically exploit the SU(3) Wigner–Eckart theorem. In analogy to the well-known SU(2) version of the Wigner–Eckart theorem, which expresses a set of matrix elements of a specific operator in terms of *one* reduced matrix element and a set of geometric Wigner functions (Clebsch–Gordan coefficients), an even larger number of individual matrix elements can be expressed in terms of *one* SU(3) reduced matrix element $\langle\,(\lambda_3\,\mu_3)\,|||\,T^{(\lambda_2\,\mu_2)}\,|||(\lambda_1\,\mu_1)\,\rangle_\rho$ and a set of SU(3) Wigner coupling coefficients [23]:

$$\begin{aligned}&\langle(\lambda_3\,\mu_3)\alpha_3|T^{(\lambda_2\,\mu_2)\alpha_2}|(\lambda_1\,\mu_1)\alpha_1\rangle\\&\quad=\sum_\rho\langle(\lambda_1\,\mu_1)\alpha_1;(\lambda_2\,\mu_2)\alpha_2|(\lambda_3\,\mu_3)\alpha_3\rangle_\rho\langle(\lambda_3\,\mu_3)|||T^{(\lambda_2\,\mu_2)}|||(\lambda_1\,\mu_1)\rangle_\rho.\end{aligned}\tag{12}$$

Here, the multiplicity label ρ enumerates multiple occurrences of the irreducible representation $(\lambda_3\,\mu_3)$ in the product $(\lambda_1\,\mu_1)\times(\lambda_2\,\mu_2)$. Note that the reduced matrix element does not depend on labels related to subgroups of SU(3), such as the angular-momentum label. This can provide computational advantages when the evaluation and/or storage of a large number of reduced matrix elements is a bottleneck in the calculation and the coupling coefficients can be efficiently generated on the fly (or expressions can be derived that eliminate or sum over the dependence on the coupling coefficients).

In order to make use of these advantages, it is necessary to express the relevant Hamiltonian, and operators associated with observables of interest, in terms of SU(3) tensors. Since the fermion creation and annihilation operators, $a^\dagger_{\eta l m \frac{1}{2}\sigma}$ and

$\tilde{a}_{\eta l m \frac{1}{2}\sigma} = (-1)^{\eta+l+m+\frac{1}{2}+\sigma} a_{\eta l -m \frac{1}{2} -\sigma}$, have SU(3) tensor character $(\eta 0)$ and (0η), respectively, a general one-plus-two-body Hamiltonian (for space $\mathbf{r}$ and spin $\mathbf{s}$ coordinates) of the form $H = \sum \langle \rho_2 | F(\mathbf{r}, \mathbf{s}) | \rho_1 \rangle a^\dagger_{\rho_2} a_{\rho_1} + \sum \langle \rho_4 \rho_3 | G(\mathbf{r}, \mathbf{r}', \mathbf{s}, \mathbf{s}') | \rho_1 \rho_2 \rangle a^\dagger_{\rho_4} a^\dagger_{\rho_3} a_{\rho_2} a_{\rho_1}$, where the sums run over all single-particle labels ρ_1, ρ_2 and $\rho_1, \rho_2, \rho_3, \rho_4$, respectively, can be re-written in an SU(3)-coupled form (with $L = S$):

$$
\begin{aligned}
H = & \sum_{\eta_1 \eta_2 (\lambda\, \mu) \kappa M S} f^{\eta_1 \eta_2 (\lambda\, \mu) \kappa M S} \left\{ a^\dagger_{\eta_1} \times \tilde{a}_{\eta_2} \right\}^{(\lambda\, \mu) S}_{\kappa L M} \\
& + \sum_{\substack{\eta_4 \eta_3 (\lambda_b\, \mu_b) S_b \\ \eta_1 \eta_2 (\lambda_a\, \mu_a) S_a \\ \rho (\lambda\, \mu) \kappa M S}} g^{\eta_4 \eta_3 (\lambda_b\, \mu_b) S_b ; \eta_1 \eta_2 (\lambda_a\, \mu_a) S_a ; \rho (\lambda\, \mu) \kappa M S} \\
& \times \{ \{ a^\dagger_{\eta_4} \times a^\dagger_{\eta_3} \}^{(\lambda_b\, \mu_b) S_b} \times \{ \tilde{a}_{\eta_2} \times \tilde{a}_{\eta_1} \}^{(\lambda_a\, \mu_a) S_a} \}^{\rho (\lambda\, \mu) S}_{\kappa L M}. \qquad (13)
\end{aligned}
$$

Here the tensor character of the creation and annihilation operators with respect to intrinsic spin transformations ($s = 1/2$) has been taken into account as well; the extension to a spin–isospin formalism is straightforward.

Tensor decompositions for relevant schematic interactions are known: the symmetry-preserving quadrupole–quadrupole interaction $Q_2 \cdot Q_2$ can be written in terms of $(\lambda\, \mu) = (0\,0), (2\,2), (2\,0)$, and $(4\,0)$ (and conjugate) tensors [17, 22], the one-body spin-orbit interaction is given in Refs. [7, 25], and expressions for the two-body pairing and tensor interactions have been derived in Refs. [7, 26, 27]. In addition, the transformation for a general two-body interaction from at JT-coupled scheme to an SU(3)-coupled expression is known (see Eq. (11) of Ref. [11]), making it possible to utilize modern, realistic interactions that are given in JT-coupled form (for an example of an SU(3) tensor analysis of the N^{3}LO interaction, see Ref. [11]).

Similarly, one- and two-body operators associated with observables of interest can be expressed in terms of SU(3) tensors. Relevant examples include quadrupole moments, electromagnetic transition strengths, and matter radii. For the electron scattering application discussed in Sec. 3.2, one needs expressions for form factors associated with longitudinal and transverse transitions in a nucleus. SU(3) tensorial expressions for the multipole operators, suitable for plane-wave Born approximation, have been derived in Ref. [28] for a momentum transfer q:

$$\hat{T}^{\chi}_{LM}(q) = \frac{Z}{A} \sum_{\substack{\eta_1\eta_2 \\ (\lambda\mu)\kappa}} C^{\chi(\lambda\mu)}_{\kappa L \eta_1 \eta_2}(bq) \{a^{\dagger}_{\eta_1} \times \tilde{a}_{\eta_2}\}^{(\lambda\mu)S=0}_{\kappa LM\Sigma=0}, \tag{14}$$

where $\chi = Coul, el, mag$ selects the character of the transition, $a^{\dagger}_{\eta}$ and $\tilde{a}_{\eta}$ are the fermion creation and annihilation operators introduced above, and the functions $C^{\chi(\lambda\mu)}_{\kappa L\eta_1\eta_2}(bq)$ involve SU(3) recoupling coefficients (here, b is the harmonic-oscillator characteristic length). This expression is representative of a typical one-body operator in an SU(3)-coupled form: the fermion character is explicit, the sums in the above SU(3) tensorial expansion extend over all possible oscillator shells, allowing for cross-shell correlations (vertical mixing) as well as symmetry breaking (horizontal mixing), and the coupling to specific SU(3) $\supset$ SO(3) and spin quantum numbers make the use of the Wigner–Eckart theorem possible.

The tools for evaluating matrix elements of general one- and two-body operators are generally available: SU(3) analogs of the well-known angular-momentum coupling and recoupling coefficients, namely, SU(3) Wigner coefficients, 6-$(\lambda\,\mu)$ and 9-$(\lambda\,\mu)$ Racah coefficients, are available numerically and in some cases analytically, as are a number of symmetry relations among them [10, 23, 29–33]. Also available are expressions that relate reduced matrix elements of SU(3) tensor products of two operators to products of reduced matrix elements.

Matrix elements of the symplectic generators $\hat{C}^{(11)}_{lm}$, $\hat{A}^{(20)}_{lm}$, and $\hat{B}^{(02)}_{lm}$ can be calculated in various ways, e.g., by using recursion formulae [4], or via vector coherent state theory [34].

Matrix elements of general one-body and two-body operators can by obtained via recursive procedures, in which the matrix element of interest is related back to valence-shell matrix elements, which can be evaluated by standard shell-model techniques. For example, in Ref. [10] the reduced matrix element of the one-body unit operator $\{a^\dagger_{\eta 1} \times \tilde{a}_{\eta 2}\}^{(\lambda\mu)S}$ between states of excitation $N\hbar\omega$ and $N'\hbar\omega$ was expressed in terms of unit matrix elements involving lower excitations, matrix elements of the symplectic generator $\hat{A}^{(20)}_{lm}$, and SU(3) Racah coupling coefficients. The resulting relation is valid for matrix elements that connect states in the same or different symplectic representations, thus it allows one to evaluate matrix elements of symmetry-breaking terms. The electron scattering form factors discussed in Sec. 3.2 were obtained using this procedure. Expressions for symplectic matrix elements of general two-body operators are given by Hecht and Suzuki [8]. Their recursive procedure reduces the matrix elements to standard shell-model matrix elements, SU(3) decoupling coefficients, and some simple expansion coefficients which are tabulated for excitations up to $10\hbar\omega$. The recursive processes of Refs. [8,10] rely on the availability of matrix elements of the standard shell-model type. These can be calculated in an SU(3)-coupled basis using a code developed by Bahri and Draayer [35].

3. Dynamical Symmetry Limit

In the previous section, the structure of the symplectic multi-shell Hilbert space was described and the ingredients needed for a full Sp(3, $\mathbb{R}$) calculation were outlined. While the various components are available, a full calculation remains a challenge, as it is equivalent to a no-core shell-model calculation. For the latter problem, sophisticated tools and techniques have been developed, and the challenges are well known [12, 14, 36]. However, if it can be demonstrated that the symplectic group provides an approximate symmetry structure for finite nuclei, it would provide a powerful motivation to recast the nuclear many-body problem in such group-theoretical framework. To determine whether such approximate symmetry is applicable, one can carry out full, multi-shell, calculations (without imposing restrictions on the Hilbert space or Hamiltonian) and extract the symmetry content of the resulting wave functions. This strategy has been pursued by Dytrych *et al.* [37–39] as will be discussed in Sec. 4. In addition, one can analyze the tensor character of state-of-the-art nuclear interactions, as has been done in Ref. [11].

A complementary approach starts with a symmetry-preserving Hamiltonian and explores to what extent nuclear phenomena can be described and which extensions are required to overcome the limitations found. Here, this latter approach is pursued.

3.1. *Dynamical Symmetry Limit: Hamiltonians and Truncation Scheme*

A major strength of the symplectic model is its ability to fully accommodate the action of the quadrupole operator, $Q_{2m} = \sqrt{\frac{16\pi}{5}} \sum_s r_s^2 Y_{2m}(\hat{r}_s)$, which can be written in terms of symplectic generators:

$$Q_{2m} = \sqrt{3}(\hat{C}_{2m}^{(11)} + \hat{A}_{2m}^{(20)} + \hat{B}_{2m}^{(02)}). \tag{15}$$

A rotationally invariant collective potential V_{coll} can be constructed from SO(3) scalars in Q_2. The quadratic term, $-\chi Q_2 \cdot Q_2$, is the well-known quadrupole–quadrupole interaction, which serves a standard ingredient in models that aim at reproducing rotational spectra and nuclear deformations. It emerges (apart from a constant) as a leading contribution in the multipole expansion of a general two-body force. Higher-order rotational scalars in Q_2 can be included in V_{coll} in order to accommodate more complicated potential forms [17]. For instance, a cubic term introduces a γ-dependence into the potential, and an effective interaction using an exponential in $Q_2 \cdot Q_2$ has recently been explored for the description of several p- and sd-shell nuclei [11, 40, 41].

The collective potential mixes states from different major shells since the collective quadrupole operator Q has non-vanishing matrix elements between shells differing by zero or two oscillator quanta [19, 20]. Thus, the potential generates coherent multi-shell admixtures in the wave functions of the system and so achieves the experimentally observed nuclear deformations and absolute $B(E2)$ rates, without the need for effective charges.

A Hamiltonian that preserves the dynamical symmetry and is relevant to the description of deformed nuclei is obtained by combining the collective potential with a harmonic-oscillator term H_0, which provides the background shell structure, and an SU(3) residual interaction:

$$H = H_0 + V_{\text{coll}} + V_{\text{res}}. \tag{16}$$

The effective residual interaction, V_{res}, is included to replace non-collective components of a more realistic Hamiltonian and the neglected effects of couplings to other Sp(3, $\mathbb{R}$) representations. It is usually expressed in terms of the SU(3) $\supset$ SO(3) integrity basis [42], i.e., it is a rotationally invariant function of the SU(3) generators. The integrity basis consists of the second- and third-order Casimir invariants, $\hat{C}_2$ and $\hat{C}_3$, of SU(3), which are just constants within any SU(3) representation, the square of the angular momentum operator, $\hat{L}^2$, and the three- and four-body scalars $\hat{X}_3 \equiv (\hat{L} \times Q^E)_{(1)} \cdot \hat{L}$ and $\hat{X}_4 \equiv (\hat{L} \times Q^E)_{(1)} \cdot (\hat{L} \times Q^E)_{(1)}$. Such a residual interaction does not couple different major oscillator shells, but it produces the observed K-band splitting, which is important for triaxial nuclei. In principle, a more general residual interaction can be constructed from U(3) scalar operators which belong to the enveloping algebra of Sp(3, $\mathbb{R}$) [42, 43].

The dynamical symmetry limit of the symplectic shell model is particularly well suited for the description of deformed nuclei. These nuclei exhibit collective behavior, that is, modes of excitation in which an appreciable fraction of the nucleons in the system participate in a coherent manner, as, for example, is the case for rotations. Symmetry-breaking single-particle spin-orbit and orbit–orbit terms, as well as pairing and other interactions, are not included in this first, approximate, treatment of the nuclear system.

Indeed, in the simplest implementation the Hilbert space of the system is truncated to one single symplectic representation. This is accomplished by selecting the vertical cone (symplectic irrep) constructed from the leading starting irrep of the $0\hbar\omega$ space. The leading irrep is defined to be the U(3) representation, $N_\sigma(\lambda_\sigma\, \mu_\sigma)$, from the lowest layer with the most symmetric spatial permutation symmetry consistent with the Pauli principle, and the maximal possible SU(3) Casimir operator value, $\langle \hat{C}_{\text{SU(3)}} \rangle [(\lambda_\sigma\, \mu_\sigma)]$, see Eq. (10). The single symplectic-irrep approximation works best for well-deformed nuclei. We will discuss such a single-irrep application next.

3.2. *Dynamical Symmetry Limit: The* ^{24}Mg *Example*

As a representative example for an application of the symplectic model in the dynamical symmetry limit, we consider the triaxially deformed nucleus ^{24}Mg. This system has been the subject of several symplectic model studies [44–46]; we summarize here the findings presented in Ref. [46] which tested the model's ability to predict electron scattering observables. The ^{24}Mg nucleus has four valence protons and neutrons in the ds-shell each and its leading Sp(3, $\mathbb{R}$) irrep is $N_\sigma(\lambda_\sigma\,\mu_\sigma) = 62.5(8\,4)$. Since both $\lambda_\sigma \neq 0$ and $\mu_\sigma \neq 0$, the symplectic Hilbert space has a rich structure. The $(8\,4)$ representation at $0\hbar\omega$ contains three rotational bands: a $K = 0$ band with $L = 0, 2, 4, 6, 8$, a $K = 2$ band with $L = 2, 3, \ldots, 10$, and a $K = 4$ band with $L = 4, 5, \ldots, 12$. At the $2\hbar\omega$ level, there are six possible SU(3) irreps, $(10\,4)$, $(8\,5)$, $(6\,6)$, $(9\,3)$, $(7\,4)$, and $(8\,2)$, which contain a total of four $K = 0$, two $K = 1$, four $K = 2$, two $K = 3$, three $K = 4$, one $K = 5$, one $K = 6$, ... bands. At the $4\hbar\omega$ level, there are 15 different SU(3) irreps, at $6\hbar\omega$, there are 25, etc. Accordingly, the number of states for a given angular momentum value L increases with the inclusion of higher excitations, but is very manageable. The symplectic model space for this case contains only 1222 states for a $(0+2+4+6)\hbar\omega$ (or $N_{\max} = 6$) calculation that includes states with $L = 0, 1, 2, \ldots, 8$.

The collective potential was chosen to be a simple quadratic, rotationally invariant, function of the microscopic quadrupole moment, and use was made of the integrity-basis operators $\hat{X}_3$ and $\hat{X}_4$ discussed above, in order to reproduce the experimentally observed K-band splitting in the spectrum of ^{24}Mg. Using the Hamiltonian

$$H_{^{24}\mathrm{Mg}} = H_0 - \chi Q_2 \cdot Q_2 + c_3\hat{X}_3 + c_4\hat{X}_4, \tag{17}$$

least squares fits were carried out to reproduce measured energies and $B(E2)$ values for $N_{\max} = 2$, 4, and 6 symplectic spaces.

Figure 3 (top) displays the energies obtained with the $N_{\max} = 6$ calculation (right part of the figure) in comparison with the experimental values [47–49] (left side). In addition to the ground

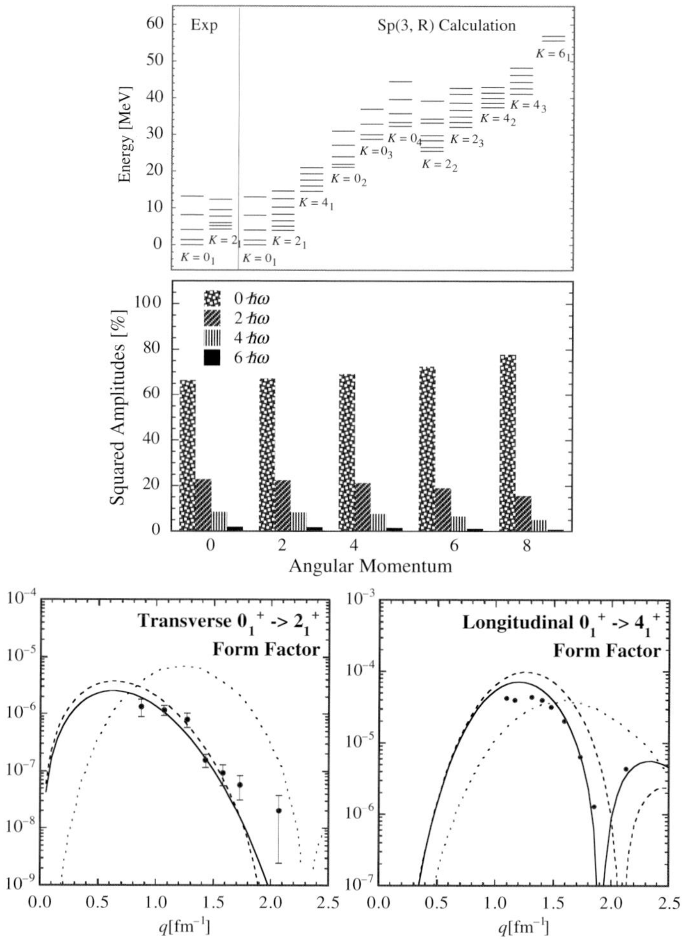

Fig. 3. Energy spectra (top), ground-band eigenstate decomposition (center) and form factor predictions (bottom) for ^{24}Mg, from a calculation with the symmetry-preserving Hamiltonian of Eq. (17). (Top) Calculated $0\hbar\omega$-dominated bands ($K = 0_1, 2_1, 4_1$) and some $2\hbar\omega$-dominated bands ($K = 0_2, 0_3, 0_4, 2_2, 2_3, 4_2, 4_3, 6_1$) are compared to experimental ground and γ-band energies [47, 49]. (Center) Contributions from $0\hbar\omega, \ldots, 6\hbar\omega$ excitations to the wave functions of members of the ground-state rotational band are shown for $L = 0, \ldots, 8$. (Bottom) Form factor predictions for transitions to selected low-lying 2^+ and 4^+ states. The dashed and solid curves correspond to $N_{\max} = 2$ and $N_{\max} = 4$ symplectic calculations, respectively. No effective charges were introduced, since multi-shell correlations are taken into account explicitly. The $0\hbar\omega$ results (dotted curves) are taken from Ref. [28]. All form factors are corrected for center-of-mass motion and the finite-size effect of the nucleon; experimental results [51, 52] are plotted at their effective momentum transfer.

($K = 0_1$) and γ ($K = 2_1$) bands, the calculated $K = 4_1$ band, which is dominated by $0\hbar\omega$ configurations, and several low-lying rotational $K = 0, 2, 4$, and 6 bands, which are dominantly $2\hbar\omega$ excitations, are shown. The decompositions of the rotational ($L = 0, 2, 4, 6, 8$) states of the ground-state band are shown in the central panel of the figure. Table 1 lists various $B(E2)$ transition rates between the low-lying states of ^{24}Mg. The results of the symplectic calculations are found to be in good agreement with the data, despite the restricted model space and the relatively simple Hamiltonian. Specifically, the γ-band is correctly located and nearly all the calculated intraband

Table 1. $B(E2)$ strengths of ^{24}Mg. Results from $N_{\rm max} = 2$, 4, and 6 symplectic calculations, from Ref. [46], are compared to experiment [47, 53]. Both intraband and interband transitions between states of the ground ($K = 0_1$) and γ ($K = 2_1$) band are given; no effective charges were employed. In the last line the static quadrupole moment of the 2_1^+ state (in units of eb) is shown, along with the experimental value by Spear *et al.* [53].

Transition				Model $B(E2)$ [W.u.]			$B(E2)$ [W.u.]
J_i	K_i	J_f	K_f	$N_{\rm max} = 2$	$N_{\rm max} = 4$	$N_{\rm max} = 6$	Exp.
2	0_1	0	0_1	17.2	20.2	20.4	20.5 ± 0.6
4	0_1	2	0_1	24.5	26.9	26.9	23 ± 4
6	0_1	4	0_1	25.2	25.5	25.2	34^{+36}_{-10}
8	0_1	6	0_1	24.4	19.4	19.2	16^{+25}_{-6}
3	2_1	2	2_1	31.6	35.6	35.3	34 ± 6
4	2_1	2	2_1	9.7	11.2	11.0	16 ± 3
5	2_1	3	2_1	15.3	17.0	16.6	28 ± 5
5	2_1	4	2_1	17.3	18.0	17.7	14 ± 6
6	2_1	4	2_1	15.3	19.4	18.3	23^{+23}_{-8}
8	2_1	6	2_1	12.4	18.0	15.9	≥ 3
2	2_1	0	0_1	1.1	1.3	1.3	1.4 ± 0.3
2	2_1	2	0_1	2.2	1.7	1.9	2.7 ± 0.4
3	2_1	2	0_1	1.9	2.4	2.3	2.1 ± 0.3
4	2_1	2	0_1	0.2	1.0	0.9	1.0 ± 0.2
4	2_1	4	0_1	2.9	2.1	2.3	1.0 ± 1.0
5	2_1	4	0_1	1.0	2.4	2.0	3.9 ± 0.8
6	2_1	4	0_1	0.2	1.2	1.0	$0.8^{+0.8}_{-0.3}$
Q [eb]				−0.171	−0.186	−0.185	-0.18 ± 0.02

and interband transition rates fall, without the use of an effective charge, within experimental uncertainties. The $N_{\max} = 4$ results are better than the $N_{\max} = 2$ results, with the $N_{\max} = 6$ calculation yielding only moderate improvements.

Also shown are electron scattering form factors (bottom panel of Fig. 3). Predictions for form factors for transitions to low-lying states of ^{24}Mg are compared to experimental values, for both the longitudinal $0_1^+ \rightarrow 4_1^+$ form factor and the transverse $0_1^+ \rightarrow 2_1^+$ form factor. The form factors are known to probe collective nuclear motion, e.g., reproducing longitudinal form factors often requires the use of effective charges. Moreover, transverse form factors are sensitive to the nature of the current flows associated with the collective motion; they have been found to be sensitive to vorticity degrees of freedom [50]. The Sp(3, $\mathbb{R}$) model accommodates different types of collective flows, including irrotational and rigid flows, and no effective charges were introduced in the calculations shown here. The $0\hbar\omega$ results are taken from an SU(3) calculation [28] that includes symmetry-breaking terms such as the spin-orbit interaction, while the $N_{\max} = 2$ and $N_{\max} = 4$ results employ the dynamical symmetry Hamiltonian of Eq. (17) and a single symplectic-irrep approximation. The calculations demonstrate good agreement with the available data when multi-shell correlations are taken into account; effective charges are not needed. Additional form factors are shown in Ref. [46].

4. Beyond the Dynamical Symmetry Limit: Describing Realistic Nuclear Systems in the Sp(3, $\mathbb{R}$) Model

The ^{24}Mg application discussed in the previous section illustrates that a schematic, symmetry-adapted Hamiltonian, evaluated in a rather restricted model space, is able to reproduce a number of experimental observables, including electron scattering form factors. This, along with other results found in symplectic model studies, provides some indication that the Sp(3, $\mathbb{R}$) symmetry, which has guided the model space construction, Hamiltonian selection, and truncation imposed, is able to capture important aspects of the nuclear many-body system. In order to better identify how the symplectic approach can be used to render large model spaces required for descriptions of realistic nuclei manageable, further work is need. It is necessary to consider realistic interactions, employ larger model spaces to assess the validity of possible truncation schemes, and describe a wider variety of nuclei and observables to expand on the capabilities of the approach and test its limits.

4.1. *Symmetry-breaking Interactions in the* Sp(3, $\mathbb{R}$) *Model*

Realistic nuclear forces are known to break SU(3) and symplectic symmetries. Important interactions, such as pairing and the spin-orbit interaction, are know to mix states belonging to different Sp(3, $\mathbb{R}$) representations. In Refs. [25–27, 54], e.g., the symmetry-breaking effects of these interaction were demonstrated for a complete model space consisting of a single major oscillator shell. In Ref. [55], Johnson *et al.* compared wave functions obtained in a highly truncated subspace of the complete $0\hbar\omega$ SU(3) model space to those obtained from a fully interacting $0\hbar\omega$ shell-model calculation. For the latter, wave functions for several *ds*- and *fp*-shell nuclei were calculated using the semi empirical Brown–Wildenthal USD and the modified Kuo–Brown interaction, respectively. The symmetry-based model space was restricted to a single SU(3) irrep, comprising the $0\hbar\omega$ subspace of an Sp(3, $\mathbb{R}$) irreducible representation. Overlaps between the resulting wave function and the "exact" wave functions calculated in the fully interacting shell model were found to be small in several cases, and this difference was attributed to the effect of the spin-orbit interaction. Even though these studies were restricted to $0\hbar\omega$ subspaces of the full symplectic model space, they illustrate the need for implementing a truncation scheme that accommodates mixing due to symmetry-breaking terms in realistic nuclear interactions.

Several studies have moved beyond the single symplectic-irrep model space: the effects of including pairing and spin-orbit effects in the Hamiltonian were investigated first in Ref. [56]. In this study of ^{20}Ne, Draayer *et al.* included the symmetry-breaking terms in the $0\hbar\omega$ subspace (schematically, this corresponds to allowing terms that connect the lowest filled circles of the cones in Fig. 2 to each other), and symmetry-preserving collective terms that generate multi-$\hbar\omega$ mixing (linking states *within* each cone of Fig. 2). Including both types of interactions was found to be essential for reproducing the rotational moment of inertia, the energy spectrum, the 2^+ quadrupole moment, and the ground-band $B(E2)$ values, without the use of effective charges. More recent studies [11, 40, 41] have followed a similar strategy for including specific symmetry-preserving multi-$\hbar\omega$ collective and symmetry-breaking $0\hbar\omega$ terms in the calculations.

The early treatments [56], as well as the more recent ones [11, 40, 41], have not yet included mixing between states that belong to different symplectic irreps and, at the same time, are excited to $N\hbar\omega$ and $N'\hbar\omega$, with $N \neq 0$ and/or $N' \neq 0$. As discussed in Sec. 2.3, all ingredients for such a calculation are now available and it will be instructive to see the effects of such couplings on calculated wave functions, moments of inertia, and transition strengths.

Some guidance for selecting an initial, horizontally, and vertically truncated, model space can be obtained from a complementary, symmetry-guided multi-shell approach. The recently developed symmetry-adapted no-core shell model (SA-NCSM) [37–39] is a full multi-shell approach that utilizes SU(3) symmetry principles to organize and label the many-body basis states in each $\hbar\omega$ subspace. Even though the SA-NCSM does not explicitly make use of the type of a vertical structure that the symplectic model imposes on the model space, it employs the Sp(3, $\mathbb{R}$) symmetry to guide model space selections. The SA-NCSM model space makes it possible to calculate the energy spectra, eigenstates, and observables for general, symmetry-breaking (one-plus-two-body) Hamiltonians. The resulting eigenstates and observables can then be investigated in terms of their SU(3) content, and symmetry-violating contributions can be identified. Analyses of the calculated eigenstates for multiple nuclei have demonstrated that results of the full no-core shell model can be reproduced in a much smaller (multi-shell) model space. It was found that the relevant subspace is dominated by those basis states which contain the lowest spin values and most deformed SU(3) irreps, with symplectic-like excitations built over these irreps (see also Chapter 8). These calculations indicate that reducing the full model space to a few (predetermined) irreps and including their symplectic multi-$\hbar\omega$ excitations may be sufficient to capture the dominant aspects of the underlying physics.

4.2. *Electron and Light-ion Scattering Reactions*

Experiments provide us with a wide range of observables that can be used to test nuclear theories and gain insights into the underlying physics. Energies, energy sequences organized in rotational bands, and electromagnetic transition rates within and between rotational bands have been the focus of many studies within the Sp(3, $\mathbb{R}$) framework. Electron scattering form factors, measured extensively in the 1970s for many stable nuclei, provide additional information on nuclear structure properties. Developments are underway to carry out electron scattering experiments with short-lived nuclei [57], thus providing new opportunities to expand our understanding of nuclear many-body systems and to test nuclear models. Few microscopic approaches are currently able to provide scattering observables for deformed nuclei without introducing effective charges (see, e.g., the discussion in Ref. [57]). A major strength of the Sp(3, $\mathbb{R}$) model is that it is able to reproduce measured electromagnetic transition strengths and electron scattering form factors without the need for effective charges [46]. Also, it accommodates a range of nuclear flows, from irrotational to rigid collective flow; this is relevant to predictions of transverse form factors (as well as moments of inertia), which depend on the internal current flows [50]. The application summarized in Sec. 3.2, as well as recent work in the framework of the SA-NCSM [58] indicate that the Sp(3, $\mathbb{R}$) structure can provide a meaningful framework for the calculation of such observables.

Since matrix elements of general one-body and two-body operators can be evaluated in the symplectic model (see Sec. 2.3), it is also possible to calculate cross-sections for elastic and inelastic scattering of neutrons and light ions from target nuclei described in the model. To this end, calculated transition densities for transitions from the ground state to various low-lying excited states (and between these states) can be folded with an effective projectile-target nucleon interaction to obtain transition potentials [59]. These potentials can be used in distorted-wave Born approximation (DWBA) or coupled-channels calculations to yield cross sections for inelastic scattering between the low-lying states of interest [60]. These types of calculations are now routinely carried out for (near-)spherical nuclei using transition densities from shell-model or (Q)RPA calculations [61–63], where (Q)RPA stands for (quasiparticle) random-phase approximation. Calculations for deformed target nuclei are more challenging [64, 65], making this an interesting area of future application for the symplectic model (and especially through its link to the cluster model, see Chapter 7).

5. Concluding Remarks

The symplectic shell model provides a valuable link between macroscopic collective descriptions of the nuclear many-body problem and microscopic shell-model approaches. We have seen that its fermionic realization and underlying group structure make microscopic calculations in physically relevant model spaces possible. Procedures and codes are available for evaluating coupling coefficients and matrix elements of symmetry-breaking operators, thus paving the way for moving beyond earlier applications.

Naturally, many challenges remain: as practitioners of traditional multi-shell shell-model calculations are keenly aware of, the efficient implementation of the calculations require careful thought and development [36]. Additional thought has to be given to the most appropriate truncation procedure: calculations in non-truncated spaces indicate that those configurations that have low spin and large deformation are dominant, but it would be desirable to develop some quantitative criteria. In addition, there is presently no computer code for including three-body forces in symplectic calculations (while such an extension is straightforward and under way), with the exception of the integrity-basis terms X_3 and X_4, which contain very specific 3- and 4-body contributions. Finally, essentially all implementations of the symplectic model employ SU(3) tensor operators and SU(3)-coupling schemes. To fully exploit the symplectic structure, the relevant operators should be expanded in terms of Sp(3, $\mathbb{R}$) tensor operators and make use of symplectic coupling and recoupling coefficients. The mathematical structures are quite challenging [18] and remain to be developed.

Acknowledgments

Insightful discussions with Michael Kruse and Vijay Sonnad are gratefully acknowledged. This work was performed under the auspices of the US Department of Energy by Lawrence Livermore National Laboratory under contract No. DE-AC52-07NA27344, with support from the LLNL LDRD program under project 15-FS-005.

References

[1] Nuclear Science Advisory Committee. Reaching for the horizon: The 2015 long-range plan for nuclear science. http://science.energy.gov/np/nsac/ (2016).
[2] G. Rosensteel and D. J. Rowe, Nuclear Sp(3,R) model, *Phys. Rev. Lett.* **38**, 10 (1977).
[3] G. Rosensteel and D. Rowe, On the algebraic formulation of collective models. I. The mass quadrupole collective model, *Ann. Phys.* **123**, 36 (1979).
[4] G. Rosensteel, A recursion formula for Sp(3,R) matrix elements, *J. Math. Phys.* **21**, 924 (1980).
[5] D. J. Rowe, Microscopic theory of the nuclear collective model, *Rep. Progr. Phys.* **48**, 1419 (1985).
[6] Y. Suzuki, Cluster and symplectic excitations in nuclei, *Nucl. Phys. A.* **448**, 395 (1986).
[7] Y. Suzuki and K. T. Hecht, Spin-orbit and tensor interactions in Sp(6,R)-model calculations, *Prog. Theor. Phys.* **77**, 190 (1987).
[8] Y. Suzuki and K. T. Hecht, Symplectic and cluster excitations in nuclei: Evaluation of interaction matrix elements, *Nucl. Phys. A.* **455**, 315 (1986).
[9] D. Rowe, Dynamical symmetries of nuclear collective models, *Prog. Part. Nucl. Phys.* **37**, 265 (1996).
[10] J. Escher and J. P. Draayer, Fermion realization of the nuclear sp(6,R) model, *J. Math. Phys.* **39**, 5123 (1998).

[11] K. D. Launey, J. P. Draayer, T. Dytrych, G.-H. Sun, and S.-H. Dong, Approximate symmetries in atomic nuclei from a large-scale shell-model perspective, *Int. J. Mod. Phys. E.* **24**, 1530005 (2015).

[12] P. Navratil, S. Quaglioni, I. Stetcu, and B. R. Barrett, Recent developments in no-core shell-model calculations, *J. Phys. G.* **36**, 083101 (2009).

[13] T. Abe, P. Maris, T. Otsuka, N. Shimizu, Y. Utsuno, and J. P. Vary, Benchmarks of the full configuration interaction, monte carlo shell model, and no-core full configuration methods, *Phys. Rev. C.* **86**, 054301 (2012).

[14] B. R. Barrett, P. Navrátil, and J. P. Vary, *Ab initio* no core shell model, *Prog. Part. Nucl. Phy.* **69**, 131 (2013).

[15] W. Leidemann and G. Orlandini, Modern *ab initio* approaches and applications in few-nucleon physics with, *Prog. Part. Nucl. Phys.* **68**, 158 (2013).

[16] G. Rosensteel and D. Rowe, On the shape of deformed nuclei, *Ann Phys.* **104**, 134 (1977). 77900483.

[17] G. Rosensteel, Irreducible u(3) tensor interactions in the symplectic collective model, *Nucl. Phys. A.* **341**, 397 (1980).

[18] G. Rosensteel. Sp(3, R) Tensors in nuclear physics. In *Proc. Int. Sym. in honor of K. T. Hecht, Sept. 1991* (1991).

[19] J. P. Elliott, Collective motion in the nuclear shell model. I. Classification schemes for states of mixed configurations, *Proc. Roy. Soc. A.* **245**, 128 (1958).

[20] J. P. Elliott, Collective motion in the nuclear shell model. II. The introduction of intrinsic wave-functions, *Proc. Roy. Soc. A.* **245**, 562 (1958).

[21] J. P. Elliott and M. Harvey, Collective motion in the nuclear shell model. III. The calculation of spectra, *Proc. Roy. Soc. A.* **272**, 557 (1962).

[22] J. Escher and A. Leviatan, Partial dynamical symmetry in the symplectic shell model, *Phys. Rev. C.* **65**, 054309 (2002).

[23] J. P. Draayer and Y. Akiyama, Wigner and Racah coefficients for SU(3), *J. Math. Phys.* **14**, 1904 (1973).

[24] J. Vergados, SU(3) $\supset$ R(3) wigner coefficients in the 2s-1d shell, *Nucl. Phys. A.* **111**, 681 (1968).

[25] J. Escher, C. Bahri, D. Troltenier, and J. Draayer, Pairing-plus-quadrupole model and nuclear deformation: A look at the spin-orbit interaction, *Nucl. Phys. A.* **633**, 662 (1998).

[26] C. Bahri, J. Escher, and J. Draayer, Monopole-pairing and deformation in atomic nuclei, *Nucl. Phys. A.* **592**, 171 (1995).

[27] C. Bahri, J. Escher, and J. Draayer, Monopole-pairing and deformation in atomic nuclei: Erratum, *Nucl. Phys. A.* **594**, 485 (1995).

[28] P. Rochford and J. P. Draayer, Electron scattering multipoles for symplectic shell model applications, *Ann. Phys.* **214**, 341 (1992).

[29] K. Hecht, SU(3) recoupling and fractional parentage in the 2s-1d shell, *Nucl. Phys.* **62**, 1 (1965).

[30] Y. Akiyama and J. P. Draayer, A user's guide to FORTRAN programs for Wigner and Racah coefficients of SU(3), *Comput. Phys. Commun.* **5**, 405 (1973).

[31] K. T. Hecht and Y. Suzuki, Some special SU(3)$\supset$R(3) Wigner coefficients and their application, *J. Math. Phys.* **24**, 785 (1982).
[32] T. Draeger, A Macsyma program for computing analytic Clebsch-Gordan coefficients of U(3)$\supset$SO(3), *Comput. Phys. Commun.* **139**, 246 (2001).
[33] C. Bahri, D. Rowe, and J. Draayer, Programs for generating Clebsch-Gordan coefficients of SU(3) in SU(2) and SO(3) bases, *Comput. Phys. Commun.* **159**, 121 (2004).
[34] D. J. Rowe, Analytical expressions for the matrix elements of the non-compact symplectic algebra, *J. Phys. A: Math. Gen.* **17**, L399 (1984).
[35] C. Bahri and J. P. Draayer, SU(3) Reduced matrix element package, *Comput. Phys. Commun.* **83**, 59 (1994).
[36] C. W. Johnson, W. E. Ormand, and P. G. Krastev, Factorization in large-scale many-body calculations, *Comput. Phys. Commun.* **184**, 2761 (2013).
[37] T. Dytrych, K. D. Sviratcheva, C. Bahri, J. P. Draayer, and J. P. Vary, Evidence for Symplectic Symmetry in *ab initio* No-core-shell-model Results for Light Nuclei, *Phys. Rev. Lett.* **98**, 162503 (2007).
[38] T. Dytrych, K. D. Sviratcheva, J. P. Draayer, C. Bahri, and J. P. Vary, *Ab initio* symplectic no-core shell model, *J. Phy. G.* **35**, 123101 (2008).
[39] T. Dytrych, K. D. Launey, J. P. Draayer, P. Maris, J. P. Vary, E. Saule, U. Catalyurek, M. Sosonkina, D. Langr, and M. A. Caprio, Collective Modes in Light Nuclei from First Principles, *Phys. Rev. Lett.* **111**, 252501 (2013).
[40] A. C. Dreyfuss, K. D. Launey, T. Dytrych, J. P. Draayer, and C. Bahri, Hoyle state and rotational features in Carbon-12 within a no-core shell model framework, *Phys. Lett. B.* **727**, 511 (2013).
[41] G. K. Tobin, M. C. Ferriss, K. D. Launey, T. Dytrych, J. P. Draayer, A. C. Dreyfuss, and C. Bahri, Symplectic no-core shell-model approach to intermediate-mass nuclei, *Phys. Rev. C.* **89**, 034312 (Mar, 2014).
[42] J. Draayer and G. Rosensteel, U(3) $\rightarrow$ R(3) integrity-basis spectroscopy, *Nucl. Phys. A.* **439**, 61 (1985).
[43] R. Gilmore and J. P. Draayer, Dynamical group chains and integrity bases, *J. Math. Phys.* **26**, 3053 (1985).
[44] G. Rosensteel, J. P. Draayer, and K. J. Weeks, Symplectic shell-model calculations for ^{24}Mg, *Nucl. Phys. A.* **419**, 1 (1984).
[45] O. Castaños and J. Draayer, Contracted symplectic model with ds-shell applications, *Nucl. Phys. A.* **491**, 349 (1989).
[46] J. Escher and J. P. Draayer, Electron scattering and multi-$\hbar\omega$ correlations, *Phys. Rev. Lett.* **82**, 5221 (1999).
[47] D. Branford, A. McGough, and I. Wright, Lifetime and branching measurements on the $K^\pi = 0^+$ and $K^\pi = 2^+$ ^{24}Mg rotational levels, *Nucl. Phys. A.* **241**, 349 (1975).
[48] M. Fewell, S. Hinds, D. Kean, and T. Zabel, Static quadrupole moment of the first excited state of ^{24}Mg, *Nucl. Phys. A.* **319**, 214 (1979).
[49] P. Endt, Strengths of gamma-ray transitions in A $= 5 - 44$ nuclei, IV, *Atomic Data and Nucl. Data Tables.* **55**, 171 (1993).

[50] M. Carvalho and D. Rowe, Transverse form factors in the collective and symplectic models, *Nucl. Phys. A.* **618**, 65 (1997).

[51] G. C. Li, M. R. Yearian, and I. Sick, High-momentum-transfer electron scattering from ^{24}Mg, ^{27}Al, ^{28}Si, and ^{32}S, *Phys. Rev. C.* **9**, 1861 (1974).

[52] A. Hotta, R. S. Hicks, R. L. Huffman, G. A. Peterson, R. J. Peterson, and J. R. Shepard, Transverse isoscalar excitations in ^{24}Mg by 180° electron scattering, *Phys. Rev. C.* **36**, 2212 (1987).

[53] R. Spear, Static quadrupole moments of first 2^+ states in the 2s-1d shell: A review of experiment and theory, *Phys. Rep.* **73**, 369 (1981).

[54] C. Vargas, J. G. Hirsch, P. O. Hess, and J. P. Draayer, Interplay between the quadrupole–quadrupole and spin-orbit interactions in nuclei, *Phys. Rev. C.* **58**, 1488 (1998).

[55] C. W. Johnson, I. Stetcu, and J. P. Draayer, SU(3) versus deformed Hartree–Fock state, *Phys. Rev. C.* **66**, 034312 (2002).

[56] J. Draayer, K. Weeks, and G. Rosensteel, Symplectic shell-model calculations for ^{20}Ne with horizontal configuration mixing, *Nucl. Phys. A.* **419**, 1 (1984).

[57] J. M. Yao, M. Bender, and P.-H. Heenen, Beyond-mean-field study of elastic and inelastic electron scattering off nuclei, *Phys. Rev. C.* **91**, 024301 (2015).

[58] T. Dytrych, A. C. Hayes, K. D. Launey, J. P. Draayer, P. Maris, J. P. Vary, D. Langr, and T. Oberhuber, Electron-scattering form factors for ^{6}Li in the *ab initio* symmetry-guided framework, *Phys. Rev. C.* **91**, 024326 (2015).

[59] G. R. Satchler, *Direct Nuclear Reactions.* Oxford University Press, Clarendon, Oxford (1983).

[60] I. J. Thompson, Coupled reaction channels calculations in nuclear physics, *Comput. Phys. Rep.* **7**, 167 (1988).

[61] G. P. A. Nobre, F. S. Dietrich, J. E. Escher, I. J. Thompson, M. Dupuis, J. Terasaki, and J. Engel, Coupled-channel calculation of nonelastic cross sections using a density-functional structure model, *Phys. Rev. Lett.* **105**, 202502 (2010).

[62] G. P. A. Nobre, F. S. Dietrich, J. E. Escher, I. J. Thompson, M. Dupuis, J. Terasaki, and J. Engel, Toward a microscopic reaction description based on energy-density-functional structure models, *Phys. Rev. C.* **84**, 064609 (2011).

[63] M. Dupuis, T. Kawano, J.-P. Delaroche, and E. Bauge, Microscopic model approach to (n,xn) pre-equilibrium reactions for medium-energy neutrons, *Phys. Rev. C.* **83**, 014602 (2011).

[64] J. Terasaki and J. Engel, Self-consistent skyrme quasiparticle random-phase approximation for use in axially symmetric nuclei of arbitrary mass, *Phys. Rev. C.* **82**, 034326 (2010).

[65] S. Péru and M. Martini, Mean field based calculations with the gogny force: Some theoretical tools to explore the nuclear structure, *Eur. Phys. J. A.* **50**, 88 (2014).

PART 3

Orderly Patterns from First Principles

Chapter 5

Lattice Quantum Chromodynamics: Symmetries and Applications

Thomas Luu* and Andrea Shindler†
Institute for Advanced Simulation (IAS-4)
Institut für Kernphysik (IKP-3)
Jülich Center for Hadron Physics
Forschungszentrum Jülich, 52425 Jülich, Germany
**t.luu@fz-juelich.de*
†a.shindler@fz-juelich.de

We give a short review of lattice quantum chromodynamics (LQCD) and its status, concentrating on its symmetries and lack thereof. We describe the formalism and motivate the use of large-scale simulations in this field. We pay particular attention to the symmetries intrinsic to the finite-volume discretized lattice, and show how clever applications of the finite-volume symmetries improve numerical simulations. We discuss LQCD calculations related to fundamental symmetries and comment on future prospects of this field from a symmetries point of view.

Contents

1. Introduction

Lattice quantum chromodynamics (LQCD) has evolved into a powerful computational tool for observables relevant to nuclear physics, high-energy physics, and more recently beyond-the-standard-model (BSM) physics. This is in large part due to the enormous advances in computer resources as well as algorithmic advances. Calculations of light nuclei, for example, are now being readily calculated and point to a rich spectrum when the masses of the quarks are larger than their physical value [1, 2]. The scattering parameters obtained from Lattice QCD calculations of multi-hadron systems provide powerful insights into the nature of the nuclear interaction [3–5], and promise to one day answer questions related to few-nucleon interactions that have not been satisfactorily constrained to date by experiment. Lattice QCD calculations are also probing hyper-nuclei [2, 6] and other exotic nuclear systems [7, 8] that existed shortly after the Big Bang, or in dense astrophysical objects. Such calculations have implications on the broader nuclear physics community since they can help determine the role of these systems in dense environments, such as during supernovae. Coupled with quantum electrodynamics (QED), recent Lattice QCD calculations have been successful in determining the neutron–proton mass difference [9] (see Refs. [10, 11] for new proposals of lattice QED calculations) for the magnetic moments of light nuclei [12]. The importance of Lattice QCD is not constrained solely to the strong and electromagnetic interactions of hadronic systems, however. In disentangling the contributions of hadronic and BSM interactions, Lattice QCD is fast becoming the "go-to" tool for such studies.

The symmetries of Lattice QCD play an important role in the algorithmic formulation of QCD. The implications of a discretized space–time lattice on various symmetries, and in particular chiral symmetry, have profound implications on the types of observables that can be assessed via Lattice QCD. In certain cases the preservation of such symmetries is of utmost importance. Not surprisingly, the demands on computational resources due to such preservation means that Lattice QCD is always hungry for more compute cycles. High-performance computing (HPC) will *always* have a resource-hungry customer: Lattice QCD.

An exhaustive review of the Lattice QCD formulation, let alone all the recent calculations that Lattice QCD has performed, is beyond the scope (and length limit) of this Chapter. For such reviews, we refer the reader to Beane *et al.* [13], for example, and references therein. Here we concentrate, instead, on the role of symmetries within QCD and discuss the impact of these symmetries (and their discrete space–time analogs) in Lattice QCD calculations. Naturally, the examples discussed in this chapter are biased toward the authors' main research areas. In turn, we discuss how Lattice QCD utilizes symmetries to extract fundamental quantities in hadronic physics. And just as important, we show how Lattice QCD can directly calculate quantities that are directly related to fundamental symmetries.

2. The Symmetries of the Quantum Chromodynamics Lagrangian

The QCD Lagrangian is based on local SU(3) color symmetry and is given by

$$\mathcal{L}_{\text{QCD}} = -\frac{1}{4} F^a_{\mu\nu} F^{\mu\nu}_a + \bar{\psi}_f \left(\mathrm{i}\not{D} - m_f \right) \psi_f, \tag{1}$$

where ψ_f is one of N_f fermion quark fields with flavor f that carries both color and spinor degrees of freedom and m_f is the quark mass of that flavor. In Eq. (1) and subsequent equations we use the Einstein convention where repeated indices are summed. We reserve Roman letters to indicate color degrees of freedom and Greek symbols for spinor degrees of freedom. The index a in this case runs as $a = 1, \ldots, 8$, while the index f runs as $f = 1, \ldots, N_F$, where the maximum value of N_F is 6.

The field tensor $F^a_{\mu\nu}$ describing the kinematics of the gluon gauge fields is

$$F^a_{\mu\nu} = \partial_\mu A^a_\nu - \partial_\nu A^a_\mu - g_s f_{abc} A^b_\mu A^c_\nu \tag{2}$$

and the covariant derivative $\not{D} = \gamma^\mu D_\mu$, which describes the coupling between the quarks and gluons with coupling strength g_s, is

$$D_\mu = \partial_\mu \mathbb{I} + \mathrm{i} g_s T^a A^a_\mu, \tag{3}$$

where $\mathbb{I}$ is the identity matrix in color space. Note that the coupling g_s is independent of the quark flavor. The generators T^a of SU(3) satisfy the following Lie algebra,

$$[T^a, T^b] = \mathrm{i} f_{abc} T^c, \tag{4}$$

where f_{abc} are the structure constants of the group and the generators T^a consist of 8 matrices of dimension 3×3 (see, e.g., Georgi [14]). The group SU(3) is non-Abelian since any two elements of the group, Λ_{α_1} and Λ_{α_2}, do not necessarily commute, $[\Lambda_{\alpha_1}, \Lambda_{\alpha_2}] \neq 0$. Inputs to the Lagrangian in Eq. (1) are the bare masses of the quark flavors and the coupling g_s. This theory is completely self-consistent once quark masses and couplings have been appropriately renormalized.

The Lagrangian in Eq. (1) is invariant under *local* SU(3) transformations,

$$\psi_f \to \Lambda_\alpha(x)\psi_f, \tag{5}$$

$$A_\mu \to \Lambda_\alpha(x)A_\mu\Lambda_\alpha^{-1}(x) + \frac{\mathrm{i}}{g_s}\left(\partial_\mu\Lambda_\alpha(x)\right)\Lambda_\alpha^{-1}(x), \tag{6}$$

where $\Lambda_\alpha(x)$ is spatially dependent and is an element of SU(3). Quarks, anti-quarks, and gluons fall under the $\mathbf{3}$, $\mathbf{3}^*$, and $\mathbf{8}$ irreducible representations of SU(3), respectively.[a] In the limit when all quark flavors are massless, $m_f \to 0$, the quark content of the Lagrangian in Eq. (1) can be written in terms of the left and right helicities of the quark fields $\psi_{f,L/R}$,

$$\bar{\psi}_{f,L}\mathrm{i}\not{D}\psi_{f,L} + \bar{\psi}_{f,R}\mathrm{i}\not{D}\psi_{f,R} + \mathcal{L}_{\text{gluons}}. \tag{7}$$

In this limit, the left and right helicities decouple and the Lagrangian is invariant under separate rotations of any $N_f \times N_f$ unitary matrix for each helicity. In other words, the Lagrangian possesses $\mathrm{U}(N_f)_L\times\mathrm{U}(N_f)_R$ *chiral* (*flavor*) symmetry. In addition to the separate left/right independent symmetries, particular linear combinations of simultaneous transformations of left and right fields also leave the Lagrangian invariant. These linear combinations correspond to vector (V) and axial-vector (A) transformations of the left/right fields, and as such the group $\mathrm{U}(N_f)_L\times\mathrm{U}(N_f)_R$ decomposes to $\mathrm{SU}(N_f)_L\times\mathrm{SU}(N_f)_R\times\mathrm{U}(1)_A\times\mathrm{U}(1)_V$.

From Noether's theorem we know that with each associated symmetry there exists a conservation law. For example, the global $\mathrm{U}(1)_V$ symmetry is the source of baryon number conservation in QCD. The symmetries and their ensuing conservation laws provide powerful tools in understanding and predicting the spectrum of the underlying theory, classically speaking. However, with quantum field theories the symmetries of the Lagrangian alone are not sufficient in this type of analysis, since certain symmetries can be broken

[a] Here we denote the irreps by their dimensionality. One could instead denote the irreps by a pair of numbers $(p\,q)$ related to the SU(3) Casimir eigenvalues [14], e.g., $\mathbf{8} = (1\,1)$.

dynamically or through *quantum fluctuations*. Probably the most famous of these cases (we are still considering the massless case) is the spontaneous breaking of chiral symmetry due to the non-zero expectation value of the quark condensate $\langle \bar{\psi}_L \psi_R \rangle$ at sufficiently low energies. This subsequently leads us to the generation of massless Nambu–Goldstone bosons (via Goldstone's theorem), which we identify as the pions and kaons we observe in nature. Central to this point is the fact that the quark condensate $\langle \bar{\psi}_L \psi_R \rangle$ does not respect chiral symmetry. This chiral symmetry of the Lagrangian is *hidden* in the ensuing spectrum. Chiral symmetry, and its violation thereof, takes on additional meaning when space–time is discretized. We will discuss this in some detail in Sec. 3.1.3. Another aspect of symmetry breaking occurs when quantum fluctuations break the chiral flavor $\mathrm{U}(1)_A$ symmetry through the chiral anomaly generated by the diagram shown in Fig. 1. A consequence of this breaking is that the η (η') meson is not identified as a Goldstone boson and as such, its mass is anomalously large compared to the pions and kaons, and the decay rate of the neutral pion π_0 to two photons is substantially

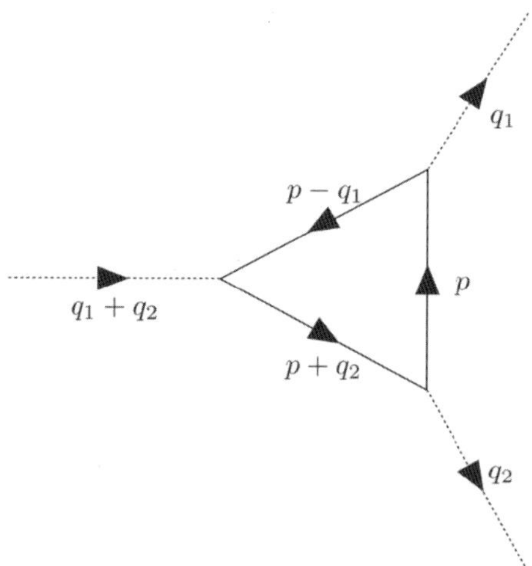

Fig. 1. Triangle diagram leading to the chiral anomaly and subsequent decay of $\pi_0 \to \gamma\gamma$. The solid black lines refer to the triangle fermion loop. The left external dotted line represents the incoming π_0, while the remaining upper and lower right external lines represent photons.

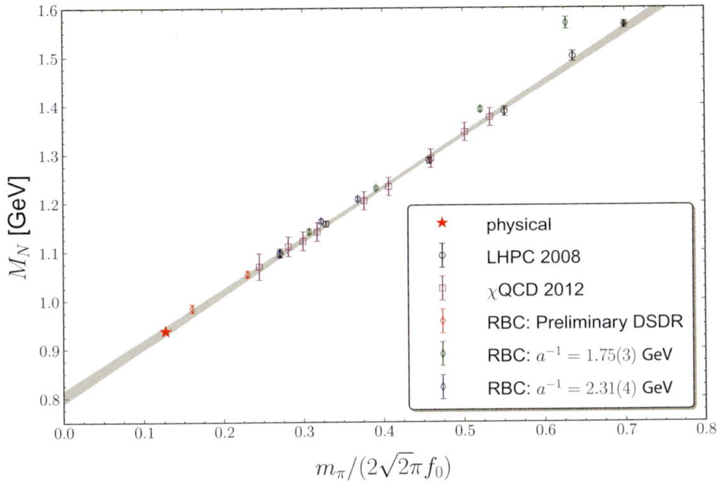

Fig. 2. Dependence of nucleon mass on pion mass calculated from Lattice QCD. Physical value is marked as a red star (figure courtesy of Walker-Loud). The linear behavior is a surprising result since chiral perturbation theory predicts a quadratic dependence.

different from what is expected using chiral symmetry arguments. We discuss the $U(1)_A$ problem in further detail in Sec. 5.1.

Even if dynamical symmetry breaking did not occur, these symmetries would be *explicitly* broken due to non-zero quark masses. If one considers only the three lightest quarks, u (up), d (down), and s (strange), the masses of these quarks are small enough such that these symmetries could still be considered approximate. For example, the near degenerate masses of the up and down quarks, compared to the QCD scale $\Lambda_{\text{QCD}} \sim 200$ MeV, means that isospin symmetry plays a strong role in the spectrum of light hadrons, such as the nucleon. In Fig. 2 we show the dependence of the nucleon mass on the pion mass (or equivalently the light-quark masses), as calculated from Lattice QCD.

3. Symmetries of Lattice Quantum Chromodynamics

The breaking of conformal symmetry leads to asymptotic freedom at high energies, while at low energies QCD exhibits confinement.

The strongly-interacting nature of QCD at such low energies precludes the use of perturbation theory in analyzing its spectrum, and one must resort to non-perturbative methods. Rather, numerical methods must be implemented. Lattice QCD offers the only procedure (to date) to rigorously calculate observables, with quantifiable uncertainties, in this low-energy regime.

Lattice QCD calculations are performed in Euclidean time and therefore there is no distinction between covariant and contravariant indices. The QCD action $S_M[\bar{\psi}, \psi, A]$ in Minkowski space, defined as the integral over space–time of the Lagrangian in Eq. (1), is obtained in Euclidean space by performing the Wick rotation $t \to -\mathrm{i}\tau$,

$$S_M \to \mathrm{i}S_M \equiv S_E = -\int dx^4 \left(\frac{1}{4}\left(F^a_{\mu\nu}\right)^2 + \bar{\psi}\left(\not{D} + m\right)\psi\right). \quad (8)$$

We have dropped the flavor index in the fermion fields and masses, so the sum over flavors is implicit. The Euclidean gamma matrices are defined as

$$\{\gamma_\mu, \gamma_\nu\} = 2\delta_{\mu\nu}. \quad (9)$$

The action in Eq. (8) plays a central role in Lattice QCD calculations since the expectation value of any operator O is given by[b]

$$\langle O \rangle = \frac{1}{\mathcal{Z}}\int \mathcal{D}A\mathcal{D}\bar{\psi}\mathcal{D}\psi \; O[A] \exp\left\{-S[\bar{\psi}, \psi, A]\right\} \quad (10)$$

$$= \frac{1}{\mathcal{Z}}\int \mathcal{D}A \; O[A] \det(M[A]) \exp\left\{-\int dx^4 \frac{1}{4}\left(F^a_{\mu\nu}\right)^2\right\} \quad (11)$$

$$\equiv \frac{1}{\mathcal{Z}}\int \mathcal{D}A \; O[A] \det(M[A]) \exp\{-S_G[A]\}, \quad (12)$$

where $M[A] \equiv \not{D} + m = \sum_\mu \gamma_\mu D_\mu + m$, det indicates a determinant over all flavors, and

$$\mathcal{Z} = \int \mathcal{D}A\mathcal{D}\bar{\psi}\mathcal{D}\psi \exp\left\{-S[\bar{\psi}, \psi, A]\right\}. \quad (13)$$

[b]Hereafter we assume that calculations will be performed in Euclidean space and therefore we drop the E subscript of the Euclidean action (8).

The symbol $\mathcal{D}$ in Eqs. (10)–(13) denotes the integration measure, and should not be confused with the covariant derivative D_μ. This is Feynman's path integral formalism applied to (Euclidean) QCD. It involves *functional* integrals over the fields A, $\bar{\psi}$, and ψ. To arrive at Eq. (11) from Eq. (10), the *Grassmann* fermion fields $\bar{\psi}$ and ψ were explicitly integrated out by taking advantage of the fermionic bilinear form of the action S, which produced the determinant in Eq. (11). Since the action S is real and bounded from below, the term $\exp\{-S\}$ in Eq. (10) can be viewed as a Boltzmann factor. Consequently $\mathcal{Z}$ can be interpreted as a partition function.

3.1. *Symmetries in Discrete Space and Time*

In practice, the functional integrals in Eq. (11) are not analytically possible, and therefore numerical methods are needed. Lattice QCD calculations make use of space–time that is approximated by a discrete hyper-cubic (four-dimensions) lattice, with lattice spacing a_s between spatial points and temporal lattice spacing a_t within the time dimension. Physical spatial dimensions L of the cube are typically of order $L \sim 4 - 6\,\text{fm}$. The typical lattice spacing sizes are $a_{t,s} \sim 0.05 - 0.1\,\text{fm}$. For the remainder of the paper we will assume that $a_s = a_t = a$. Periodic boundary conditions (PBCs) are typically employed at the faces of the spatial cube, while anti-periodic boundary conditions (aPBCs) are used in the temporal directions to account for the antisymmetry of the fermions.

Lattice QCD calculations are equivalent to simulating (Euclidean) QCD at finite temperature, with the inverse temperature β defined as the length of the time direction T. Lattice QCD thermodynamic simulations therefore typically have $T < L$, whereas "cold" QCD lattice calculations use $T \gtrsim L$. Because of the finite volume, the allowed momentum modes are discrete and given by $\vec{p}_n = 2\pi\vec{n}/L$, where $\vec{n} = (n_x, n_y, n_z)$ are a triplet of integers. The discretization further restricts the momenta to reside within the first Brillouin zone, for example, $-\pi/a \leq p_x < \pi/a$, and similarly for the other directions. Thus, the discretization acts as an ultraviolet cutoff.

3.1.1. *Preserving gauge symmetry on the lattice with the Wilson action*

Wilson [15] showed that it was possible to maintain gauge symmetry on the lattice by re-expressing the QCD action in terms of *parallel transporters* $U(x, y)$, where

$$U(x, x + a\hat{\mu}) = \exp\left(-ig \int_x^{x+a\hat{\mu}} ds A_\mu(s)\right) \tag{14}$$

$$\equiv U_\mu(x). \tag{15}$$

Note that $U^\dagger_\mu(x) = U_{-\mu}(x)$. $U_\mu(x)$ is an SU(3) unitary matrix and as such transforms as

$$U_\mu(x) \to \Lambda_\alpha(x) U_\mu(x) \Lambda^\dagger_\alpha(x + a\hat{\mu}). \tag{16}$$

Maintaining gauge invariance is essential to reducing the number of needed parameters in numerical simulations, as many renormalizations are protected due to the preserved symmetry.

The covariant derivative D_μ acting on ψ can be approximated as

$$\begin{aligned} D_\mu \psi(x) &= \frac{1}{2}(\nabla_\mu + \nabla^*_\mu)\psi(x) \\ &= \frac{1}{2a}[U_\mu(x)\psi(x + a\hat{\mu}) - U_{-\mu}(x)\psi(x - a\hat{\mu})] \end{aligned} \tag{17}$$

where

$$\begin{aligned} \nabla_\mu \psi(x) &= \frac{1}{a}[U_\mu(x)\psi(x + a\hat{\mu}) - \psi(x)], \\ \nabla^*_\mu \psi(x) &= \frac{1}{a}[\psi(x) - U_{-\mu}(x)\psi(x - a\hat{\mu})]. \end{aligned} \tag{18}$$

The "symmetric" form of the derivative, together with thermal boundary conditions, ensures charge-conjugation symmetry. The gauge action can be expressed as

$$S_G[U] = \frac{6}{g^2} \sum_{x,\mu,\nu} \left(1 - \frac{1}{3} \Re \, \mathrm{Tr} U_{\mu\nu}(x)\right) \tag{19}$$

$$= \int dx^4 \frac{1}{4} (F_{\mu\nu})^2 + O(a^2), \tag{20}$$

where $U_{\mu\nu}(x)$ is called a *plaquette* and is given by

$$U_{\mu\nu}(x) = U_\mu(x) U_\nu(x + a\hat{\mu}) U_\mu^\dagger(x + a\hat{\mu} + a\hat{\nu}) U_\nu^\dagger(x + a\hat{\nu}). \tag{21}$$

Under the trace, the plaquette, and any other objects built from it, is invariant under gauge transformations. In the classical limit that $a \to 0$, one has that $S_G[U] \to S_G[A]$ up to an irrelevant (infinite) constant. The group elements $U_\mu(x)$ can be interpreted as "gauge links" since they reside between the nodes of the lattice, whereas the fermions themselves reside on the lattice nodes. The Wilson action,

$$S_W[\bar{\psi}, \psi, U] = S_G[U] + S_F[\bar{\psi}, \psi, U] - \frac{a}{2} \bar{\psi} \nabla_\mu \nabla_\mu^* \psi, \tag{22}$$

combines the gauge action $S_G[U]$ built from links in Eq. (19), with the fermion action $S_F[\bar{\psi}, \psi, U]$ using the covariant derivative defined in Eq. (17), as well as an additional *Wilson* term $-\frac{a}{2} \bar{\psi} \nabla_\mu \nabla_\mu^* \psi$ which removes fermion doublers in the continuum $a \to 0$ limit [15]. We discuss the Wilson term and its consequences in more detail in the following sections.

To compute an observable O, one first generates ensembles $\{U\}$ of gauge links with probability distribution

$$P(U) = \frac{1}{\mathcal{Z}} \det(M[U]) \exp\{-S[U]\}. \tag{23}$$

The expectation value of O is then estimated by

$$\langle O \rangle \approx \frac{1}{N_{\mathrm{cf}}} \sum_i O[U_i] \quad \text{for } U_i \in \{U\}, \tag{24}$$

where N_{cf} is the number of configurations within the ensemble $\{U\}$.

3.1.2. *Discretization errors and the continuum limit*

To relate Lattice QCD numerical calculations with experimental results the *continuum limit*, $a \to 0$, of Lattice QCD is of fundamental importance. From a practical point of view performing the continuum limit amounts to calculating at several values of the lattice spacing a the quantity of interest, and subsequently performing an extrapolation to $a = 0$. The rate of the discretization errors is highly constrained by the symmetries of the Lattice QCD action and a symmetry analysis can be used to improve, and in many cases remove, the leading lattice artifacts. A very important role in this respect is played by chiral symmetry.

To understand the form of the leading discretization errors one writes down the most general effective theory that describes Lattice QCD for energy lower than the inverse lattice spacing. This effective theory, the so-called Symanzik effective action [16], is highly constrained by the symmetries of the lattice action. The simplest realization of Lattice QCD, the so-called Wilson action introduced in Sec. 3.1.1, has leading $O(a)$ discretization errors. The underlying reason is the breaking of chiral symmetry induced by the Wilson term

$$\frac{a}{2}\overline{\psi}\nabla_\mu\nabla_\mu^*\psi. \tag{25}$$

The lack of chiral symmetry even for vanishing bare quark mass allows the presence of a dimension-5 operator in the effective theory

$$a\overline{\psi}\sigma_{\mu\nu}F_{\mu\nu}\psi, \tag{26}$$

the so-called clover term [17], proportional to the lattice spacing a.

To remove the leading discretization error it is rather standard procedure to add a counterterm to the lattice action proportional to the clover term with a properly tuned coefficient, $c_{\rm sw}$ [18]. The observation that the clover term breaks chiral symmetry can be used to fix $c_{\rm sw}$ imposing the restoration of the chiral Ward identities [19] that are themselves violated at finite lattice spacing by terms of $O(a)$. Chiral symmetry plays thus a key role in improving the rate with which the continuum limit is approached.

3.1.3. *The Nielsen–Ninomiya theorem*

The observation that the leading $O(a)$ dimension-5 operators in the Symanzik effective action break chiral symmetry suggests that a solution to improve the rate of the continuum limit would be to construct a lattice D that preserves chiral symmetry, which would be formally equivalent to

$$\gamma_5 D + D\gamma_5 = 0. \tag{27}$$

The Nielsen–Ninomiya theorem [20] states that chiral symmetry for a single flavor lattice action is incompatible with locality (i.e., bounded by an exponential with vanishing rate in the continuum limit), absence of massless spurious fermion states (doublers) and proper QCD continuum limit. This theorem can be circumvented by relaxing the chiral symmetry condition in Eq. (27) with the milder Ginsparg–Wilson (GW) relation [21]

$$\gamma_5 D + D\gamma_5 = aDR\gamma_5 D, \tag{28}$$

where R is a local operator. Because of the locality of R the GW relation guarantees that at distances larger than the range of R the states propagating are effectively chiral (see Ref. [22] for an excellent review).

Solutions to the GW relation can be found: the perfect action of Hasenfratz and Niedermayer [23], the overlap operator of Neuberger [24, 25], and the domain-walls construction of Kaplan and Shamir [26, 27]. The fact that the GW relation implies the existence of chiral Ward identities [28] and of the index theorem [29], indicates the presence of a lattice symmetry. In fact Lüscher observed that every massless lattice operator satisfying the GW relation possesses an exact lattice chiral symmetry [30]. The consequence of Lüscher symmetry is the absence of leading $O(a)$ discretization errors. The clover term of Eq. (26) breaks the lattice chiral symmetry, so it cannot appear in the Symanzik effective theory describing GW fermions. The only $O(am)$ discretization effects could rise due to the presence of a mass term, i.e., a chirally breaking term, in the lattice action. The mass term can be introduced though in the lattice action in a way that it flips sign under the lattice chiral symmetry, forbidding the presence of such $O(am)$ terms. This implies that all discretization effects are of $O(a^2)$. Numerically GW fermions are challenging and it would be numerically advantageous to have "cheaper" lattice fermions that preserve all or most of their nice properties.

3.1.4. *Wilson twisted mass formulation*

An important observation comes from understanding the pattern of chiral symmetry breaking induced by the Wilson term and the mass term. By analyzing the impact of these two chiral symmetry breaking sources in the discretization effects, we can try to understand if there is a way to automatically remove the leading $O(a)$ effects in physical quantities without adding proper counterterms to the lattice action and local operators.

In this respect, an interesting alternative to the Wilson action and the non-perturbatively $O(a)$ improved clover fermions is given by the Wilson twisted mass (Wtm) fermions. The Wtm lattice action was initially introduced to cure the problem of exceptional configurations in the quenched approximation [31]. The Wtm lattice action is the Wilson action with a mass term that is chirally rotated with respect to the Wilson term in Eq. (25). For $N_f = 2$ degenerate quarks $\chi^T = (\chi_u, \chi_d)$ the mass term reads

$$m\overline{\chi}(x)\chi(x) + \mathrm{i}\mu\overline{\chi}(x)\gamma_5\tau^3\chi(x), \tag{29}$$

where τ^3 is the Pauli matrix in flavor space and μ is the twisted mass. This lattice action is particularly interesting when the untwisted bare quark mass m is tuned in such a way that the renormalized untwisted quark mass, $m_R = 0$, vanishes in the continuum limit. In this particular setup the twist angle $\omega = \arctan(\mu_R/m_R) \simeq \pi/2$ and the lattice theory is said to be maximally twisted. We stress that even if the mass term has an unusual form, $\mathrm{i}\mu\overline{\chi}(x)\gamma_5\tau^3\chi(x)$ in the continuum limit, it can be rotated back to the standard mass form with a non-anomalous chiral rotation. This also shows that the breaking of parity and isospin symmetry in Eq. (29) is only apparent in the continuum limit and simply induces different lattice artifacts in the observables with respect to standard Wilson fermions. This particular choice of twist angle is useful because the lattice theory now enjoys the property of *automatic* $O(a)$ *improvement* [32]. This property ensures that physical quantities have the $O(a)$ discretization effects automatically removed with a single non-perturbative tuning, i.e., the untwisted quark mass. The first numerical evidence of the property of automatic $O(a)$ was shown in Ref. [33]. A thorough discussion about Wilson twisted mass and automatic $O(a)$ improvement can be found in Refs. [34, 35] and references therein. Here, we briefly discuss the tree-level of the theory. The propagator of a doublet of Wilson twisted mass fermions is

$$\widetilde{G}(p) = \frac{-\mathrm{i}\gamma_\mu \mathring{p}_\mu + \mathcal{M}(p) - \mathrm{i}\mu\gamma_5\tau^3}{\mathring{p}_\mu^2 + \mathcal{M}(p)^2 + \mu^2}, \tag{30}$$

where

$$\mathring{p}_\mu = \frac{1}{a}\sin(ap_\mu), \quad \mathcal{M}(p) = m + \frac{1}{2}a\hat{p}_\mu^2, \quad \hat{p}_\mu = \frac{2}{a}\sin\left(\frac{ap_\mu}{2}\right). \tag{31}$$

If we now expand the denominator in powers of the lattice spacing a we obtain

$$p^2 + m_0^2 + am_0p^2 + \mu^2 + O(a^2), \quad \text{where } p^2 = p_\mu p_\mu. \tag{32}$$

The leading $O(a)$ discretization effects of the dispersion relation are given by the term am_0p^2.

From Eq. (32) we see that if we set $m_0 = 0$ we have a massive theory with mass μ without $O(a)$ discretization errors. To understand how this can happen it is good to understand the origin of the $O(a)$ term. The $O(a)$ term, am_0p^2, originates from the cross term of the Wilson term and the untwisted mass term in $\mathcal{M}(p)$. There is no cross term between Wilson and twisted mass because they point in different "directions" in the chiral-flavor space. In other words, it is possible to find a chiral transformation that leaves invariant the twisted mass term but not the Wilson term. The final lattice theory with vanishing untwisted quark mass is thus automatically $O(a)$ improved.

In the interacting case the tuning of quark mass can be done using chiral Ward identities [36,37], but the conclusions are the same. There is a chiral symmetry transformation that leaves invariant the continuum theory but it flips the sign of the clover term (26). In this case then the insertion of the clover term in the Symanzik expansion of a generic correlator vanishes and so do the lattice artifacts linear in a.

Wilson twisted mass fermions has been used in the recent years in large-scale Lattice QCD calculation by the European Twisted Mass Collaboration for important phenomenological applications [38–42].

3.2. *Symmetries of the Finite-volume Point Groups*

In most Lattice QCD calculations, PBCs are imposed on the quark and gluons fields in the spatial directions of the lattice volume. Since calculations are performed within a finite cubic volume, and if we further restrict ourselves to observables with zero center-of-mass motion (CM), it is the irreducible representations (irreps) of the cubic, or octahedral, group O that determine the degeneracies of the eigenstates in the lattice volume. This has direct implications to extracting angular-momentum dependent quantities, such as phase shifts δ_l (discussed below). In the infinite-volume limit, angular momenta fall under SO(3) symmetry, which has an infinite number of irreps, each labeled by l and has dimension $2l + 1$. In a cubic volume, the octahedral group has five irreps labeled A_1, A_2, T_1, T_2, and E, with dimensions 1, 1, 3, 3, and 2, respectively. The cube is also invariant under parity operations, so each irrep can be further divided as even or odd under parity operations (e.g., $A_1 \rightarrow A_1^{\pm}$). The spatial component of all lattice-calculated quantities falls under these irreps. The decomposition of cubic groups in terms of SO(3) irreps, some of which are given in Table 1, is infinite in dimension. More importantly, the decomposition shows that different angular momenta mix in a cubic volume. If the center of the group is enlarged to Z_2, one has the *double cover* of the cubic group O_h [43]. Odd-half

Table 1. Cubic volume decomposition in terms of SO(3) irreps.

Cubic irrep Γ	SO(3) irrep decomposition l	Γ	l
A_1^+	$l = 0, 4, 6, 8, \ldots$	A_1^-	$l = 9, 13, 15, \ldots$
A_2^+	$l = 6, 10, 12, \ldots$	A_2^-	$l = 3, 7, 9, \ldots$
T_1^+	$l = 4, 6, 8, \ldots$	T_1^-	$l = 1, 3, 5, 7, 9, \ldots$
T_2^+	$l = 2, 4, 6, 8, \ldots$	T_2^-	$l = 3, 5, 7, 9 \ldots$
E^+	$l = 2, 4, 6, 8, \ldots$	E^-	$l = 5, 7, 9, \ldots$

spin angular momentum can now be represented by three additional irreps, $G_1^{\pm}$, $G_2^{\pm}$, and H. As before, the decomposition of these irreps in terms of SU(2) irreps is infinite and mixes SU(2) irreps.

For observables with non-zero CM motion, only the symmetry operations that leave both the cube and the CM momentum vector invariant are allowed. Such operations can only be a subset of those that define O, and so the symmetry group in this case is a subgroup of O. Examples of such subgroups are the C_{nv} point groups, where n is an integer. Similar behavior occurs for finite volumes that are asymmetric (e.g., rectangular), or through the implementation of twisted boundary conditions (TBCs) (which in effect induces non-zero CM). We will discuss the latter in more detail in Sec. 4.2.

4. "Scattering" within a Finite Volume Using Lattice Quantum Chromodynamics

Maiani and Testa [44] showed that an immediate consequence of working in an (infinite) Euclidean space is that information about scattering processes, which are intrinsically defined in Minkowski space, cannot be assessed from infinite-volume Euclidean correlators except at kinematic thresholds. Their "No-Go" theorem implied that physically relevant scattering information would always be masked by Euclidean continuum states.

A "loop-hole" to the "No-Go" theorem comes from the fact that Lattice QCD calculations are performed within *finite* Euclidean volumes. Lüscher [45–47] exploited this fact when he showed, from a field-theoretical point of view, how scattering parameters could be related to the discrete energy eigenvalues within a finite volume[c] (see also Chapter 6). His relations are valid up to inelastic thresholds.

4.1. *Implications of Reduced Symmetry Due to Finite Volume*

Given a positive eigenenergy ϵ of two interacting particles with A_1 symmetry (here we assume the rest masses of both particles are subtracted from ϵ), the interacting momentum q associated with this energy is defined through $\epsilon = \frac{q^2}{2\mu L^2}$, where μ is the reduced mass. Lüscher's formula relates this energy to the S-wave phase shift at this momentum, $\delta_0(q)$,

$$q \cot(\delta_0(q)) = \frac{2}{\sqrt{\pi}L} \mathcal{Z}_{00}\left(1; \frac{\mu L^2 \epsilon}{2\pi^2}\right), \tag{33}$$

where

$$\mathcal{Z}_{lm}(s; x) = \sum_{\vec{n}} \frac{|n|^l \, Y_{lm}(\Omega_{\vec{n}})}{[|\vec{n}|^2 - x]^s}, \tag{34}$$

is the generalized zeta function. The sum is over all triplets of integers $\vec{n}$. Note that for $(l, m) = (0, 0)$, the sum in Eq. (34) is divergent and must be regulated.

Given the discussion in Sec. 3.2, it is apparent that Eq. (33) is an approximation since the energy eigenvalue ϵ includes not only S-wave information, but also $l = 4$, 6, and higher partial waves due to mixing. Fortunately, Lüscher showed that for large volumes, the higher partial waves are suppressed by inverse powers of L. Luu and Savage [50] give explicit expressions for these higher-order terms, as well as expressions that relate partial waves to the other cubic irrep eigenenergies. In Fig. 3 the extracted scattering length for the isospin-$\frac{3}{2}$ kaon-pion system is shown as a function of the reduced mass of the system. Also shown is the leading order (LO) chiral prediction for the system. As somewhat of a mystery, the calculated

---

[c]The relationship between scattering parameters and finite-volume energies within a non-relativistic formulation has been well documented, particularly within the condensed matter community [48, 49].

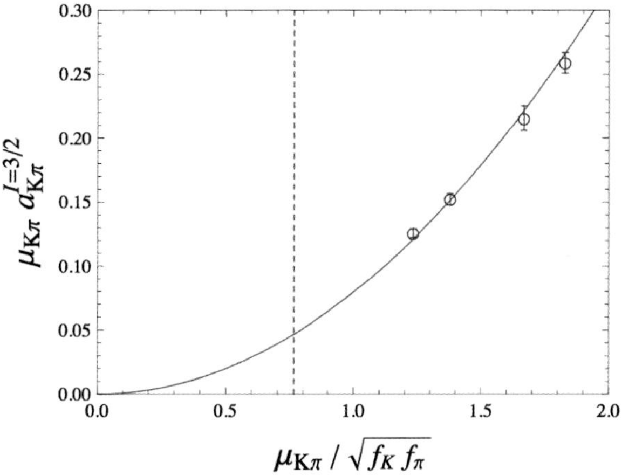

Fig. 3. The isospin-$\frac{3}{2}$ kaon-pion scattering length dependence on reduced pion-kaon mass and decay constants, compared to LO prediction (solid line). The physical point occurs at the vertical dashed line.

results (which are fully non-perturbative) show remarkably good agreement with the LO chiral prediction for very large values of the reduced mass.

4.2. *Calculations with Twisted Boundary Conditions*

By modifying the symmetries of the system, numerical improvements can be made. For example, the spatial symmetries within Lattice QCD calculations can be modified by appropriate applications of TBCs,

$$\psi(\vec{r} + \vec{n}L) = e^{i\vec{n}\cdot\vec{\theta}}\psi(\vec{r}), \tag{35}$$

where $\vec{n}$ is a triplet of integers and $\vec{\theta}$ represents a triplet of twist angles for the three different spatial directions. When $\vec{\theta} = (0, 0, 0)$ one has PBCs in the spatial directions, which is most commonly used in Lattice QCD. On the other hand, aPBCs can be imposed when $\vec{\theta} = (\pi, \pi, \pi)$. In Lattice QCD, TBCs are equivalent to introducing a background U(1) gauge field imposed on the quarks, subsequently endowing them with a momentum dependent on the twist angle [51, 52]. As mentioned above, an induced momentum on the quarks can reduce the spatial symmetry to one of the subgroups of the O_h. Briceno *et al.* [53] investigated the behavior of the reduced symmetries due to twists and their impact on finite-volume corrections for the two-body system. It was found that specific twists angles, $\vec{\theta} = (\frac{\pi}{2}, \frac{\pi}{2}, \frac{\pi}{2})$, greatly reduced the finite-volume shifts. Recently Körber and Luu [54] derived the solution for twist angles that eliminate the LO finite-volume shifts for three particles within a finite volume,

$$\sum_{j=1}^{3}(\cos(\vec{\theta}_1 \cdot \mathbf{e}_j) + \cos(\vec{\theta}_2 \cdot \mathbf{e}_j) + \cos(\vec{\theta}_3 \cdot \mathbf{e}_j)) = 0, \qquad (36)$$

where $\vec{\theta}_i$ is the vector of twist angles for the ith particle and $\mathbf{e}_j$ is a unit vector in the jth direction. Figure 4 shows an example contour of twist angles that satisfy this condition.

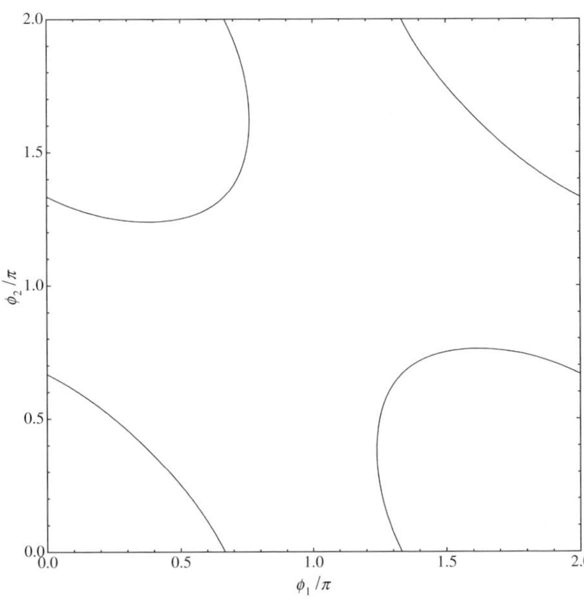

Fig. 4. Example of contours of twist angles θ_1 and θ_2 that eliminate LO finite-volume shifts in three-body calculations. Here the twist angle for the third particle is $\vec{\theta}_3 = -(\vec{\theta}_1 + \vec{\theta}_2)$ and $\vec{\theta}_i = (\theta_i, \theta_i, \theta_i)$ (figure courtesy of C. Körber).

5. Lattice Quantum Chromodynamics Calculations of Fundamental Symmetries

The Lagrangian in Eq. (1) has discrete Poincare symmetries in charge conjugation C, parity P, and time reversal T. As such, the Lagrangian is invariant under simultaneous combinations of all three combinations of C, P, and T. However, there is no reason to demand that the Lagrangian is symmetric under individual or combinations of any two of these symmetries, as long as the overall CPT symmetry is respected. Here, we discuss recent developments in the field of Lattice QCD as applied to these fundamental symmetries.

5.1. *The* $\mathbf{U(1)_A}$ *Anomaly*

The $U(1)_A$ problem (see Fig. 1 and associated text in Sec. 2) has a very ingenious solution based on the realization [55, 56] that the QCD vacuum has a rather complicated structure, making $U(1)_A$ not a true symmetry of the QCD Lagrangian. The anomalous nature of the $U(1)_A$ symmetry transformation effectively adds an extra term (so-called θ term) to the QCD Lagrangian

$$i\theta \frac{g_s^2}{32\pi^2} F^a_{\mu\nu} \widetilde{F}^a_{\mu\nu}, \tag{37}$$

where

$$\widetilde{F}^a_{\mu\nu} = \frac{1}{2}\varepsilon_{\mu\nu\rho\sigma} F^a_{\rho\sigma}, \tag{38}$$

is the dual field tensor. This term violates both parity P and time reversal T, but preserves charge conjugation C, and therefore violates CP, thus potentially inducing an electric dipole moment (EDM) in particles with non-vanishing spin. The experimental bound on the neutron EDM, $|d_n| < 3 \times 10^{-26} e\,\text{cm}$ [57] implies that the angle $|\theta| < 10^{-9}$. The smallness of the θ parameter is usually denoted as the *strong CP problem*. There are several mechanism that might explain why the θ angle is so small that we cannot cover in this chapter.

The problem is emphasized in the Standard Model, because the diagonalization of the mass matrix in the weak sector effectively adds an additional θ term to the QCD Lagrangian. The neutron EDM has also a contribution form the CP violation induced by the Cabibbo–Kobayashi–Maskawa (CKM) matrix. It turns out that the leading contribution to the quarks EDM is a three-loops effect [58] and it is currently around 5 orders of magnitude smaller than the experimental bound of the neutron EDM. There are other flavor diagonal CP-violating sources originating from BSM physics that can induce a non-vanishing neutron EDM. At lower energies they are usually parametrized by gluonic and fermionic higher dimensional operators.

Thus, a measurement of any non-vanishing neutron EDM between the current bound and the very small CKM contribution is a direct evidence of physics beyond the Standard Model or a small and non-vanishing θ angle. The contribution of the strong dynamics to the EDM, induced by the θ angle or other BSM sources, can be determined using Lattice QCD. In the next section we briefly review this topic that has recently seen a renewed interest especially thanks to new techniques and larger computational resources.

5.2. *Neutron Electric Dipole Moment from the θ Term*

To determine the θ-term contribution to the neutron EDM one has to calculate the electromagnetic current matrix element J_μ between two nucleon states in a θ vacuum. This amounts to inserting the topological charge operator on all space–time points to the three-point correlation function, as described below.

Generally there are two main reasons why a calculation of an EDM within Lattice QCD is highly non-trivial. The first one is that the signal-to-noise ratio for baryonic correlation functions degrades exponentially in Euclidean time for light quark masses. The second reason is the non-trivial renormalization of the CP-odd operator. Furthermore the θ angle precludes the use of standard stochastic techniques because it introduces an imaginary term in the real Euclidean QCD action. First pioneering works with Lattice QCD [59–61] treated θ in a perturbative manner allowing a direct determination of the nucleon EDM contribution proportional to θ. The same methods have been used on a recent neutron EDM determination [62]. This method is very reasonable considering that $|\theta|$ is known to be very small from current experimental bounds. Ref. [63] tries to circumvent this problem by performing a calculation with imaginary θ and then rotating back to a real value of θ in the final result.

Here, we want to briefly review a new method that we have proposed in Refs. [64, 65], still based on a perturbative expansion in powers of θ, but that resolves the renormalization ambiguities using the gradient flow for the gauge fields [66] and that allows a theoretically sound continuum limit. Recently the same method has been also adopted by the European Twisted Mass Collaboration [67].

To obtain the nucleon EDMs one computes the matrix element of the electromagnetic current J_μ between nucleon states in the θ vacuum, with mass M_N, spins s, s' and momentum transfer $\mathbf{q} = \mathbf{p}' - \mathbf{p}$

$$\langle N^\theta(\mathbf{p}', s')|J_\mu|N^\theta(\mathbf{p}, s)\rangle = \bar{u}_N^\theta(\mathbf{p}', s')\Gamma_\mu(Q^2)u_N^\theta(\mathbf{p}, s), \tag{39}$$

where $Q^2 = -q^2$. The form factor $\Gamma_\mu(Q^2)$, parametrizing the matrix element, has the most general structure consistent with the symmetries of the action such as gauge, Lorentz, CPT and C invariance. $\Gamma_\mu(Q^2)$ can be decomposed in a linear combination of CP-even and CP-odd form factors and the EDM is proportional to the CP-odd form factor $F_3(Q^2)$ at zero momentum transfer

$$d_N \simeq \theta \frac{F_3^N(0)}{2M_N} + O(\theta^3). \tag{40}$$

Although the action with a non-vanishing θ angle is not amenable to Monte Carlo methods, the desired result for EDM can still be obtained by expanding the relevant correlation function around $\theta = 0$ and calculating only the linear term in it. For a generic correlation function $\mathcal{O}$ we can write

$$\langle \mathcal{O} \rangle_\theta \simeq \langle \mathcal{O} \rangle_{\theta=0} + \mathrm{i}\theta \left\langle \mathcal{O} \int d^4x\, q(x) \right\rangle_{\theta=0} + O(\theta^2), \tag{41}$$

where $q(x)$ is the topological charge density

$$q(x) = \frac{1}{64\pi^2} \epsilon_{\mu\nu\rho\sigma} F^a_{\mu\nu}(x) F^a_{\rho\sigma}(x). \tag{42}$$

To have a theoretically sound definition of the correlation function

$$\left\langle \mathcal{O} \int d^4x \; q(x) \right\rangle_{\theta=0}, \tag{43}$$

eliminating any possible renormalization ambiguities and thus having a well-defined continuum limit, we have proposed [64, 65] to directly compute the matrix element in Eq. (43) defining the topological charge density with the gradient flow [66].

We do not intend to go into the details of the gradient flow for gauge fields in this chapter and we refer to the original works [66, 68, 69] for additional details. Relevant for the EDM calculation is to recognize that the flow equation represents the steepest descent equations in the space of gauge fields. The flow drives the gauge configurations toward local minima of the action and the topological charge defined at non-vanishing flow-time as

$$\mathcal{Q}(t) = \int d^4x \; q(x,t) \tag{44}$$

with

$$q(x,t) = \frac{1}{64\pi^2} \epsilon_{\mu\nu\rho\sigma} G^a_{\mu\nu}(x,t) G^a_{\rho\sigma}(x,t) \tag{45}$$

quickly reaches close-to-integer values saturating the corresponding instanton bound. The renormalization properties of the gauge fields at non-vanishing flow time guarantee that the topological charge defined with the gradient flow is free from renormalization ambiguities and has a renormalization constant equal to 1 [66, 70]. This method has been successfully tested in a proof-of-principle calculation [65] where for the first time the continuum limit for the EDM has been taken. In Fig. 5 we plot the CP-odd form factor $F_3(Q^2)$ and the corresponding extrapolated value to zero momentum transfer, i.e., the values of the neutron and proton EDM. These results, even if obtained at a rather high value of the pion mass and in the quenched approximation, are very encouraging and they indicate that Lattice QCD is mature for an accurate and precise determination of the EDM for nucleons and light nuclei. We conclude by mentioning that the method we have proposed for the

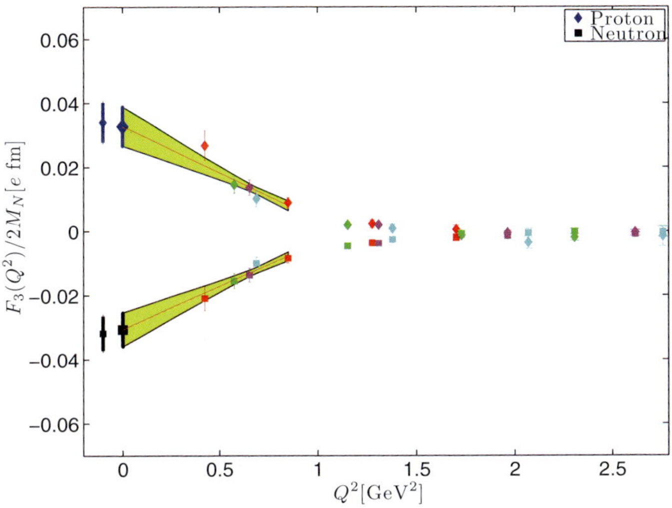

Fig. 5. The CP-odd form factor $F_3(Q^2)/2M_N$ of the proton and the neutron as a function of Q^2. The linear extrapolation in Q^2, shown with a yellow band, is consistent with a chiral perturbation theory calculation at next-to-leading-order [71]. Different colors represent different lattice spacings from $a \simeq 0.1$ fm (cyan) to $a \simeq 0.05$ fm (green). The lightly displaced extrapolated form factor uses only the three finest lattice spacings.

determination of the EDM coming from the θ angle can be extended to fermionic and gluonic higher dimensional operators parameterizing physics BSM, and work in this direction is in progress.

6. Heavy Quark Symmetry and B-physics on the Lattice

In this last section, we consider systems which are composed by a heavy quark and light quarks. In these systems the heavy quark is surrounded by a strongly interacting cloud of light QCD degrees of freedom exchanging momenta of the order of Λ_{QCD}. Despite the non-perturbative nature of these interactions the fact that the heavy quark mass m_Q is much larger than Λ_{QCD} leads to symmetry considerations and to the construction of an effective field theory.

The quantum numbers of the heavy quark can be probed only for momenta larger than the heavy quark mass, while the light degrees of freedom interact with the heavy quark mostly via soft gluons with momenta of the order of $\Lambda_{\rm QCD}$. This means that the light degrees of freedom are mostly blind to the flavor or the spin of the heavy quark. The only interaction is via the exchange of soft gluons, thus to a very good approximation the heavy quark is just a color source for the light degrees of freedom.

In the limit $m_Q \to \infty$ heavy–light hadrons with the same flavor or spin quantum numbers have very similar physical properties, such as the hadron masses. These considerations can be put on a firmer ground realizing that they are all realizations of new symmetries for the strong interactions of heavy quarks at low energies. If the masses of the N_h heavy quarks in the heavy–light system are large compared with $\Lambda_{\rm QCD}$, then there is an SU($2N_h$) spin-flavor symmetry group. This heavy-quark symmetry is not an exact symmetry of the Lagrangian even for infinite quark mass. It is rather a symmetry of the effective interaction of the heavy and light quarks for large quark masses. The results obtained using heavy-quark symmetry are not model dependent but rather QCD results in a given kinematical domain. The symmetry breaking corrections of order $\Lambda_{\rm QCD}/m_Q$ and higher can be studied in a systematic way by building an effective field theory, the heavy quark effective field theory (HQET), from a suitable infinite quark mass limit of QCD. The construction of HQET is covered by many review articles (see, for example, Ref. [72]). The leading term of the HQET is the static quark action that following Eichten and Hill [73] can be written on the lattice as

$$S_h = a^4 \sum_x \overline{\psi}_h(x) \nabla_0^* \psi_h(x), \tag{46}$$

where the four-component spinors $\psi_h, \overline{\psi}_h$ satisfy the constraints

$$P_+ \psi_h = \psi_h, \quad \overline{\psi}_h P_+ = \overline{\psi}_h, \quad \text{with } P_+ = \frac{1}{2}(1+\gamma_0), \tag{47}$$

effectively leaving only two degrees of freedom. The lattice backward derivative ∇_0^* indicates that static quarks propagate only forward in time. The same action can be written down for the static anti-quark with a forward lattice derivative. If we define

$$\Gamma_i = \frac{1}{2}\gamma_5\gamma_0\gamma_i, \tag{48}$$

then the Eichten–Hill (EH) action exhibits the SU(2) spin symmetry

$$\psi_h \to V\psi_h, \quad \overline{\psi}_h \to \overline{\psi}_h V^{-1}, \quad \text{with } V = \mathrm{e}^{\mathrm{i}\phi_i\Gamma_i}. \tag{49}$$

The static lattice action possesses an additional local symmetry, which is the conservation of heavy quark flavor number. If we define an arbitrary space dependent function $\alpha(\mathbf{x})$ then the action is invariant under

$$\psi_h \to \exp^{\mathrm{i}\alpha(\mathbf{x})}\psi_h, \quad \overline{\psi}_h \to \overline{\psi}_h \exp^{-\mathrm{i}\alpha(\mathbf{x})}. \tag{50}$$

The discussion of the renormalization and discretization errors of the lattice static action simplifies using these symmetries [74].

Static-light correlation functions are the LO terms of any HQET expansion and the static propagator is the building block of any calculation within HQET. Unfortunately, the näive discretization of the EH action (46) induces an exponential growth in Euclidean time, $\exp(\delta x_0)$ of the noise-to-signal ratio in any static-light correlator. The rate $\delta \sim 1/a$ diverges in the continuum limit. By changing the way the static quark propagates in time it is possible to reduce the value of δ, solving the signal-to-noise problem for this system. The different lattice actions proposed in Ref. [75] effectively reduce δ allowing a safe statistical and systematic determination of static-light correlation functions for large Euclidean times [76]. All the actions proposed in Ref. [75] share the same symmetry properties, mentioned earlier, as the EH action. In this way the renormalization and the discussion of the $O(a)$ terms is the same as for the EH action.

The Alpha collaboration has pioneered the determination of the HQET expansion parameters with finite-volume matching techniques [77, 78]. Many results have been obtained by the Alpha collaboration using the static lattice action proposed in Ref. [75] and the HQET expansion such as a non-perturbative determination of the b-quark mass [79], m_b, and the B-mesons decay constants [80], f_B and f_{B_s}.

Other methods can be used for B-physics lattice calculations, which for lack of space we cannot discuss in detail here. The one proposed by the ETMC collaboration, for example, uses relativistic heavy quarks but heavily relies on the HQET expansion of physical quantities. It is based on the observation that ratios of heavy–light quantities computed at slightly different heavy quark masses have an exact static limit and a smoother inverse heavy quark mass expansion. For this reason it is sometimes called the "*ratio method*" [81, 82]. Results obtained with the ratio method include the b-quark mass, the B-mesons decay constants, f_B and f_{B_s}, and the bag-parameter of the neutral B-meson oscillation, including contributions from BSM physics [83]. Other methods have been used as well to address B-physics quantities with Lattice QCD and we refer to the annual lattice symposium contributions for updated references.

7. Conclusions

In this chapter we have emphasized how symmetries play a fundamental role in QCD, and in numerical studies of non-perturbative QCD on the lattice. They (and their violations thereof) constrain the pattern of the low-energy hadronic spectrum. In Lattice QCD, symmetries take on additional meaning since their preservation helps reduce the number of fitted constants needed to describe the discretized action. Not surprisingly, preserving symmetries on the lattice usually goes hand-in-hand with increased computational requirements. At the same time, the reduced symmetries of the lattice, and their modifications at the boundaries, can be utilized to improve numerical studies of QCD, as we have argued in this chapter. Indeed, it is because of the reduced symmetry in a finite volume that enables the extraction of infinite-volume scattering parameters via Lüscher's method. Finally, the precision of current Lattice QCD calculations is enabling a new class of calculations related to fundamental symmetries (and violations thereof) coming from operators that have BSM origins. Such calculations help disentangle the strong hadronic contributions from BSM physics.

Acknowledgements

We thank C. Körber and A. Walker-Loud for permission to use their figures in this chapter.

References

[1] K. Orginos, A. Parreno, M. J. Savage, S. R. Beane, E. Chang, and W. Detmold, Two nucleon systems at $m_\pi \sim 450$ MeV from lattice QCD, *Phys. Rev. D.* **92**, 114512 (2015).

[2] S. R. Beane, E. Chang, S. D. Cohen, W. Detmold, H. W. Lin, T. C. Luu, K. Orginos, A. Parreno, M. J. Savage, and A. Walker-Loud, Light nuclei and hypernuclei from quantum chromodynamics in the limit of SU(3) flavor symmetry, *Phys. Rev. D.* **87**, 034506 (2013).

[3] S. R. Beane, W. Detmold, T. C. Luu, K. Orginos, M. J. Savage, and A. Torok, Multi-pion systems in lattice QCD and the three-pion interaction, *Phys. Rev. Lett.* **100**, 082004 (2008).

[4] S. R. Beane, E. Chang, W. Detmold, H. W. Lin, T. C. Luu, K. Orginos, A. Parreno, M. J. Savage, A. Torok, and A. Walker-Loud, The $I = 2$ $\pi\pi$ S-wave scattering phase shift from lattice QCD, *Phys. Rev. D.* **85**, 034505 (2012).

[5] S. R. Beane, W. Detmold, P. M. Junnarkar, T. C. Luu, K. Orginos, A. Parreno, M. J. Savage, A. Torok, and A. Walker-Loud, SU(2) Low-energy constants from mixed-action lattice QCD, *Phys. Rev. D.* **86**, 094509 (2012).

[6] S. R. Beane, E. Chang, S. D. Cohen, W. Detmold, H. W. Lin, T. C. Luu, K. Orginos, A. Parreno, M. J. Savage, and A. Walker-Loud, Hyperon–nucleon interactions and the composition of dense nuclear matter from quantum chromodynamics, *Phys. Rev. Lett.* **109**, 172001 (2012).

[7] S. R. Beane, E. Chang, W. Detmold, H. W. Lin, T. C. Luu, K. Orginos, A. Parreno, M. J. Savage, A. Torok, and A. Walker-Loud, The deuteron and exotic two-body bound states from lattice QCD, *Phys. Rev. D.* **85**, 054511 (2012).

[8] S. R. Beane, E. Chang, S. D. Cohen, W. Detmold, H. W. Lin, K. Orginos, A. Parreo, and M. J. Savage, Quarkonium-nucleus bound states from lattice QCD, *Phys. Rev. D.* **91**, 114503 (2015).

[9] S. Borsanyi *et al.*, *Ab initio* calculation of the neutron-proton mass difference, *Science.* **347**, 1452 (2015).

[10] M. G. Endres, A. Shindler, B. C. Tiburzi, and A. Walker-Loud, Massive photons: An infrared regularization scheme for lattice QCD+QED (2015). arXiv:1507.08916.

[11] B. Lucini, A. Patella, A. Ramos, and N. Tantalo, Charged hadrons in local finite-volume QED+QCD with C* boundary conditions, *J. High Energy Phys.* **02**, 076 (2016).

[12] S. R. Beane, E. Chang, S. Cohen, W. Detmold, H. W. Lin, K. Orginos, A. Parreno, M. J. Savage, and B. C. Tiburzi, Magnetic moments of light

nuclei from lattice quantum chromodynamics, *Phys. Rev. Lett.* **113**, 252001 (2014).
[13] S. R. Beane, W. Detmold, K. Orginos, and M. J. Savage, Nuclear physics from lattice QCD, *Prog. Part. Nucl. Phys.* **66**, 1 (2011).
[14] H. Georgi, *Lie Algebras in Particle Physics: From Isospin to Unified Theories*. Frontiers in Physics Series, *Westview Press* is a Member of the Perseus Books Group (1999).
[15] K. G. Wilson, Confinement of quarks, *Phys. Rev. D.* **10**, 2445 (1974).
[16] K. Symanzik, Continuum limit and improved action in lattice theories: (I). Principles and ϕ^4 theory, *Nucl. Phys. B.* **226**, 187 (1983).
[17] B. Sheikholeslami and R. Wohlert, Improved continuum limit lattice action for QCD with Wilson fermions, *Nucl. Phys. B.* **259**, 572 (1985).
[18] M. Luscher, S. Sint, R. Sommer, P. Weisz, and U. Wolff, Nonperturbative O(a) improvement of lattice QCD, *Nucl. Phys. B.* **491**, 323 (1997).
[19] M. Luscher, S. Sint, R. Sommer, and P. Weisz, Chiral symmetry and O(a) improvement in lattice QCD, *Nucl. Phys. B.* **478**, 365 (1996).
[20] H. B. Nielsen and M. Ninomiya, A no-go theorem for regularizing chiral fermions, *Phys. Lett. B.* **105**, 219 (1981).
[21] P. H. Ginsparg and K. G. Wilson, A remnant of chiral symmetry on the lattice, *Phys. Rev. D.* **25**, 2649 (1982).
[22] F. Niedermayer, Exact chiral symmetry, topological charge and related topics, *Nucl. Phys. Proc. Suppl.* **73**, 105 (1999).
[23] P. Hasenfratz and F. Niedermayer, Perfect lattice action for asymptotically free theories, *Nucl. Phys. B.* **414**, 785 (1994).
[24] H. Neuberger, Exactly massless quarks on the lattice, *Phys. Lett. B.* **417**, 141 (1998).
[25] H. Neuberger, More about exactly massless quarks on the lattice, *Phys. Lett. B.* **427**, 353 (1998).
[26] D. B. Kaplan, A Method for simulating chiral fermions on the lattice, *Phys. Lett. B.* **288**, 342 (1992).
[27] Y. Shamir, Chiral fermions from lattice boundaries, *Nucl. Phys. B.* **406**, 90 (1993).
[28] P. Hasenfratz, Lattice QCD without tuning, mixing and current renormalization, *Nucl. Phys. B.* **525**, 401 (1998).
[29] P. Hasenfratz, V. Laliena, and F. Niedermayer, The Index theorem in QCD with a finite cutoff, *Phys. Lett. B.* **427**, 125 (1998).
[30] M. Luscher, Exact chiral symmetry on the lattice and the Ginsparg–Wilson relation, *Phys. Lett. B.* **428**, 342 (1998).
[31] R. Frezzotti, P. A. Grassi, S. Sint, and P. Weisz, Lattice QCD with a chirally twisted mass term, *J. High Energy Phys.* **08**, 058 (2001).
[32] R. Frezzotti and G. C. Rossi, Chirally improving Wilson fermions. 1. O(a) improvement, *J. High Energy Phys.* **08**, 007 (2004).
[33] K. Jansen, A. Shindler, C. Urbach, and I. Wetzorke, Scaling test for Wilson twisted mass QCD, *Phys. Lett. B.* **586**, 432 (2004).
[34] A. Shindler, Twisted mass lattice QCD, *Phys. Rep.* **461**, 37–110 (2008).

[35] S. Sint. Lattice QCD with a chiral twist. In *Workshop on Perspectives in Lattice QCD Nara, Japan, November, 2005* (2007).
[36] K. Jansen, M. Papinutto, A. Shindler, C. Urbach, and I. Wetzorke, Light quarks with twisted mass fermions, *Phys. Lett. B.* **619**, 184 (2005).
[37] K. Jansen, M. Papinutto, A. Shindler, C. Urbach, and I. Wetzorke, Quenched scaling of Wilson twisted mass fermions, *J. High Energy Phys.* **09**, 071 (2005).
[38] P. Boucaud *et al.*, Dynamical twisted mass fermions with light quarks, *Phys. Lett. B.* **650**, 304 (2007).
[39] B. Blossier *et al.*, Light quark masses and pseudoscalar decay constants from $N_f = 2$ Lattice QCD with twisted mass fermions, *J. High Energy Phys.* **04**, 020 (2008).
[40] C. Alexandrou *et al.*, Light baryon masses with dynamical twisted mass fermions, *Phys. Rev. D.* **78**, 014509 (2008).
[41] B. Blossier *et al.*, Pseudoscalar decay constants of kaon and D-mesons from $N_f = 2$ twisted mass Lattice QCD, *J. High Energy Phys.* **07**, 043 (2009).
[42] R. Baron *et al.*, Light meson physics from maximally twisted mass lattice QCD, *J. High Energy Phys.* **08**, 097 (2010).
[43] M. Tinkham, *Group Theory and Quantum Mechanics.* Dover Books on Chemistry and Earth Sciences, Dover Publications, Mineola, NY (2003).
[44] L. Maiani and M. Testa, Final state interactions from Euclidean correlation functions, *Phys. Lett. B.* **245**, 585 (1990).
[45] H. W. Hamber, E. Marinari, G. Parisi, and C. Rebbi, Considerations on Numerical Analysis of QCD, *Nucl. Phys. B.* **225**, 475 (1983).
[46] M. Luscher, Volume dependence of the energy spectrum in massive quantum field theories. II. Scattering states, *Commun. Math. Phys.* **105**, 153 (1986).
[47] M. Luscher, Two particle states on a torus and their relation to the scattering matrix, *Nucl. Phys. B.* **354**, 531 (1991).
[48] T. D. Lee, K. Huang, and C. N. Yang, Eigenvalues and eigenfunctions of a Bose system of hard spheres and its low-temperature properties, *Phys. Rev.* **106**, 1135 (1957).
[49] T. D. Lee and C. N. Yang, Many-body problem in quantum mechanics and quantum statistical mechanics, *Phys. Rev.* **105**, 1119 (1957).
[50] T. Luu and M. J. Savage, Extracting scattering phase-shifts in higher partial-waves from lattice QCD calculations, *Phys. Rev. D.* **83**, 114508 (2011).
[51] M. Guagnelli, K. Jansen, F. Palombi, R. Petronzio, A. Shindler, and I. Wetzorke, Continuous external momenta in nonperturbative lattice simulations: A Computation of renormalization factors, *Nucl. Phys. B.* **664**, 276 (2003).
[52] P. F. Bedaque, Aharonov–Bohm effect and nucleon nucleon phase shifts on the lattice, *Phys. Lett. B.* **593**, 82 (2004).
[53] R. A. Briceno, Z. Davoudi, T. C. Luu, and M. J. Savage, Two-baryon systems with twisted boundary conditions, *Phys. Rev. D.* **89**, 074509 (2014).
[54] C. Körber and T. Luu, Applying twisted boundary conditions for few-body nuclear systems, *Phys. Rev. C.* **93**, 054002 (2016).

[55] G. 't Hooft, Symmetry Breaking through Bell–Jackiw anomalies, *Phys. Rev. Lett.* **37**, 8 (1976).
[56] G. 't Hooft, Computation of the quantum effects due to a four-dimensional pseudoparticle, *Phys. Rev. D.* **14**, 3432 (1976). [Erratum: *Phys. Rev. D.* **18**, 2199 (1978)].
[57] C. A. Baker *et al.*, An improved experimental limit on the electric dipole moment of the neutron, *Phys. Rev. Lett.* **97**, 131801 (2006).
[58] A. Czarnecki and B. Krause, Neutron electric dipole moment in the standard model: Valence quark contributions, *Phys. Rev. Lett.* **78**, 4339 (1997).
[59] E. Shintani, S. Aoki, N. Ishizuka, K. Kanaya, Y. Kikukawa, Y. Kuramashi, M. Okawa, Y. Tanigchi, A. Ukawa, and T. Yoshie, Neutron electric dipole moment from lattice QCD, *Phys. Rev. D.* **72**, 014504 (2005).
[60] F. Berruto, T. Blum, K. Orginos, and A. Soni, Calculation of the neutron electric dipole moment with two dynamical flavors of domain wall fermions, *Phys. Rev. D.* **73**, 054509 (2006).
[61] E. Shintani, S. Aoki, and Y. Kuramashi, Full QCD calculation of neutron electric dipole moment with the external electric field method, *Phys. Rev. D.* **78**, 014503 (2008).
[62] E. Shintani, T. Blum, T. Izubuchi, and A. Soni, Neutron and proton electric dipole moments from $N_f = 2 + 1$ domain-wall fermion lattice QCD, *Phys. Rev. D.* **93**, 094503 (2016).
[63] F. K. Guo, R. Horsley, U. G. Meissner, Y. Nakamura, H. Perlt, P. E. L. Rakow, G. Schierholz, A. Schiller, and J. M. Zanotti, The electric dipole moment of the neutron from 2+1 flavor lattice QCD, *Phys. Rev. Lett.* **115**, 062001 (2015).
[64] A. Shindler, J. de Vries, and T. Luu. Beyond-the-standard-model matrix elements with the gradient flow. In *Proc. 32nd Int. Sym. on Lattice Field Theory*, PoS(LATTICE2014), (2014), New York, NY, p. 251.
[65] A. Shindler, T. Luu, and J. de Vries, Nucleon electric dipole moment with the gradient flow: The θ-term contribution, *Phys. Rev. D.* **92**, 094518 (2015).
[66] M. Lüscher, Properties and uses of the Wilson flow in lattice QCD, *J. High Energy Phys.* **08**, 071 (2010). [Erratum: *J. High Energy Phys.* **03**, 092 (2014)].
[67] C. Alexandrou, A. Athenodorou, M. Constantinou, K. Hadjiyiannakou, K. Jansen, G. Koutsou, K. Ottnad, and M. Petschlies, Neutron electric dipole moment using N_f=2+1+1 twisted mass fermions, *Phys. Rev. D.* **93**, 074503 (2016).
[68] M. Luscher and P. Weisz, Perturbative analysis of the gradient flow in non-abelian gauge theories, *J. High Energy Phys.* **02**, 051 (2011).
[69] M. Luscher, Chiral symmetry and the Yang–Mills gradient flow, *J. High Energy Phys.* **04**, 123 (2013).
[70] M. Cé, C. Consonni, G. P. Engel, and L. Giusti, Non-Gaussianities in the topological charge distribution of the SU(3) Yang–Mills theory, *Phys. Rev. D.* **92**, 074502 (2015).

[71] J. de Vries, R. Higa, C. P. Liu, E. Mereghetti, I. Stetcu, R. G. E. Timmermans, and U. van Kolck, Electric dipole moments of light nuclei from chiral effective field theory, *Phys. Rev. C.* **84**, 065501 (2011).
[72] M. Neubert, Heavy quark symmetry, *Phys. Rep.* **245**, 259 (1994).
[73] E. Eichten and B. R. Hill, An effective field theory for the calculation of matrix elements involving heavy quarks, *Phys. Lett. B.* **234**, 511 (1990).
[74] M. Kurth and R. Sommer, Renormalization and O(a) improvement of the static axial current, *Nucl. Phys. B.* **597**, 488 (2001).
[75] M. Della Morte, A. Shindler, and R. Sommer, On lattice actions for static quarks, *J. High Energy Phys.* **08**, 051 (2005).
[76] M. Della Morte, S. Durr, J. Heitger, H. Molke, J. Rolf, A. Shindler, and R. Sommer, Lattice HQET with exponentially improved statistical precision, *Phys. Lett. B.* **581**, 93 (2004); Erratum: *Phys. Lett. B.* **612**, 313 (2005).
[77] J. Heitger and R. Sommer, Nonperturbative heavy quark effective theory, *J. High Energy Phys.* **02**, 022 (2004).
[78] B. Blossier, M. Della Morte, P. Fritzsch, N. Garron, J. Heitger, H. Simma, R. Sommer, and N. Tantalo, Parameters of heavy quark effective theory from $N_f = 2$ lattice QCD, *J. High Energy Phys.* **09**, 132 (2012).
[79] F. Bernardoni *et al.*, The b-quark mass from non-perturbative $N_f = 2$ Heavy Quark Effective Theory at $O(1/m_h)$, *Phys. Lett. B.* **730**, 171 (2014).
[80] F. Bernardoni *et al.*, Decay constants of B-mesons from non-perturbative HQET with two light dynamical quarks, *Phys. Lett. B.* **735**, 349 (2014).
[81] B. Blossier *et al.*, A Proposal for B-physics on current lattices, *J. High Energy Phys.* **04**, 049 (2010).
[82] P. Dimopoulos *et al.*, Lattice QCD determination of m_b, f_B and f_{B_s} with twisted mass Wilson fermions, *J. High Energy Phys.* **01**, 046 (2012).
[83] N. Carrasco *et al.*, B-physics from $N_f = 2$ tmQCD: The Standard model and beyond, *J. High Energy Phys.* **03**, 016 (2014).

Chapter 6

Ab initio Lattice Effective Field Theory and Wigner's SU(4) Symmetry

Dean Lee
Department of Physics, North Carolina State University, Box 8202
Raleigh, North Carolina, USA
dean_lee@ncsu.edu

I review *ab initio* lattice simulations for nuclear systems using chiral effective field theory. In this discussion we explore the importance of an approximate symmetry of the low-energy nuclear interactions where the four nucleon degrees of freedom can be regarded as four components with an underlying SU(4) symmetry group.

Contents

1. Introduction

Effective field theory (EFT) is a theoretical framework for arranging the low-energy interactions of a system in powers of particle momenta. Chiral effective field theory applies this framework to the low-energy interactions of protons and neutrons while explicitly including the interactions of pions [1–9]. Pions are treated differently from other mesons since they become massless in the limit of massless quarks, thereby inducing long-range exchange interactions. The low-energy expansion of chiral EFT is organized in powers of Q, where Q denotes the typical momentum of the nucleons as well as explicit factors of the mass of the pion. The most important interactions are called leading order (LO) or $O(Q^0)$. The next most important contributions are next-to-leading order (NLO) or $O(Q^2)$. The terms after this are next-to-next-to-leading order (NNLO) or $O(Q^3)$, and so on.

Lattice effective field theory refers generally to lattice simulations based upon the framework of effective field theory. There are some reviews in the literature which discuss current methods used in lattice effective field theory [10, 11]. This is one of several different *ab initio* approaches being used to calculate the properties of few- and many-nucleon systems. Some others include the no-core shell model [12–14], constrained-path [15–18] and fixed-node [19, 20] Green's function Monte Carlo, auxiliary-field diffusion Monte Carlo [21–23], self-consistent Green's functions [24], in-medium similarity renormalization group [25, 26], and coupled cluster methods [27–29]. Lattice effective field theory differs from other *ab initio* approaches in several respects. One difference is that many different phenomena can be studied in lattice EFT using the same lattice action. In principle, all systematic errors are introduced up front when defining the low-energy effective theory, as opposed to the particular computational scheme used to calculate observables. Once the low-energy effective lattice action is determined, it may be used to calculate bound nuclei, transition matrix elements, scattering, the ground state of neutron matter, or bulk thermodynamic properties at non-zero temperature.

On the computational side, lattice EFT is helped by efficient lattice methods developed for lattice QCD and condensed matter

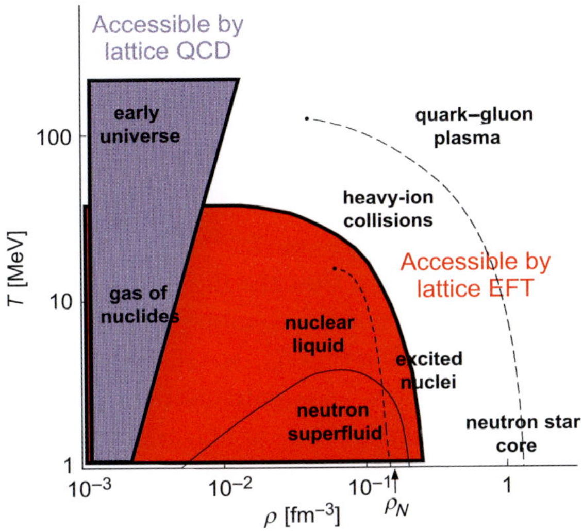

Fig. 1. Phase diagram showing regions accessible by lattice QCD simulations and regions accessible by nuclear lattice EFT simulations.

applications. This includes Markov Chain Monte Carlo techniques, auxiliary fields [30, 31], pseudofermion methods [32], and non-local updating schemes such as hybrid Monte Carlo [33–35]. Lattice EFT was first used in studies of infinite nuclear matter [36] and infinite neutron matter with and without explicit pion interactions [37–40]. The method has also been used to study light nuclei in pionless EFT [41] and chiral EFT at LO [42]. Dilute neutrons have been simulated at NLO in chiral EFT [43, 44] as well as light nuclei at NNLO [45, 46]. As indicated in Fig. 1, lattice EFT is able to explore regions of the QCD phase diagram not computationally accessible to lattice QCD. It also provides a test of the duality between quark and hadronic descriptions in the region where the two methods overlap.

2. Scattering on the Lattice

At any given order in the chiral EFT expansion, there will in general be short-range interaction coefficients which depend on the chosen regularization of ultraviolet physics. On the lattice this ultraviolet regularization is provided by the lattice spacing, unless some additional regularization is applied to the lattice interactions. In order to determine the values of the short-range two-nucleon interaction coefficients, we make a comparison of nucleon–nucleon scattering on the lattice with experimental scattering data. The extension to three-nucleon interaction coefficients is also required at NNLO, and that procedure on the lattice has been discussed in Epelbaum *et al.* [45]

Lüscher [47–49] has shown that the finite-volume energy levels for a two-body system in a periodic cubic box are directly connected to the infinite-volume scattering matrix (see also Chapter 5). While the method is very useful at low momenta, it becomes less accurate at higher momenta and higher orbital angular momenta. Also spin-orbit coupling and partial-wave mixing are difficult to measure accurately using Lüscher's method due to scattering artifacts produced by the periodic cubic boundary. A simple alternative approach has been developed to measure phase shifts for non-relativistic point particles on the lattice using a spherical wall boundary [50, 51].

In this approach, a hard spherical wall boundary is imposed on the relative separation between the two particles. This wall is placed at some chosen radius R_{wall} and removes copies of the interactions produced by the periodic lattice. Working in the center-of-mass frame, we solve the time-independent Schrödinger equation as a function of the relative separation between particles and find spherical standing waves which vanish at $r = R_{\text{wall}}$. At values of r beyond the range of the interaction, the spherical standing wave can be written as a superposition of products of spherical harmonics and spherical Bessel functions. We find

$$[j_l(kr)\cos\delta_l - y_l(kr)\sin\delta_l]\, Y_{l,l_z}(\theta,\phi), \tag{1}$$

where the energy is

$$E = 2\frac{k^2}{2m} = \frac{k^2}{m} \tag{2}$$

and the phase shift for partial wave l is δ_l. We can extract k from the energy E of the standing wave, and the phase shift δ_l is determined by setting the wave function in Eq. (1) to zero at the wall boundary.

When the total intrinsic spin of the two nucleons is non-zero, spin-orbit coupling generates mixing between partial waves. In this case the standing wave at the wall boundary is decomposed into spherical harmonics and coupled-channels equations are solved to extract the phase shifts and mixing angles. The spherical wall method was used to calculate phase shifts and mixing angle for low-energy nucleon–nucleon scattering [43]. More recently, the spherical wall approach has been improved in accuracy and computational efficiency [52]. In the improved approach we project onto spherical harmonics Y_{ℓ,ℓ_z} with angular momentum quantum numbers ℓ, ℓ_z. This allows us to construct radial position states for a given partial wave,

$$|r\rangle^{\ell,\ell_z} = \sum_{\vec{r}'} Y_{\ell,\ell_z}(\hat{r}')\delta_{r,|\vec{r}'|}|\vec{r}'\rangle. \tag{3}$$

We require that r is less than half the box length $L/2$. Using this technique we can construct a radial lattice Hamiltonian for each partial wave.

It is also useful to introduce auxiliary potentials in the region lying just in front of the spherical wall boundary [52]. We use one auxiliary potential in order to tune the energy of the scattering wave. This auxiliary potential is a spherical attractive well positioned in front of the spherical wall boundary. We can tune to any scattering energy by adjusting the depth of the well. For systems with partial wave mixing due to spin-orbit coupling, we also include a Hermitian but imaginary off-diagonal auxiliary potential for the two coupled channels. This breaks time reversal symmetry, and the resulting standing wave solutions now have both real and imaginary parts that are linearly independent. From the real and imaginary solutions we can now determine the scattering phase shifts and mixing angle at any given value of the scattering energy with much greater accuracy than could be done previously.

3. Auxiliary-field Projection Monte Carlo

The interactions of nucleons are simulated with auxiliary-field projection Monte Carlo. One can view the nucleons as interacting with each other while diffusing in space as a function of Euclidean time,

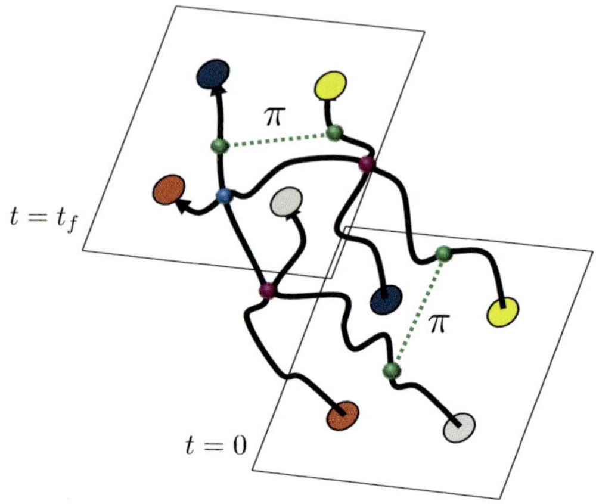

Fig. 2. A sketch showing worldlines of nucleons which evolve as a function of Euclidean time with contact interactions and the exchange of pions.

as indicated in Fig. 2. In the actual lattice calculations, however, the interactions are treated using auxiliary fields. The auxiliary field formalism is based upon the Gaussian integral identity [30, 31],

$$\exp\left[-\frac{C}{2}(N_i^\dagger B_{ij} N_j)^2\right] = \sqrt{\frac{1}{2\pi}} \int_{-\infty}^{\infty} ds \times \exp\left[-\frac{1}{2}s^2 + \sqrt{-C}\, s(N_i^\dagger B_{ij} N_j)\right], \tag{4}$$

where B_{ij} is a Hermitian matrix, C is an interaction coefficient, N_i and $N_i^\dagger$ are the nucleon fields, and $i(j)$ are indices labeling the nucleon degrees of freedom. We imply summation over repeated indices. Instead of the two-body interaction $(N_i^\dagger B_{ij} N_j)^2$, we have a one-body density $N_i^\dagger B_{ij} N_j$ coupled to the auxiliary variable s. We introduce an auxiliary variable in the same manner at every point in space and time on the lattice. As a result of this transformation, the particles completely decouple from one another and interact only with the auxiliary field. The original interactions are recovered exactly upon integrating over all possible configurations of the auxiliary field. Pion exchange potentials are reproduced in the same way using pion fields with gradient couplings to the nucleon fields.

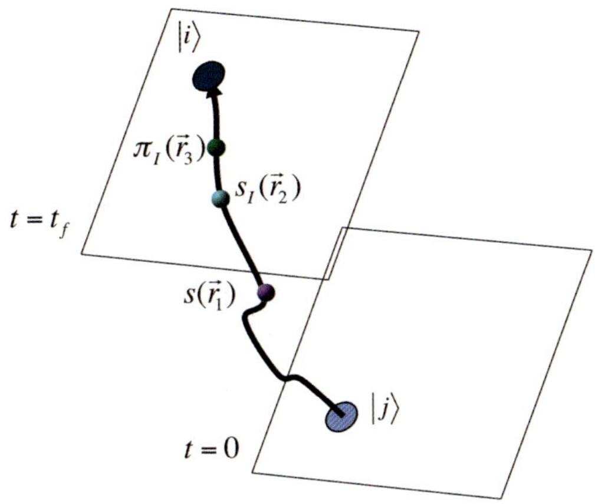

Fig. 3. A sketch showing a worldline for a single nucleon evolving in a background of pion fields and auxiliary fields. The subscript on the auxiliary field s_I and pion field π_I is a vector isospin index.

Each nucleon evolves as a single particle in a fluctuating background of pion fields and auxiliary fields. The original interactions are reproduced after integrating over the fluctuating pion and auxiliary fields. For a simulation with A nucleons, the amplitude for a given configuration of pion and auxiliary fields is proportional to the determinant of the $A \times A$ matrix M_{ij}. The entries of M_{ij} are single nucleon worldline amplitudes for a nucleon starting at state j at $t = 0$ and ending at state i at $t = t_f$. This is shown in Fig. 3. (See also Chapter 9 for applications of the auxiliary field formalism to medium-mass and heavy nuclei.)

4. SU(4) Symmetry

In order to improve the efficiency of the nuclear lattice simulations, it is useful to define a lattice Hamiltonian $H_{\mathrm{SU}(4)}$ which is an SU(4)-symmetric approximation to the leading-order Hamiltonian H_{LO}. This SU(4) symmetry was first investigated by Wigner [53]. It refers to a symmetry group which rotates the four nucleon degrees of freedom as four components of an SU(4) multiplet (see also Chapter 2). The low-energy nuclear interactions show an approximate SU(4) symmetry in the S-wave scattering channels.

While the Coulomb interaction and one-pion exchange interaction break this SU(4) symmetry, the short-distance part of the S-wave nucleon–nucleon interactions obey the symmetry rather well [54]. In the limit of large number of colors, one can view the symmetry as arising from a combinatorial enhancement of interaction terms which are spin and flavor independent [55, 56]. Furthermore, recent lattice QCD simulations have shown that SU(4) symmetry becomes increasingly accurate at heavier quark masses [57, 58].

There is experimental evidence that some predictions that one can derive from SU(4) symmetry are well satisfied by the spectrum of light nuclei [59, 60]. In Ref. [59], I have shown that any fermionic system with an SU($2N$)-symmetric interaction potential with negative-definite Fourier transform is in an $2N$ clustering phase and obeys a set of spectral convexity bounds, and for example, for SU(4), it implies an alpha-clustering phase. This is displayed in Fig. 4, where we have plotted the ground-state energy vs. particle number, A. We draw all lines connecting points with even A within the interval

$$2NK \leq A \leq 2N(K+1) \tag{5}$$

for any integer K. The statement of the spectral convexity bounds is that all of the points must lie on or above the drawn lines. This was proved [60] by applying the Hölder inequality to Euclidean-time

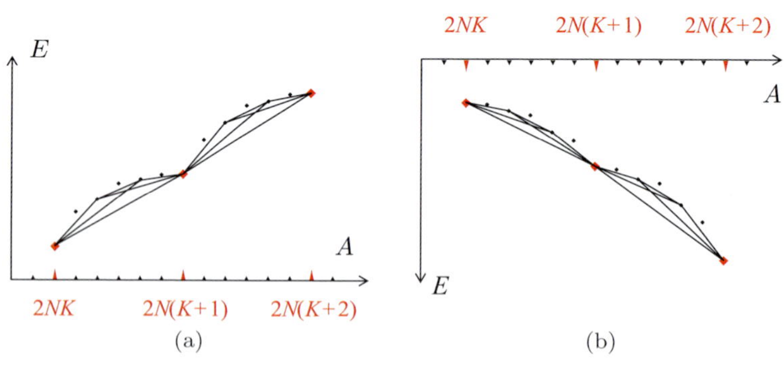

Fig. 4. Plot of the ground-state energy vs. number of particles A for (a) a confined weakly-attractive system, and (b) a self-bound strongly-attractive system. The lines show the corresponding SU($2N$) convexity bounds [59].

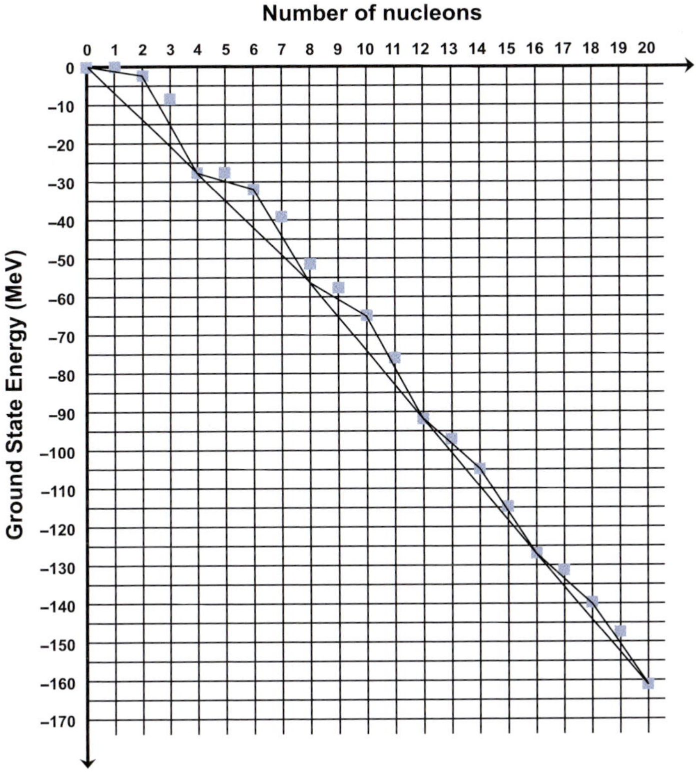

Fig. 5. Ground-state energy of nuclei vs. number of nucleons up to $A = 20$. The lines show the corresponding SU(4) convexity bounds [59].

correlation functions corresponding to different particle-number sectors. An immediate corollary of the theorem is that the system can be simulated using a single auxiliary field without any sign oscillations.

The low-energy nucleon–nucleon interaction are only approximately SU(4) symmetric. However one can see in Fig. 5 that all of the SU(4) convexity constraints are satisfied for nuclei up to $A = 20$. Furthermore, recent lattice simulations have also shown alpha-cluster structures consistent with an SU(4)-clustering phase in carbon-12 [61–64], oxygen-16 [65], and other alpha-like nuclei [66]. In Fig. 6 we show the alpha-cluster substructures seen in the low-lying even-parity

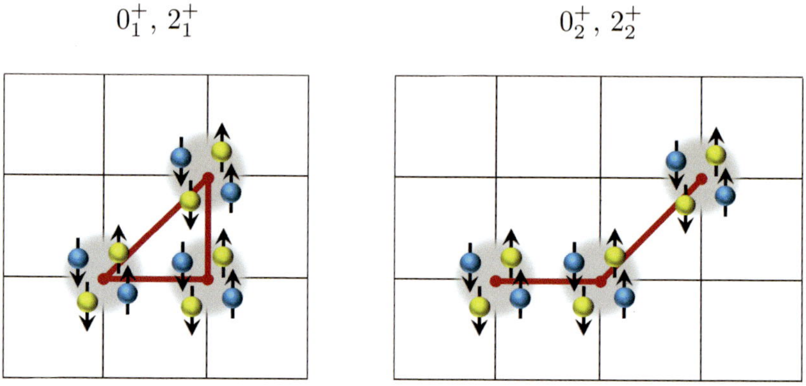

Fig. 6. Illustrations of the alpha-cluster substructures of carbon-12. (Left) The 0_1^+ ground state and 2_1^+ state dominated by a compact triangle configuration. (Right) The 0_2^+ Hoyle state and 2_2^+ state predominantly in an obtuse triangle or bent-arm configuration.

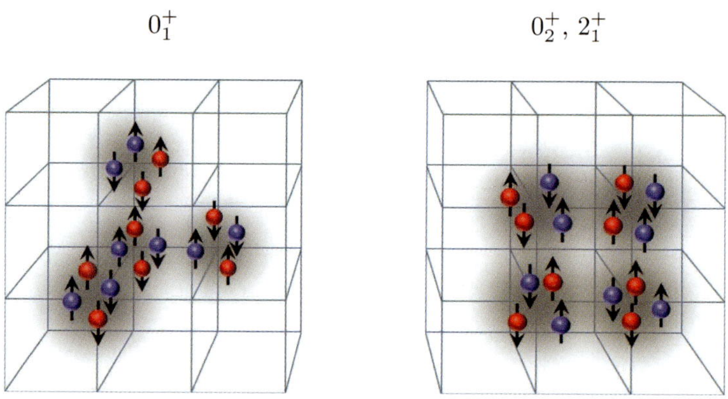

Fig. 7. Illustrations of the alpha-cluster substructures of oxygen-16. (Left) The 0_1^+ ground state mostly in a tetrahedral configuration. (Right) The 0_2^+ and 2_1^+ states predominantly in a square or planar configuration.

states of carbon-12. On the left panel we show the 0_1^+ ground state and 2_1^+ state dominated by a compact triangle configuration. On the right panel we show the 0_2^+ Hoyle state and 2_2^+ state dominated by an obtuse triangle or bent-arm configuration. In Fig. 7 we show the alpha-cluster substructures seen in oxygen-16. The left panel shows the 0_1^+ ground state mostly in a tetrahedral configuration, and the right panel shows the 0_2^+ and 2_1^+ states mostly in a square or planar configuration.

5. Leading Order Lattice Calculations and $H_{\mathrm{SU}(4)}$

We now discuss some details of the LO lattice calculations. Let, $|\Psi^{\mathrm{init}}\rangle$ be an A-body Slater determinant of single-nucleon standing waves in a periodic cube. Let H_{LO} denote the leading-order Hamiltonian, which includes instantaneous one-pion exchange and contact interactions. Let us define a trial wave function using Euclidean time projection with the SU(4)-symmetric Hamiltonian $H_{\mathrm{SU}(4)}$ previously discussed,

$$|\Psi(t')\rangle = \exp\left[-H_{\mathrm{SU}(4)}t'\right]|\Psi^{\mathrm{init}}\rangle. \tag{6}$$

Using this trial wave function we define the amplitude

$$Z(t) = \langle\Psi(t')|\exp\left[-H_{\mathrm{LO}}t\right]|\Psi(t')\rangle, \tag{7}$$

as well as an estimate of the energy,

$$E(t) = -\frac{\partial}{\partial t}[\ln Z(t)]. \tag{8}$$

The amplitude in Eq. (7) is computed in the auxiliary field formalism as integrals over the pion and auxiliary fields,

$$Z(t) = \int ds ds_S ds_I ds_{SI} d\pi_I \det M_{ij}(s, s_S, s_I, s_{SI}, \pi_I, t), \tag{9}$$

where the matrix M_{ij} consists of single nucleon amplitudes,

$$M_{ij}(s, s_S, s_I, s_{SI}, \pi_I, t) = \langle i|T\exp\left[-H_{\mathrm{LO}}(s, s_S, s_I, s_{SI}, \pi_I, t)\right]|j\rangle. \tag{10}$$

The symbol T denotes time ordering. In this lattice calculation there are 3 pion fields and 16 auxiliary fields. The auxiliary field s couples to total nucleon density, s_S couples to spin density with $S = 1, 2, 3$, while s_I couples to isospin density with $I = 1, 2, 3$, and s_{SI} couples to spin-isospin density with $S, I = 1, 2, 3$.

In the limit of large t we find

$$\lim_{t\to\infty} E(t) = E_0, \tag{11}$$

where E_0 is the energy of the lowest eigenstate $|\Psi_0\rangle$ of H_{LO} with a non-zero inner product with $|\Psi(t')\rangle$. To compute the expectation value of some operator O, we define

$$Z_O(t) = \langle \Psi(t')| \exp[-H_{\text{LO}}t/2]\, O\, \exp[-H_{\text{LO}}t/2]|\Psi(t')\rangle. \tag{12}$$

The expectation value of O for $|\Psi_0\rangle$ can be calculated in the large t limit as

$$\lim_{t\to\infty} \frac{Z_O(t)}{Z(t)} = \langle \Psi_0|O|\Psi_0\rangle. \tag{13}$$

We can view $\exp\left[-H_{\text{SU}(4)}t'\right]$ as an approximate inexpensive low-energy filter and $\exp\left[-H_{\text{LO}}t\right]$ as an exact low-energy filter. The projection $\exp\left[-H_{\text{SU}(4)}t'\right]$ is computationally inexpensive because it requires only one auxiliary field and, more importantly, the path integral is strictly positive for any even number of nucleons [59]. Although there is no positivity guarantee for odd numbers of nucleons, sign oscillations are suppressed in odd-A systems when they are only one particle or one hole away from an even system with no sign oscillations. Higher-order contributions in the chiral expansion, Coulomb interactions, and isospin-breaking effects are computed as perturbative corrections to the LO results.

5.1. *Symmetry-sign Extrapolation*

We conclude with one additional application of SU(4) symmetry to nuclear lattice simulations using effective field theory. Since the leading-order Hamiltonian H_{LO} is not SU(4) invariant, the auxiliary-field projection Monte Carlo calculations suffer from sign oscillations to a degree that depends on the number of protons and neutrons. For this reason most studies have considered nuclei with equal and even numbers of protons and neutrons where the sign oscillations are smallest. Recently a technique called symmetry-sign extrapolation has been developed which uses the approximate SU(4) symmetry of the nuclear interaction to control the sign oscillations without introducing uncontrolled systematic errors [67].

The basic idea of symmetry-sign extrapolation method is to smoothly connect the leading-order lattice Hamiltonian, $H_{\rm LO}$, to an SU(4)-symmetric Hamiltonian, $H_{\rm SU(4)}$, which has no sign problem and which is specifically tuned to produce approximately the same radius and binding energy for the system of interest. This can be the same SU(4)-symmetric Hamiltonian as used in Eq. (6) or a different Hamiltonian. We then define a one-parameter family of Hamiltonians $H(d_h)$, interpolating from $H_{\rm SU(4)}$ at $d_h = 0$ to $H_{\rm LO}$ at $d_h = 1$,

$$H(d_h) = (1 - d_h)H_{\rm SU(4)} + d_h H_{\rm LO}. \tag{14}$$

We then perform calculations using $H(d_h)$ for several values of d_h and extrapolate to the limit $d_h = 1$.

An extrapolation technique similar to symmetric-sign extrapolation has been used in shell model Monte Carlo calculations [68, 69] (cf. Chapter 9). In that case, the extrapolation is implemented by decomposing the Hamiltonian into "good" and "bad" sign parts, H_G and H_B, respectively. The calculations proceed by multiplying the coefficients of H_B by a parameter g and extrapolating from $g < 0$, where the simulations are free from sign oscillations, to the physical point $g = 1$. For symmetry-sign extrapolation and nuclear lattice effective field theory simulations, the analysis in terms of "good" and "bad" signs is not the entire story.

Most of the interactions can be divided into two groups which are sign-free by themselves. But a large portion of the sign oscillations in the nuclear lattice simulations are due to interference between the different underlying symmetries for the various different interactions. Since this effect is quadratic in the interfering interaction coefficients, the growth of the sign problem is somewhat slow. In many cases, it is possible to extrapolate from values not so far away from the physical point $d_h = 1$, and the method shows great promise for improving the quality of lattice results for proton-rich and neutron-rich nuclei.

Acknowledgements

I would like to thank my excellent collaborators on the projects cited in this Chapter from whom I have learned much. These include Serdar Elhatisari, Evgeny Epelbaum, Hermann Krebs, Timo Lähde, Bingnan Lu, Thomas Luu, Ulf-G. Meißner, and Gautam Rupak. Partial financial support was provided by the U.S. Department of Energy (DE-FG02-03ER41260). Computational resources were provided by the Jülich Supercomputing Centre.

References

[1] S. Weinberg, Nuclear forces from chiral lagrangians, *Phys. Lett. B* **251**, 288 (1990).
[2] S. Weinberg, Effective chiral lagrangians for nucleon–pion interactions and nuclear forces, *Nucl. Phys. B* **363**, 3 (1991).
[3] C. Ordonez and U. van Kolck, Chiral lagrangians and nuclear forces, *Phys. Lett. B* **291**, 459 (1992).
[4] C. Ordonez, L. Ray, and U. van Kolck, Nucleon–nucleon potential from an effective chiral lagrangian, *Phys. Rev. Lett.* **72**, 1982 (1994).
[5] U. van Kolck, Few nucleon forces from chiral lagrangians, *Phys. Rev. C* **49**, 2932 (1994).
[6] E. Epelbaum, W. Glockle, and U.-G. Meißner, Low-momentum effective theory for nucleons, *Phys. Lett. B* **439**, 1 (1998).
[7] E. Epelbaum, W. Glockle, and U.-G. Meißner, Nuclear forces from chiral lagrangians using the method of unitary transformation. I: Formalism, *Nucl. Phys. A* **637**, 107 (1998).
[8] P. F. Bedaque and U. van Kolck, Effective field theory for few-nucleon systems, *Ann. Rev. Nucl. Part. Sci.* **52**, 339 (2002).
[9] E. Epelbaum, H.-W. Hammer, and U.-G. Meißner, Modern theory of nuclear forces, *Rev. Mod. Phys.* **81**, 1773 (2009).
[10] D. Lee, Lattice simulations for few- and many-body systems, *Prog. Part. Nucl. Phys.* **63**, 117 (2009).
[11] J. E. Drut and A. N. Nicholson, Lattice methods for strongly interacting many-body systems, *J. Phys. G* **40**, 043101 (2013).
[12] P. Navratil, V. G. Gueorguiev, J. P. Vary, W. E. Ormand, and A. Nogga, Structure of $A = 10 - 13$ nuclei with two-plus three-nucleon interactions from chiral effective field theory, *Phys. Rev. Lett.* **99**, 042501 (2007).
[13] P. Maris, J. P. Vary, and A. M. Shirokov, *Ab initio* no-core full configuration calculations of light nuclei, *Phys. Rev. C* **79**, 014308 (2009).
[14] T. Dytrych, K. Launey, J. Draayer, P. Maris, J. Vary, *et al.*, Collective Modes in Light Nuclei from First Principles, *Phys. Rev. Lett.* **111**, 252501 (2013).
[15] S. C. Pieper, K. Varga, and R. B. Wiringa, Quantum monte carlo calculations of $A = 9, 10$ nuclei, *Phys. Rev. C* **66**, 044310 (2002).

[16] S. C. Pieper, Quantum Monte Carlo calculations of light nuclei, *Nucl. Phys. A* **751**, 516 (2005).
[17] L. Marcucci, M. Pervin, S. C. Pieper, R. Schiavilla, and R. Wiringa, Quantum Monte Carlo calculations of magnetic moments and M1 transitions in $A = 7$ nuclei including meson-exchange currents (2008).
[18] S. C. Pieper, *Ab initio* calculations of ^{12}C and neutron drops, *B. Am. Phys. Soc.* **54**, 70 (2009).
[19] S. Y. Chang *et al.*, Neutron matter: A superfluid gas, *Nucl. Phys. A* **746**, 215 (2004).
[20] A. Gezerlis and J. Carlson, Strongly-paired fermions: Cold atoms and neutron matter, *Phys. Rev. C* **77**, 032801 (2008).
[21] S. Gandolfi, F. Pederiva, S. Fantoni, and K. E. Schmidt, Auxiliary field Diffusion Monte Carlo calculation of nuclei with A<40 with tensor interactions, *Phys. Rev. Lett.* **99**, 022507 (2007).
[22] S. Gandolfi, A. Y. Illarionov, K. E. Schmidt, F. Pederiva, and S. Fantoni, Quantum Monte Carlo calculation of the equation of state of neutron matter, *Phys. Rev. C* **79**, 054005 (2009).
[23] S. Gandolfi, A. Illarionov, S. Fantoni, J. Miller, F. Pederiva, *et al.*, Microscopic calculation of the equation of state of nuclear matter and neutron star structure, *Mon. Not. Roy. Astron. Soc.* **404**, L35 (2010).
[24] V. Soma, C. Barbieri, and T. Duguet, *Ab-initio* Gorkov-Green's function calculations of open-shell nuclei, *Phys. Rev. C* **87**, 011303 (2013).
[25] K. Tsukiyama, S. K. Bogner, and A. Schwenk, In-medium similarity renormalization group for nuclei, *Phys. Rev. Lett.* **106**, 222502 (2011).
[26] H. Hergert, S. Binder, A. Calci, J. Langhammer, and R. Roth, *Ab Initio* Calculations of even oxygen isotopes with chiral two-plus three-nucleon interactions, *Phys. Rev. Lett.* **110**, 242501 (2013).
[27] M. Wloch *et al.*, *Ab initio* coupled-cluster study of ^{16}O, *Phys. Rev. Lett.* **94**, 212501 (2005).
[28] G. Hagen, D. J. Dean, M. Hjorth-Jensen, T. Papenbrock, and A. Schwenk, Benchmark calculations for ^{3}H, ^{4}He, ^{16}O and ^{40}Ca with *ab initio* coupled-cluster theory, *Phys. Rev. C* **76**, 044305 (2007).
[29] G. Hagen, T. Papenbrock, D. J. Dean, and M. Hjorth-Jensen, Medium-mass nuclei from chiral nucleon–nucleon interactions, *Phys. Rev. Lett.* **101**, 092502 (2008).
[30] J. Hubbard, Calculation of partition functions, *Phys. Rev. Lett.* **3**, 77 (1959).
[31] R. L. Stratonovich, On a method of calculating quantum distribution functions, *Soviet Phys. Doklady.* **2**, 416 (1958).
[32] D. H. Weingarten and D. N. Petcher, Monte Carlo integration for lattice gauge theories with fermions, *Phys. Lett. B* **99**, 333 (1981).
[33] R. T. Scalettar, D. J. Scalapino, and R. L. Sugar, New algorithm for the numerical simulation of fermions, *Phys. Rev. B* **34**, 7911 (1986).
[34] S. Gottlieb, W. Liu, D. Toussaint, R. L. Renken, and R. L. Sugar, Hybrid molecular dynamics algorithms for the numerical simulation of quantum chromodynamics, *Phys. Rev. D* **35**, 2531 (1987).

[35] S. Duane, A. D. Kennedy, B. J. Pendleton, and D. Roweth, Hybrid Monte Carlo, *Phys. Lett. B* **195**, 216 (1987).
[36] H. M. Müller, S. E. Koonin, R. Seki, and U. van Kolck, Nuclear matter on a lattice, *Phys. Rev. C* **61**, 044320 (2000).
[37] D. Lee, B. Borasoy, and T. Schäfer, Nuclear lattice simulations with chiral effective field theory, *Phys. Rev. C* **70**, 014007 (2004).
[38] D. Lee and T. Schäfer, Neutron matter on the lattice with pionless effective field theory, *Phys. Rev. C* **72**, 024006 (2005).
[39] D. Lee and T. Schäfer, Cold dilute neutron matter on the lattice I: Lattice virial coefficients and large scattering lengths, *Phys. Rev. C* **73**, 015201 (2006).
[40] D. Lee and T. Schäfer, Cold dilute neutron matter on the lattice II: Results in the unitary limit, *Phys. Rev. C* **73**, 015202 (2006).
[41] B. Borasoy, H. Krebs, D. Lee, and U.-G. Meißner, The triton and three-nucleon force in nuclear lattice simulations, *Nucl. Phys. A* **768**, 179 (2006).
[42] B. Borasoy, E. Epelbaum, H. Krebs, D. Lee, and U.-G. Meißner, Lattice simulations for light nuclei: Chiral effective field theory at leading order, *Eur. Phys. J. A* **31**, 105 (2007).
[43] B. Borasoy, E. Epelbaum, H. Krebs, D. Lee, and U.-G. Meißner, Chiral effective field theory on the lattice at next-to- leading order, *Eur. Phys. J. A* **35**, 343 (2008).
[44] B. Borasoy, E. Epelbaum, H. Krebs, D. Lee, and U.-G. Meißner, Dilute neutron matter on the lattice at next-to-leading order in chiral effective field theory, *Eur. Phys. J. A* **35**, 357 (2008).
[45] E. Epelbaum, H. Krebs, D. Lee, and U.-G. Meißner, Lattice chiral effective field theory with three-body interactions at next-to-next-to-leading order, *Eur. Phys. J. A* **41**, 125 (2009).
[46] E. Epelbaum, H. Krebs, D. Lee, and U.-G. Meißner, Lattice effective field theory calculations for A = 3,4,6,12 nuclei, *Phys. Rev. Lett.* **104**, 142501 (2010).
[47] M. Lüscher, Volume dependence of the energy spectrum in massive quantum field theories. 1. stable particle states, *Commun. Math. Phys.* **104**, 177 (1986).
[48] M. Lüscher, Volume dependence of the energy spectrum in massive quantum field theories. 2. scattering states, *Commun. Math. Phys.* **105**, 153 (1986).
[49] M. Lüscher, Two particle states on a torus and their relation to the scattering matrix, *Nucl. Phys. B* **354**, 531 (1991).
[50] B. Borasoy, E. Epelbaum, H. Krebs, D. Lee, and U.-G. Meißner, Two-particle scattering on the lattice: Phase shifts, spin- orbit coupling, and mixing angles, *Eur. Phys. J. A.* **34**, 185 (2007).
[51] J. Carlson, V. Pandharipande, and R. Wiringa, Variational calculations of resonant states in ^{4}He, *Nucl. Phys. A.* **424**, 47 (1984).
[52] B.-N. Lu, T. A. Lähde, D. Lee, and U.-G. Meißner, Precise determination of lattice phase shifts and mixing angles (2015). arxiv:1506.05652.
[53] E. Wigner, On the consequences of the symmetry of the nuclear hamiltonian on the spectroscopy of nuclei, *Phys. Rev.* **51**, 106 (1937).

[54] T. Mehen, I. W. Stewart, and M. B. Wise, Wigner symmetry in the limit of large scattering lengths, *Phys. Rev. Lett.* **83**, 931 (1999).
[55] D. B. Kaplan and M. J. Savage, The spin-flavor dependence of nuclear forces from large-N QCD, *Phys. Lett. B* **365**, 244 (1996).
[56] A. Calle Cordon and E. Ruiz Arriola, Wigner symmetry, large N_c and renormalized one-boson exchange potential, *Phys. Rev. C* **78**, 054002 (2008).
[57] S. Beane *et al.*, Nucleon–nucleon scattering parameters in the limit of SU(3) flavor symmetry, *Phys. Rev. C* **88**, 024003 (2013).
[58] S. R. Beane, E. Chang, W. Detmold, K. Orginos, A. Parreño, M. J. Savage, and B. C. Tiburzi, *Ab initio* calculation of the $np \to d\gamma$ radiative capture process, *Phys. Rev. Lett.* **115**, 132001 (2015).
[59] J.-W. Chen, D. Lee, and T. Schäfer, Inequalities for light nuclei in the Wigner symmetry limit, *Phys. Rev. Lett.* **93**, 242302 (2004).
[60] D. Lee, Spectral convexity for attractive SU($2N$) fermions, *Phys. Rev. Lett.* **98**, 182501 (2007).
[61] E. Epelbaum, H. Krebs, D. Lee, and U.-G. Meißner, *Ab initio* calculation of the Hoyle state, *Phys. Rev. Lett.* **106**, 192501 (2011).
[62] E. Epelbaum, H. Krebs, T. Lähde, D. Lee, and U.-G. Meißner, Structure and rotations of the Hoyle state, *Phys. Rev. Lett.* **109**, 252501 (2012).
[63] E. Epelbaum, H. Krebs, T. A. Lähde, D. Lee, and U.-G. Meißner, Viability of carbon-based life as a function of the light quark mass, *Phys. Rev. Lett.* **110**, 112502 (2013).
[64] E. Epelbaum, H. Krebs, T. A. Lähde, D. Lee, and U.-G. Meißner, Dependence of the triple-alpha process on the fundamental constants of nature, *Eur. Phys. J. A.* **49**, 82 (2013).
[65] E. Epelbaum, H. Krebs, T. A. Lähde, D. Lee, U.-G. Meißner, and G. Rupak, *Ab initio* calculation of the spectrum and structure of ^{16}O, *Phys. Rev. Lett.* **112**, 102501 (2014).
[66] T. A. Lähde, E. Epelbaum, H. Krebs, D. Lee, U.-G. Meißner, and G. Rupak, Lattice effective field theory for medium-mass nuclei, *Phys. Lett. B* **732**, 110 (2014).
[67] T. A. Lähde, T. Luu, D. Lee, U.-G. Meißner, E. Epelbaum, H. Krebs, and G. Rupak, Nuclear lattice simulations using symmetry-sign extrapolation, *Eur. Phys. J. A.* **51**, 92 (2015).
[68] Y. Alhassid, D. Dean, S. Koonin, G. Lang, and W. Ormand, Practical solution to the Monte Carlo sign problem: Realistic calculations of ^{54}Fe, *Phys. Rev. Lett.* **72**, 613 (1994).
[69] S. E. Koonin, D. J. Dean, and K. Langanke, Shell model monte carlo methods, *Phys. Rept.* **278**, 1 (1997).

Chapter 7

Clustering in Light Nuclei with the Correlated Gaussian Approach

Yasuyuki Suzuki[*,†,§] and Wataru Horiuchi[‡,¶]
[*]*Department of Physics, Niigata University*
Niigata 950-2181, Japan
[†]*RIKEN Nishina Center*
Wako 351-0198, Japan
[‡]*Department of Physics, Hokkaido University*
Sapporo 060-0810, Japan
[§]*suzuki@nt.sc.niigata-u.ac.jp*
[¶]*whoriuchi@nucl.sci.hokudai.ac.jp*

We focus on clustering, especially α-clustering in light nuclei. The nuclear wave function of the cluster model consists of cluster-internal and cluster-relative parts. As the relative motion between the clusters is spatially extended, a shell-model representation of clustering becomes very hard even with present-day large-scale calculations. In contrast to the shell-model basis truncation, the stochastic variational method using a correlated Gaussian basis makes it possible to describe, with no cluster ansatz *a priori*, both localized cluster states and shell-model-like states as exemplified in ^{4}He and ^{16}O. We also take advantage of the cluster model to calculate electroweak response functions of ^{4,6}He that involve continuum states.

Contents

1. Introduction

Nuclear force belongs to the strong interaction. Its action in nuclei is, however, strongly modified by the Pauli principle acting among nucleons, and the so-called effective two-nucleon interaction becomes much weaker than that in free space. This supports the assumption of a mean field for nuclei, a basis of the shell model. The residual nucleon–nucleon interaction is then amenable to large-scale shell-model calculations that successfully reproduce many nuclear properties.

In contrast to the shell model, an α-cluster model — or a cluster model in general — assumes a formation of subunit systems. Its physical ground is that e.g., the α-particle is tightly bound or its binding energy per particle is considerably large. Historically the α-cluster model was employed to account for α-decay phenomena. In nuclear reactions initial and final channels always contain some clusters (fragment nuclei) and while they are in the interaction region they exchange energy, momentum and nucleons, etc. In this way the cluster model is not necessarily a structure theory but also a unified theory of structure and reactions [1].

Though the cluster model was mostly applied to the two-cluster case, the situation has considerably changed with the developments in experimental techniques using radioactive nuclear beams [2]. The nuclear structure study has largely been extended to cover unstable nuclei near neutron and proton driplines, where especially in light nuclei, exotic structure has been recognized including halo, Borromean, and thick neutron skin, etc. The breakdown of the shell structure has been also discussed. Since the neutrons near the Fermi level are weakly bound in neutron-rich nuclei, the spatial extension of their motion reaches up to large distances. The nuclei near the dripline have just one or very few particle-bound states, and most of the states can be observed as a resonance, suggesting the important role of continuum states. These features have motivated us to develop a theory that, by going beyond the mean-field theory, can describe both the asymptotics of a weakly bound system and few-body dynamics, or the correlation among the constituents. Extending the cluster model to a microscopic multicluster model that incorporates more than two clusters, many successful examples have been shown for the light exotic nuclei [3–11]. Recent developments in cluster physics can be seen in Ref. [12].

The organization of this Chapter is the following. Section 2 consists of three parts: first, we briefly introduce the basic elements of the cluster model with an emphasis on a correlated Gaussian (CG) basis. Second, we remind readers of the microscopic symplectic model to point out its interesting and complementary relationship to the cluster model, and third, we demonstrate that representing a spatially localized α-cluster in single-particle basis costs a huge configuration space. Section 3 presents some specific applications to bound- and unbound-state problems that we have recently worked out. We show that both of the ground and first excited states in ^{16}O and ^{4}He are reproduced by calculations of *ab initio* type and stress that the structure of their first excited states is most simply understood in the cluster concept. We then show a useful application of the cluster model to calculating electroweak strength (response) functions in ^{4,6}He. Section 4 draws a summary.

2. Cluster Model

2.1. *Microscopic Cluster Model with Correlated Gaussians*

The wave function of an A-nucleon system is obtained by solving the Schrödinger equation with a translationally invariant Hamiltonian

$$H = \sum_{i=1}^{A} T_i - T_{\rm cm} + \sum_{j>i=1}^{A} V_{ij} + \sum_{k>j>i=1}^{A} V_{ijk}, \tag{1}$$

where T_i is the kinetic energy of nucleon i, $T_{\rm cm}$ is the kinetic energy of the center-of-mass, and V_{ij} and V_{ijk} are two-body and three-body interactions, respectively. The Coulomb potential between the protons is included. For a large A system, the two-body interaction is an effective force and approximations on the wave functions are necessary. In contrast, a few-nucleon system is described with the use of realistic forces and the wave function can be obtained by accurately solving the Schrödinger equation.

Suppose a cluster model consisting of N subsystems in which the ith cluster contains A_i nucleons ($\sum_{i=1}^{N} A_i = A$). The ansatz of the cluster model assumes a basis in a form

$$\Psi_{JM\pi}^{NIL} = \mathcal{A}\Big\{[\phi_I^N(\text{int})\psi_L^N(\text{rel})]_{JM}\Big\}, \tag{2}$$

where $JM\pi$ denote the total angular momentum, its z-projection, and the parity of the system, and $\mathcal{A}$ is an antisymmetrizer. The cluster-intrinsic wave function $\phi_I^N(\text{int})$ is given by successively coupling each cluster wave function ϕ_{I_i} as $[\cdots[[\phi_{I_1}\phi_{I_2}]_{I_{12}} \phi_{I_3}]_{I_{123}}\cdots\phi_{I_N}]_I$, while $\psi_L^N(\text{rel})$ stands for the cluster relative-motion function with total orbital angular momentum L. The angular momentum coupling is indicated by the square bracket. A few remarks should be made here. The wave function ϕ_{I_i} usually includes the ground-state and/or excited-state solutions of the ith cluster Hamiltonian but they are often approximated by appropriate model wave functions. Some quantum numbers other than the spin I_i, e.g., the isospin, may be necessary to specify ϕ_{I_i}, but they are suppressed here to simplify the notation. One or few of A_i can be unity, which means that the corresponding ϕ_{I_i} is nothing but the spin–isospin function of the nucleon cluster.

Once ϕ_{I_i}'s are set, a main effort in the cluster model is to determine $\psi_L^N(\text{rel})$. The quality of the solution is influenced by how many clusters and what kinds of clusters are employed. For low-energy problems, N usually corresponds to the number of fragments of the low-lying thresholds. Approximating the total wave function as

$$\Psi_{JM\pi} = \sum_{NIL} \Psi_{JM\pi}^{NIL}, \tag{3}$$

we obtain the equation of motion for $\psi_L^N(\text{rel})$:

$$\sum_{NIL} \langle \Psi_{JM\pi}^{N'I'L'} | H - E | \Psi_{JM\pi}^{NIL} \rangle = 0, \tag{4}$$

which should be satisfied for all $N'I'L'$. With this approach both the ground state of ^{4}He and the reactions of astrophysical interest, $^2\text{H}(^2\text{H},\gamma)^4\text{He}$ and $^2\text{H}(^2\text{H},p)^3\text{He}$, are successfully described with a realistic force [13–15].

The dynamics of the cluster relative motion becomes complicated with increasing N. For its realistic description, a good trial function for $\psi_L^N(\text{rel})$ has to be chosen. We have demonstrated that the trial function composed of combinations of the CG leads to virtually exact solutions for relatively small N cases. Let $(\boldsymbol{x}_1, \boldsymbol{x}_2, \ldots, \boldsymbol{x}_{N-1})$ denote $N-1$ relative coordinates among the clusters. The CG is defined as

$$\psi_{LM_L}^N(\text{rel}) = \exp(-\tilde{\boldsymbol{x}}B\boldsymbol{x})\theta_{LM_L}(\boldsymbol{x}), \tag{5}$$

where $\tilde{\boldsymbol{x}}B\boldsymbol{x} = \sum_{i,j=1}^{N-1} B_{ij}\boldsymbol{x}_i \cdot \boldsymbol{x}_j$ with B being a positive-definite, symmetric matrix. The elements of B are variational parameters and the non-diagonal elements explicitly couple the different coordinates, which is useful to represent the correlated motion between the clusters. The function $\theta_{LM_L}(\boldsymbol{x})$ describes the angular motion of the clusters. Its most elegant representation is made by introducing a global vector, $\tilde{u}\boldsymbol{x} = \sum_{i=1}^{N-1} u_i\boldsymbol{x}_i$, where u_i's are variational parameters. A remarkable feature of Eq. (5) is that its functional form is invariant for any choice of $\boldsymbol{x}$, which is a key for describing a variety of structure in a unified manner. See Refs. [14, 16, 18–20] for details and applications. In order to make this CG approach practical, the stochastic variational method (SVM) [3, 4, 9, 16–18] has been developed to select most important basis functions among a large number of candidate bases. As shown in Sec. 3.1, this approach makes it possible to reproduce those states which have quite different character, e.g., localized cluster states and shell-model-like states.

The CG cluster model is not based on an antisymmetrized product of single-particle orbits. Applying it to a large A system is limited by $A!$ permutations of $\mathcal{A}$. To handle this problem, some efforts have been made to express the CG cluster wave functions by an integral transformation of the Slater determinants built from single-particle Gaussian wave packets (GWP) [16, 18]:

$$\phi_\kappa(\boldsymbol{s}_1, \dots, \boldsymbol{s}_A) = \mathcal{A}\Big\{ \prod_{i=1}^{A} \varphi^{\nu}_{\boldsymbol{s}_i}(\boldsymbol{r}_i)\phi_{\frac{1}{2}\sigma_i}\phi_{\frac{1}{2}\tau_i} \Big\}, \tag{6}$$

where $\kappa = (\sigma_1\tau_1, \dots, \sigma_A\tau_A)$ is the set of the spin–isospin quantum numbers of the nucleons, and $\varphi^{\nu}_{\boldsymbol{s}}(\boldsymbol{r})$ is the GWP centered at $\boldsymbol{s}$ (with a width parameter ν),

$$\varphi^{\nu}_{\boldsymbol{s}}(\boldsymbol{r}) = \Big(\frac{2\nu}{\pi}\Big)^{3/4} \exp(-\nu(\boldsymbol{r} - \boldsymbol{s})^2). \tag{7}$$

This approach has been successfully applied to a multicluster description of light nuclei [3–11]. Especially ^{11}Li, the most well-known halo nucleus, was described as a six-cluster model ($A = 11$, $N = 6$) of $\alpha + {}^3\mathrm{H} + n + n + n + n$ [21, 22]. In this way, possible excitations or modifications of the ^{9}Li ($\alpha + {}^3\mathrm{H} + n + n$) core due to the two halo neutrons were examined.

Use of the GWP (7) in the cluster model was made popular by Ref. [23], the main purpose of which was to examine the possibility of describing many-particle many-hole configurations with the α-cluster model. In recent molecular dynamics calculations [24, 25], the nuclear wave function of type (6) with a projection onto parity and angular momentum has been adopted to search for an optimum set of the positions $\boldsymbol{s}_1, \ldots, \boldsymbol{s}_A$. Though no cluster ansatz is put *a priori* in those calculations, the intrinsic density calculated as a snapshot often exhibits clustering. It should be noted, however, that such clustering may be blurred by the angular momentum projection and the correlated motion of the nucleons that can not be represented by one or few combinations of the Slater determinants.

2.2. *Cluster vs. Symplectic Model*

The nucleons in the clusters are subject to the Pauli principle. When the clusters are closely located and overlap strongly, it is meaningless to attempt at identifying which cluster the nucleons belong in. Such cluster states often have large overlap with particular shell-model configurations. It is of considerable interest and importance to know what configurations are covered by the cluster model. This problem has been studied mostly for two-cluster cases assuming that the cluster wave functions ϕ_{I_i} are given by the harmonic-oscillator (HO) shell-model functions with a common oscillator constant. The Elliott SU(3) model [26, 27] is useful to make a link between the cluster model and the HO shell model [28, 29]. For example, the α+^{12}C cluster model for ^{16}O, where the wave functions of the α-particle and ^{12}C are respectively described by the $|(0s)^4[4](00)\rangle$ and $|(0s)^4(0p)^8[444](04)\rangle$ shell-model configurations (with $[f_1 f_2 \cdots]$ the space symmetry labels), contains the states labeled by the SU(3) $(\lambda\mu)$ representations

$$(00) \times (04) \times (Q0) = \sum_{k=0}^{4} (Q + k - 4, k). \tag{8}$$

Here Q is the number of HO quanta of the α-^{12}C relative motion. Because of the Pauli principle, Q must be $Q \geqq 4$ and in the case of $4 \leqq Q \leqq 7$ some $(\lambda\mu)$ states do not appear. The α+^{12}C cluster model is found to contain the important shell-model configurations needed to describe low-lying spectrum of ^{16}O [30,31]. As the clusters develop and have little overlap, the corresponding cluster state strongly deviates from the usual shell-model configurations.

As mentioned above, some of the most important shell-model configurations can be expressed by exciting the relative motion degree of freedom of the clusters. A careful comparison with experiment, however, indicates that the cluster model, if the clusters are frozen to their ground states, tends to miss some states with simple core excitations, overestimates cluster decay widths, and underestimates $E2$ transition rates. With regard to this problem we pay attention to the non-compact real symplectic group Sp$(6,\mathbb{R})$, often denoted as Sp$(3,\mathbb{R})$ [32–36] (see also Chapter 3).

The Sp$(6,\mathbb{R})$ model has been developed as a microscopic theory of the nuclear collective motion with monopole and quadrupole collectivity. As an extension of the U(3) group generated by $C_{\alpha\beta} = \sum_{i=1}^{A}\frac{1}{2}\big(b_\alpha^\dagger(i)b_\beta(i) + b_\alpha(i)b_\beta^\dagger(i)\big)$ $(\alpha,\ \beta = x, y, z)$, the Sp(6, $\mathbb{R}$) group also contains spectrum-generating operators

$$A_{\alpha\beta}^\dagger = \sum_{i=1}^{A} b_\alpha^\dagger(i)b_\beta^\dagger(i), \quad A_{\alpha\beta} = \sum_{i=1}^{A} b_\alpha(i)b_\beta(i), \tag{9}$$

where $b_\alpha^\dagger(i)$ and $b_\alpha(i)$ are the (center-of-mass excitation free) creation and annihilation operators for HO quanta. The kinetic energy and the isoscalars (IS) monopole and quadrupole operators are all expressed by the Sp$(6,\mathbb{R})$ generators. Thanks to this property, the Sp$(6,\mathbb{R})$ group provides a means for selecting shell-model configurations needed to build up monopole and quadrupole collectivity. The Sp$(6,\mathbb{R})$ shell-model space is decomposable into infinite but countable collection of subspaces, each of which itself is an infinite dimensional irreducible representation subspace of the Sp(6, $\mathbb{R}$) group. That subspace, called a symplectic band, is characterized by its bandhead.

Both microscopic cluster and Sp(6, $\mathbb{R}$) models enable us to select specific core excited states from the huge space of possible shell-model excitations. Although they overlap partly, they have their own advantages, and are expected to play a complementary role [37, 38]. The overlap between the two states is efficiently calculated using a recursion relation if the cluster wave functions ϕ_{I_i} are given by the lowest possible Pauli-allowed configurations [39].

Table 1. Overlap between the α+^{12}C cluster and Sp(6, $\mathbb{R}$) states of SU(3) symmetry $(\lambda_0 + \mathcal{N}, \mu_0)$, where $(\lambda_0\, \mu_0)$ is the SU(3) label of the Sp(6, $\mathbb{R}$) bandhead and $\mathcal{N}$ denotes the number of HO quanta excited from the bandhead. Values are taken from Ref. [39].

	$\mathcal{N}$						
$(\lambda_0\, \mu_0)$	0	2	4	6	8	10	12
(0 0)	1	0.808	0.558	0.376	0.263	0.191	0.143
(2 1)	1	0.816	0.581	0.405	0.290	0.213	0.161
(4 2)	1	0.818	0.602	0.437	0.322	0.242	0.185
(6 3)	1	0.822	0.629	0.476	0.364	0.281	0.220
(8 4)	1	0.832	0.668	0.531	0.421	0.335	0.267

Table 1 lists an example of the overlap between the α+^{12}C cluster and Sp(6, $\mathbb{R}$) states for ^{16}O. The large overlaps between the first excitation ($\mathcal{N} = 2$) indicate that the α+^{12}C cluster basis significantly contains the quadrupole collectivity. Incorporating both excitations into a unified framework has been made possible by developing a method for evaluating interaction matrix elements [40, 41] and such calculations were applied to study the monopole and quadrupole strengths in light nuclei [42–44]. This type of calculations should be further extended to include higher $\mathcal{N}$ excitations since the monopole strength carried by the cluster model attracts renewed interest as the experimental information on the monopole strength reaches higher in the excitation energy [45].

2.3. *Spatially Localized Clusters in Single-particle Basis*

Describing a localized cluster near the nuclear surface is a challenge for the shell model. To examine how many HO quanta are needed to represent such cluster states, let us consider a specific $p + p + n + n$ configuration

$$\Psi^{\nu}_{\boldsymbol{s}}(\alpha) = \left[\prod_{i=1}^{4} \phi^{\nu}_{\boldsymbol{s}}(\boldsymbol{r}_i)\right] \phi_{S=0,T=0}(1, 2, 3, 4), \tag{10}$$

where $\phi_{S=0,T=0}(1,2,3,4)$ is the totally antisymmetric spin-isospin function. The four nucleons occupy the same GWP. The wave function (10) represents the $(0s)^4$ α-cluster located at $\boldsymbol{s}$ whose magnitude is set to $s = r_0 A^{1/3}$ ($r_0 = 1.1\,\text{fm}$). The size of the α-cluster increases as $\nu^{-1/2}$.

The single-particle content of $\phi^{\nu}_{\boldsymbol{s}}(\boldsymbol{r})$ controls the possibility of a shell-model description of the cluster state. The probability that $\phi^{\nu}_{\boldsymbol{s}}(\boldsymbol{r})$ contains orbits with the orbital angular momentum l is given by

$$P_\eta(l) = \sum_{m=-l}^{l} \frac{1}{4\pi} \int d\hat{\boldsymbol{s}} \int_0^\infty dr\, r^2 \left|\langle Y_{lm}(\hat{\boldsymbol{r}})|\phi^{\nu}_{\boldsymbol{s}}(\boldsymbol{r})\rangle\right|^2, \tag{11}$$

which results in $(2l+1)\mathrm{i}_l(\eta)e^{-\eta}$, where

$$\eta = \nu s^2 \tag{12}$$

is a dimensionless quantity and $\mathrm{i}_l(z)$ is the modified spherical Bessel function, $\sqrt{\frac{\pi}{2z}} I_{l+1/2}(z)$. The square root of η indicates the ratio of the position of the cluster to its size, so that it is a measure of spatial localization of the cluster. In case that the cluster size is as small as the free α-particle ($\nu = 0.26\,\text{fm}^{-2}$), η quickly grows from 2.0 ($A = 16$) to 11 ($A = 208$) as a function of A. On the other hand, if ν is the standard HO width parameter ($\nu = 0.48\,A^{-1/3}\,\text{fm}^{-2}$), η gradually increases from 1.5 ($A = 16$) to 3.5 ($A = 208$). According to the uncertainty relation of the angular momentum and the angular position, the orbit $\phi^{\nu}_{\boldsymbol{s}}(\boldsymbol{r})$ with large η contains many high l components. For example, the cumulative sum, $\sum_{l=0}^{l_{\text{max}}} P_\eta(l)$, exceeds 80% at $l_{\text{max}} = 1, 4, 5, 6$ for $\eta = 1, 5, 10, 15$, respectively. Applying this analysis to the α-cluster wave function (10) enables us to obtain its partial-wave contents. Figure 1 plots the probability of finding the configurations with partial waves l_1, l_2, l_3, l_4 ($l_i \leq l_{\text{max}}$) as a function of l_{max}. The contribution of the high partial-waves becomes important and spreads over many configurations as η increases. In case that ν is chosen to be the same as that of the nucleus, it is possible to fairly well describe the surface α-clustering even in heavy nuclei provided the shell-model space includes the single-particle

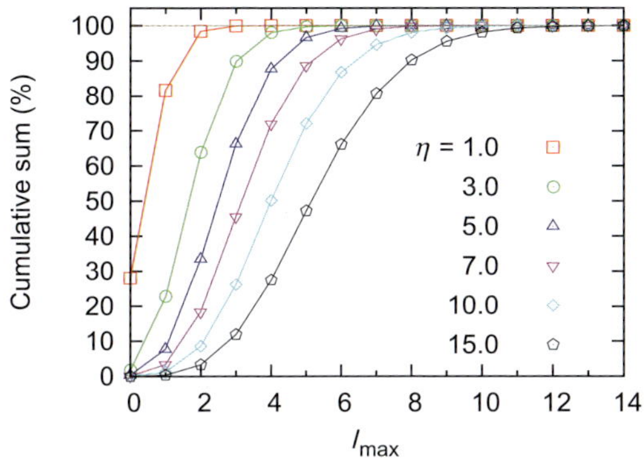

Fig. 1. Cumulative sum of the probabilities of finding the components with each partial wave $l_i \leq l_{\max}$ in the α-cluster wave function, which has a center-of-mass motion characterized by η, Eq. (12), and has zero orbital angular momentum.

orbits up to $l_{\max} = 4$. However, if ν is chosen to be that of the free α-particle, we need even more high l values, say, up to $l_{\max} = 10$, which is certainly beyond present-day shell-model calculations.

The extent to which the surface α-clustering develops is quantified by an α spectroscopic amplitude (SA) that is crucial to determine α-decay rate. The α-decay amplitude of the nuclear state $\Psi_{JM}(A)$ to the daughter nucleus $\phi_I(A-4)$ is defined by

$$y_{(IL)J}(r) = \sqrt{\binom{A}{4}} \langle [[\phi_0(\alpha)\phi_I(A-4)]_I Y_L(\hat{\boldsymbol{R}})]_{JM} \times \frac{\delta(R-r)}{Rr} | \Psi_{JM}(A) \rangle, \tag{13}$$

where $\boldsymbol{R}$ is the relative distance vector between the α-particle and the daughter nucleus. If $\phi_I(A-4)$ is identical with $\phi_0(\alpha)$, the right hand side of Eq. (13) has to be multiplied by a factor $1/\sqrt{2}$. The amplitude $y_{(IL)J}(r)$ near the nuclear surface is of vital importance for the α decay and α clustering [46], and the one calculated by the shell model is usually too small to reproduce the observed α width [47]. The integrated quantity $S_{(IL)J} = \int_0^\infty dr r^2 [y_{(IL)J}(r)]^2$ is called α spectroscopic factor (SF). Even though $S_{(IL)J}$ is large, it does not always mean that the probability of finding the α-particle near the surface is enhanced. We show examples of $y_{(IL)J}(r)$ in Secs. 3.1 and 3.2.

3. Specific Applications to Bound- and Unbound-state Problems

One of the characteristic features in light nuclei is the coexistence and competition of the shell and cluster structure, that is, the mean-field dominating motion and the clustering. Since each structure is usually described by the respectively tailored models, a unified description of both structure has been challenging in nuclear theory. Since the CG approach accounts for various correlations among the nucleons and in addition, is flexible enough to describe the tail of the wave function, we apply it to four- to six-body problems where both types of structure may show up. Especially we ask questions whether the clusters appear naturally without assuming them *a priori* and how the effect of cluster distortion is observed.

3.1. *Core Plus Few-nucleon Model with Application to* ^{16}O

We adopt a core plus few-nucleon model to focus on the coexistence of the cluster and shell structure. As noted in Sec. 2.1, a fully microscopic calculation for this model is extremely hard for a large A system. We instead treat the core as a point particle though its microscopic structure is taken into consideration as explained below. In this way a many-body problem reduces to the few-body problem, which can be accurately solved without any further assumption. The correlation among the few nucleons called valence nucleons has to be fully taken into account together with the one between the core.

One important remark should be made concerning the Pauli principle due to the microscopic structure of the core. The valence nucleon can not occupy the orbits in the core that are called forbidden states and to enforce this requirement an explicit anti-symmetrization of the core and valence nucleons is in principle necessary. However, this is extremely hard, as detailed in Ref. [47], which provides cluster-shell mixed basis calculations for a cluster decay. A simpler version, we adopt here, is to eliminate the forbidden states from the Hilbert space for the valence-nucleon motion, which is a basic idea of the orthogonality condition model (OCM) [52]. In this semi-microscopic approach, the Hamiltonian of the N-body system consisting of the core and $N-1$ nucleons takes a form

$$H = H_{\text{core}} + \sum_{i=1}^{N} T_i - T_{\text{cm}} + \sum_{i=1}^{N-1} U_i + \sum_{j>i=1}^{N-1} V_{ij} + \lambda \sum_{i=1}^{N-1} \Gamma_i, \tag{14}$$

where H_{core} is the core Hamiltonian and the core-nucleon potential U_i is chosen to be consistent with available data. The Γ_i of Eq. (14) is a projector onto the forbidden orbits, and a positive constant λ is taken to be large enough to ensure that the forbidden components are practically eliminated from the solution [53]. This model combined with the CG approach has successfully been applied to, for example, ^{6}He [48], ^{16}C [49], ^{22}C [50], and ^{31}Ne [51].

We test a $^{12}\text{C} + n + n + p + p$ five-body model for ^{16}O [54]. The ground state of ^{16}O is doubly closed, but contrary to the shell structure its first excited state is 0^+ with the unexpectedly low excitation energy, 6.05 MeV. This issue has been a long-standing puzzle and many theoretical works have been devoted to resolve it. The 0_2^+ state has not been reproduced even in modern large-scale calculations [55, 56], and only recently the first attempt has been made to understand this state within an *ab initio* framework [57] (see also Chapter 6). A recent shell-model calculation [58] reproduces most of the levels of ^{16}O but the energy of the 0_2^+ state is still 1 MeV higher than experiment. One successful classic is the $^{12}\text{C}+\alpha$ OCM calculation [30,31], which reproduces all the low-lying levels but the 0^- (10.95 MeV) state, including the 0_2^+ state as well as its parity-inverted partner [59]. One important reason for the success is that the $\alpha+^{12}\text{C}$ cluster model contains the most important shell-model configurations as noted in Sec. 2.2. The success of this calculation also indicates that the mechanism of the low excitation energy is the α-clustering. To confirm whether this is really the case or not, it is crucially important to perform the five-body calculation for $J^\pi = 0^+$ without assuming the α-cluster.

The input of the ^{12}C $+ n + n + p + p$ five-body model is detailed in Ref. [54]. The ^{12}C core is described by $0s_{1/2}$ and $0p_{3/2}$ HO wave functions ($\hbar\omega = 16.0\,\mathrm{MeV}$). The U_i consists of central and spin-orbit terms and it is set to reproduce the low-lying $1/2^-$, $1/2^+$, and $5/2^+$ spectrum of ^{13}C. The valence nucleons interact with the Minnesota (MN) potential [60] that fairly well reproduces the binding energies and sizes of d, ^{3}H, and α-particle. The five-body wave function is expressed in terms of a combination of many CG of Eq. (5). No spurious center-of-mass motion is involved. Each basis contains many variational parameters, and therefore preparing the configuration space by discretizing the parameters on grids results in an enormous dimension of at least 10^{10}. Instead we test a number of candidate bases with the SVM [3, 4, 9, 16–18], choose the best one among them and increase the basis dimension one by one until a convergence is reached. This procedure is computer-time expensive, especially because the forbidden states have to be eliminated but no other viable methods are at hand to get converged solutions for this problem. Convergence is slow and more than 9,000 basis states are required [54]. The most basis states are used to eliminate the forbidden states.

After the convergence two 0^+ states appear below ^{12}C$+\alpha$ threshold and their energies are both remarkably close to experiment [54], that is, the excitation energy of the 0_2^+ state is very well reproduced. The present five-body model has thus resolved the puzzle. In order to quantify the extent to which the four valence nucleons behave as an α-cluster, we display in the left panel of Fig. 2 the ^{12}C $+ \alpha$ SA of ^{16}O defined in Eq. (13). The SA results of the ^{12}C $+ \alpha$ OCM calculation [30, 31] are also plotted for comparison. For the 0_1^+ state, the highest peak is located at around 4 fm, which is shorter than the touching distance of ^{12}C$+\alpha$ ($\approx$4.9 fm). In that case, the four valence nucleons have significant components distorted from the α-particle configuration because of the interaction from the core, and the SF is not that large (0.105). Compared to this, the OCM SA has a peak at a shorter distance and much larger inner oscillation, leading to a

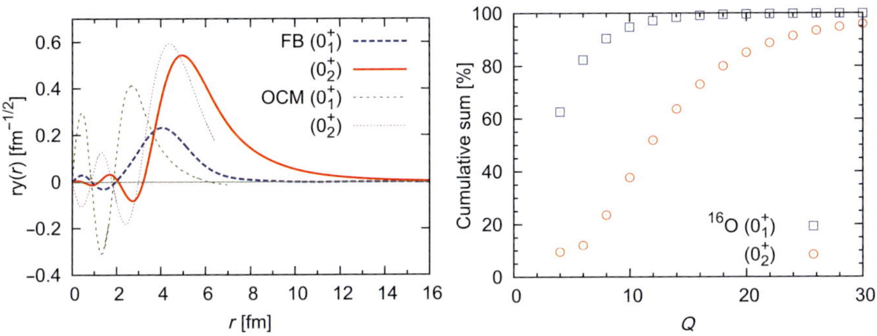

Fig. 2. (Left) α+^{12}C SAs of the ground and first excited 0^+ states of ^{16}O calculated in the five-body model (labeled "FB") [54]. Results with the ^{12}C + α orthogonality condition model (labeled "OCM") [30, 31] are also plotted for comparison. (Right) Cumulative sum of occupation probabilities as a function of the number of HO quanta, Q, of the ground and first excited states of ^{16}O ($\hbar\omega = 16.0$ MeV).

larger value (0.300) for the OCM SF. The difference between the two results suggests that the four valence nucleons tend to gain energy by balancing between the two factors, the formation of an α-cluster and the single-particle motion around the core. The latter causes the distortion of the α-cluster. Its effect is taken into account in the five-body calculation, whereas it is neglected in the OCM, leading to the apparent large SF value. It should be noted that in spite of these differences the nodes of oscillation of the SA appear almost at the same positions, which is a consequence of the Pauli principle. On the other hand, the SA of the 0_2^+ state is much larger than that of the 0_1^+ state and has a peak at the touching distance with a very long tail, which is a characteristic feature of developed cluster structure. The shape of the calculated SA is similar to that of the OCM calculation. The SF of the five-body model is 0.680, in agreement with the OCM value, 0.679. It is not an overstatement that the 0_2^+ state is an α+^{12}C cluster state and the α-clustering is a key to resolve the puzzle.

Why is it so difficult to describe the 0_2^+ state with shell-model calculations? From the SA of the 0_2^+ state, the degree of localization measure η, defined in Eq. (12), is estimated to be about 6.2. Figure 1 clearly indicates that a fair description of the α-cluster with $\eta \approx 6$ requires $l_{\max} = 4$–5 for the partial wave of each nucleon. It would be impossible for present-day computers to perform a shell-model calculation including all the configurations that satisfy this condition. To substantiate this argument, we calculate the probability distribution of the number Q of HO quanta contained in the five-body wave functions that have no spurious center-of-mass components. This is easily done by applying the method of Ref. [61]. The right panel of Fig. 2 plots the cumulative sum of the probabilities for the ground and first excited states, respectively. The HO frequency is set the same as the occupied states in the ^{12}C core. The distribution of the ground state starts with $Q = 4$ corresponding to the closed-shell configuration of $(0p_{1/2})^2_\nu(0p_{1/2})^2_\pi$, and rapidly converges with increasing Q. It is interesting to note, however, that the ground state contains the closed-shell configuration only at the 60% level and $2\hbar\omega$ components at the 20% level. The distribution of the 0^+_2 state is quite different from that of the ground state, as already predicted by the cluster model [30, 31]. The closed-shell component is only 10%. The probability is peaked at $Q = 10$–12 with about 14% and distributes up to large Q values. The cumulative sum up to $Q = 12$ is only 47%, and exceeds 90% only when all the components up to $Q = 26$ are included. Therefore, it is practically impossible to describe such a state by the HO basis with the usual major-shell truncation. Developing an innovative method will be indispensable to cope with clustering states.

It is interesting to extend the five-body model to include the core excitation, as formulated in Ref. [62]. It is of great interest to apply the $^{12}\mathrm{C} + n + n + p + p$ model to the low-lying 1^- and 2^+ levels of ^{16}O as well because they are key levels for the radiative capture reaction of $^{12}\mathrm{C}(\alpha, \gamma)^{16}\mathrm{O}$. It is desirable to study those states in such a framework that can describe both the shell-model and cluster states in a unified way as shown above.

3.2. *Ab initio Approach to the* ^{4}He *Structure*

The α-particle is tightly bound as the ground state of the doubly magic nucleus, ^{4}He. Contrary to its shell structure, the first excited state of ^{4}He is not of negative parity but 0^+, which is in exactly the same situation as ^{16}O. Above the 0_2^+ state of ^{4}He, seven negative-parity states are observed below $E_x = 26\,\text{MeV}$ [63]. It is of great interest to clarify what mechanism is responsible for lowering the 0_2^+ state. An answer to this question is obtained by *ab initio* calculations with use of realistic nuclear forces for various J^π states of ^{4}He [19, 64, 65].

Similarly to the case of ^{16}O, the competition and coexistence of the shell and cluster structure plays a key role in governing the low-lying spectrum of ^{4}He [64]. The clustering in ^{4}He is brought about by two clusters of $3N + N$ ($^3\text{H} + p$ and $^3\text{He} + n$), that is, the dominant mode of the 0_2^+ state of ^{4}He is the relative motion between $3N$ and N. It is also found that the tensor force plays a decisive role to reproduce the correct level ordering of the negative-parity states. In particular, the reason why the excitation energy of the $J^\pi T = 0^-0$ state gets low is due to a strong coupling between different angular momentum channels, $(LS) = (11)$ and (22), caused by the tensor force. If one uses an effective interaction that contains only a central term, the negative-parity levels would be almost degenerate [65].

The importance of the clustering in the low-lying $T = 0$ excited states close to the 0_2^+ state is made clear by the $^3\text{He} + n$ SA, which is defined analogously to Eq. (13). The left panel of Fig. 3 displays the SA with $(IL) = (00)$ for the ground and first excited states

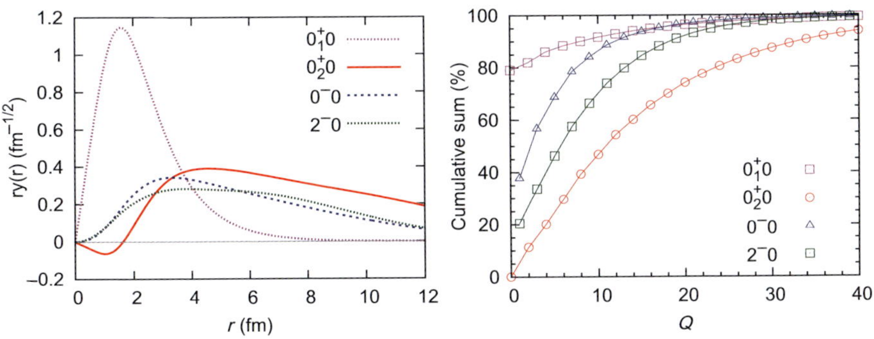

Fig. 3. (Left) $^3\text{He} + n$ SAs of the ground and excited states of ^{4}He [64]. (Right) Cumulative sum of occupation probabilities as a function of the number of HO quanta for the ground and first excited 0^+ states and the two negative-parity states, 0^-0 and 2^-0.

of ^{4}He [64]. The realistic G3RS nucleon–nucleon interaction [66] is employed. The SA of the ground state is peaked at around 1.8 fm and its relative distance is considerably shorter than the radius of ^{3}He (∼2.2 fm). That is, the neutron has a strong overlap with the ^{3}He cluster. On the contrary, the SA of the 0_2^+ state has a peak beyond the surface region of ^{3}He with a very long tail. The calculated SF is very large, 1.03. The SAs of the 0^-0 and 2^-0 states for $(IL) = (11)$ also exhibit surface peaked structure with a long tail, though the P-wave centrifugal barrier makes their peak positions shift to the origin. The SFs of those states are also quite large, 0.58 and 0.52, respectively. These two negative-parity states can be interpreted as a parity-inverted partner of the 0_2^+ state since they all share the $3N + N$ intrinsic structure.

The 0_2^+ state is a good example of the Feshbach resonance. It appears between the $^3\text{H} + p$ and $^3\text{He} + n$ thresholds that have small spacing. Because of the isospin conservation the 0_2^+ state contains almost equal mixing of the $^3\text{H} + p$ and $^3\text{He} + n$ components and it is stabilized thanks to the $^3\text{H} + p$ Coulomb potential despite that it is above the $^3\text{H} + p$ threshold. In the case of the 0^- and 2^- states, both channels are open but the P-wave centrifugal barrier plays a role for stabilizing them. The decay widths of these S- and P-wave resonant states are estimated using an R-matrix type formula, $\Gamma = 2P_L(kr)(\hbar^2 r/2\mu)[y_{(IL)J}(r)]^2$ with an appropriate channel radius r, where P_L is the penetrability, k is the wave number of the decaying energy, and μ is the reduced mass. The decay widths obtained are all consistent with empirical data [64].

The degree of clustering is also seen by analyzing the occupation probability of HO quanta. The right panel of Fig. 3 displays the cumulative sum of the occupation probabilities of the number of HO quanta [67], where the Argonne $v8'$ [68] (AV8$'$) plus a central three-body force [69] is employed. The HO frequency $\hbar\omega$ is set to reproduce the ground-state matter radius with the $(0s)^4$ configuration ($\hbar\omega$ = 21.6 MeV). For the ground state the $Q = 0$ contribution is approximately 80% and higher Q components are due to tensor and short-range repulsive forces. Since the ground state of ^{4}He is strongly correlated as shown in, e.g., Refs. [70–72] and references therein, the HO quanta distribution is strongly affected by the features of the interaction employed [67]. Compared to the ground state, the convergence of the distribution for the excited states, especially for the 0_2^+ state, is extremely slow. The high Q component is definitely needed to describe the S-wave relative motion function reaching far distances (see the left panel of Fig. 3). The inversion-doublet partners, especially the $J^\pi T = 2^- 0$ state, also need high Q components in order to represent the asymptotic behavior of the P-wave relative motion function.

3.3. *Strength Functions of* ^{4}He

The electroweak strength function of a nucleus, which connects its ground state with the excited states by a perturbation due to the electroweak interaction, is an important quantity that carries information on the resonant and continuum structure. In the previous subsection the structure of the ^{4}He spectrum was discussed in a bound-state approximation. This is reasonable when the state has a small decay width. Recent computational advances allow us to study the four-nucleon system by taking a proper account of continuum coupling. In fact ^{4}He is a testing ground of investigating nucleon dynamics, correlations as well as nucleon–nucleon interactions. For example, Ref. [73] discussed the effect of different three-body interactions on the inelastic (e, e') form factor to the 0_2^+ state, which is nothing but the IS monopole excitation. Similarly the seven negative-parity states can be excited by the spin-dipole (SD) and electric dipole ($E1$) operators [74, 75]. The SD operator belongs to the first forbidden transition operator of the weak interaction and will play an important role in neutrino–nucleus reactions that are expected to occur in the scenario of a core collapse supernova [76]. The operator can be a leading order for $N = Z$ nuclei where the Fermi and Gamow–Teller type transitions are suppressed. We show results of the above-mentioned response functions of ^{4}He obtained in the four-body calculations using realistic nuclear forces.

The response function at the excitation energy E for the transition of the ground state Ψ_0 is defined by

$$S(p, \lambda, E) = \mathcal{S}_{f\mu} |\langle \Psi_f | \sum_{i=1}^{4} \mathcal{O}_{\lambda\mu}(i) T_i^p | \Psi_0 \rangle|^2 \delta(E_f - E_0 - E), \quad (15)$$

where $\mathcal{O}_{\lambda\mu}(i) = [(\boldsymbol{r}_i - \boldsymbol{x}_4)\boldsymbol{\sigma}_i]_{\lambda\mu}$ and $(\boldsymbol{r}_i - \boldsymbol{x}_4)_\mu$ for the SD and $E1$ operators specified by angular momentum and its projection, $\lambda\mu$, respectively. Here, $\boldsymbol{x}_4$ is the center-of-mass coordinate of ^{4}He, and T_i^p distinguishes different types of isospin operators, 1 ($p =$ IS, isoscalar), τ_{3i} ($p =$ IV0, isovector), and $\frac{1}{2}(1 - \tau_{3i})$ ($p = E1$). $\mathcal{S}_{f\mu}$ denotes a sum over f and μ and indicates that all the final states Ψ_f satisfying the energy conservation are included.

All of the excited states of ^{4}He are in the continuum, but this does not pose any problem in calculating $S(p, \lambda, E)$ because the sum over f in Eq. (15) is replaced by the imaginary part of the Green's function, $-\frac{1}{\pi}\mathrm{Im}[1/(E + E_0 - H + \mathrm{i}\epsilon)]$. The calculation of the Green's function involves the inversion of the Hamiltonian and we use the complex scaling method (CSM) that is widely used for calculating the strength function [77, 78]. With the CSM we can adopt square-integrable ($\mathcal{L}^2$) basis functions for the continuum wave functions. The accuracy of CSM crucially depends on how completely the $\mathcal{L}^2$ basis functions are prepared to simulate the continuum. We prepare them by paying attention to the sum rule of the electroweak strength functions (i), and the decay channels of the continuum, (ii) and (iii), as follows [74, 75]: (i) configurations of single-particle-like excitations that are obtained by acting $\mathcal{O}_{\lambda\mu}$ on Ψ_0, (ii) configurations of $3N + N$ (^{3}H $+ p$ and ^{3}He $+ n$) two-body disintegration, and (iii) configurations of ^{2}H $+ p + n$ three-body disintegration. These disintegration configurations are much better described using appropriate relevant coordinates but not the single-particle coordinates. The CG basis is so powerful that all the above configurations can be expressed without any difficulty, and the needed matrix elements can be evaluated easily [19].

Figure 4 plots the SD strength functions of IV0 and IS types calculated with the AV8$'$+3NF potential. For the sake of comparison,

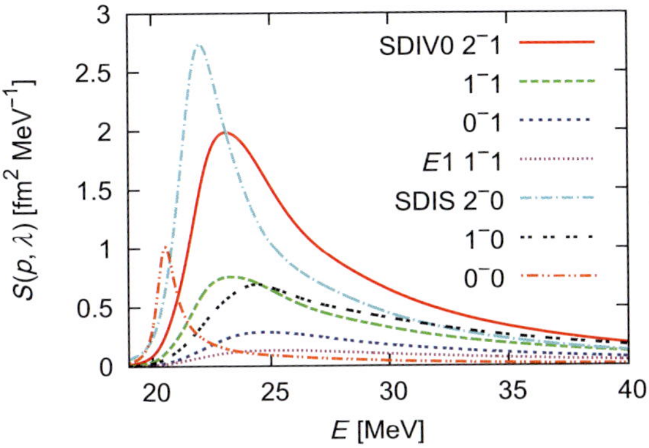

Fig. 4. Isoscalar (IS) and isovector (IV0) spin-dipole (SD) and electric dipole ($E1$) strength functions for ^{4}He as a function of the excitation energy. The AV8$'$ + 3NF potential is used. This figure is adapted from figures in Ref. [75].

the $E1$ strength function is also plotted. As seen in the figure, the three SD strength functions of IS type show narrower widths than the $E1$ strength function. These IS SD strength functions, especially for the 0^-0 and 2^-0 cases, give much narrower distribution than the IV0 strength functions. It should be noted that the peak positions of the IS and IV0 strength functions for the 0^- states become more largely separated than those for the 1^- and 2^- states. The reason for this difference is understood by analyzing the role played by the tensor force [19, 65]. All the peak positions of the strength functions well correspond to the observed excitation energies of the seven $T = 0$ and 1 negative-parity states of ^{4}He [63]. Table 2 compares the resonance parameters extracted from the strength functions to experiment [75]. The agreement between them is reasonable considering the fact that most of the widths are larger than 5 MeV.

The non-energy-weighted sum rules (NEWSR) for the $E1$ and SD operators are fully satisfied as expected [74, 75]. The NEWSR of the SD operator is found to be sensitive to the tensor correlation of the ground state of ^{4}He. It is also found that the SD transition between the inversion-doublet partners is very large, exhausting almost all the strengths from the 0_2^+ state.

Table 2. Resonance energies E_R and widths Γ, given in MeV, of the negative-parity levels of ^{4}He [75]. Experimental data are taken from Ref. [63].

	E_R		Γ	
$J^\pi T$	Cal.	Exp.	Cal.	Exp.
0^-0	20.54	21.01	1.06	0.84
2^-0	22.03	21.84	3.10	2.01
2^-1	23.11	23.33	5.58	5.01
1_1^-1	23.34	23.64	7.17	6.20
1^-0	24.44	24.25	9.57	6.10
0^-1	24.71	25.28	9.98	7.97
1_2^-1	25.36	25.95	13.24	12.66

3.4. *Electric Dipole Strength Function of* ^{6}He

As the last application, we discuss the so-called soft dipole mode [79–81] of ^{6}He. As a typical two-neutron halo nucleus, ^{6}He is well understood with $\alpha+n+n$ structure, but we go beyond that by treating ^{6}He as a full six-nucleon system [82] without assuming the frozen α-cluster. This is especially important because different results are obtained under the frozen-core assumption [83–85].

We employ the MN two-body potential [60] tuned to reproduce the two-neutron separation energy of ^{6}He. Calculated matter and charge radii of ^{6}He are consistent with measurement, indicating the $\alpha+n+n$ three-body structure. To be more exact, however, we obtain 5% core swelling in ^{6}He, which is never obtained in the three-body model with the ^{4}He core being fixed. The final states reached by the $E1$ operator are expressed in terms of many CG basis states similarly to the ^{4}He case: (i) the single-particle excitations from the ground state of ^{6}He, (ii) $\alpha+n+n$ three-body disintegration, and (iii) $^3\mathrm{H}+^2\mathrm{H}+n$ three-body disintegration channels.

Two peaks are found in the $E1$ strength function at the excitation energy $E_x = 3.0$ and 32.9 MeV. Figure 5 displays the proton and neutron transition densities at those energies. The proton and neutron transition densities for the peak at 3.0 MeV exhibit in-phase oscillation in the inner region of ^{6}He and out-of-phase

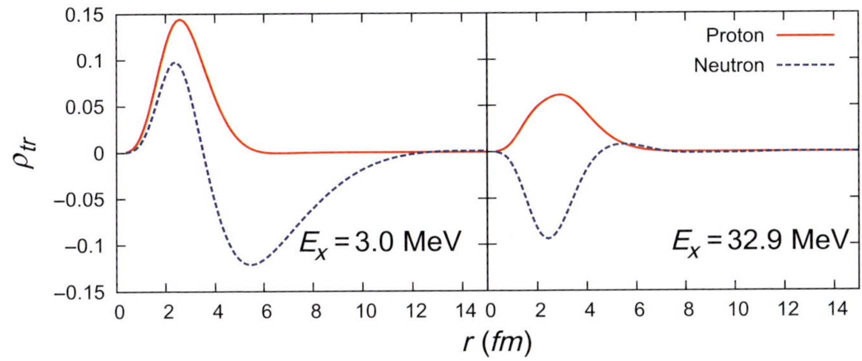

Fig. 5. *E1* transition densities (ρ_{tr}) for protons and neutrons of ^{6}He. The figures are plotted based on the data in Ref. [82].

oscillation in the outer region of $r > 3.5$ fm, which corresponds to a macroscopic picture of the soft dipole mode. In the case of the peak at 32.9 MeV, out-of-phase transition is found in the whole region, which is characteristic of the macroscopic picture of the giant dipole mode. By evaluating the *p*-*p* distance as a function of E_x and comparing it to that of the α-particle, we can clearly monitor the distortion of the core in ^{6}He [82]. Up to $E_x = 20$ MeV, the α-core does not change, and only the valence neutrons contribute to the strength function, which is consistent with the soft dipole picture. At about $E_x = 30$ MeV where the giant dipole resonance occurs, the α-cluster is strongly distorted from the one in the ground state. In this way the precise few-body approach succeeds to examine the validity of the soft dipole mode in ^{6}He.

4. Summary

The nuclear cluster model takes into account an important feature, that is, the nuclear system often dissociates into a few subsystems. The basic ansatz of microscopic cluster models is that the wave function of the system can be well approximated by an antisymmetrized product of the cluster-internal and cluster relative-motion parts. In this Chapter, we have focused on the use of the CG basis to represent the latter part as it accurately describes various types of correlated motion of the subsystems.

The cluster-model wave function overlaps with ordinary shell-model wave functions especially when the clusters are located close to each other, but strongly deviates from them as the spatial extension of the clusters develops. As a consequence of these two factors, the coexistence and competition of the shell and cluster structure can occur in light nuclei. We have demonstrated this phenomenon in the ground and first excited 0^+ states of ^{4}He and ^{16}O with the four- and five-body calculations of *ab initio* type, in which no cluster assumption is invoked *a priori*. The evidence for clear clustering in both excited states has been given by analyzing the SA for the decay into the cluster partitions as well as the probability distribution of the number of HO quanta. We have argued that describing the surface clustering in terms of a product of single-particle functions costs enormous effort that is far beyond the limit of present-day computers.

The cluster model can properly describe the asymptotics of the nuclear wave function. This advantage has been utilized in the calculation of the electroweak response functions of ^{4,6}He, where the relevant continuum states are taken into account according to the cluster model. With the *ab initio* calculation, we have succeeded to reproduce the resonance parameters of the negative-parity states of ^{4}He and to clarify the mechanism of the low-lying soft dipole mode in ^{6}He.

It is very hard to apply the cluster model presented here to larger systems. Two approaches will be interesting as future works. One is to develop a mixed basis model of cluster and symplectic excitations because the latter is a microscopic theory for collective nuclear motion and both are expected to play a complementary role. The other is to apply the core plus few nucleon model to medium and heavy nuclei to study how the four nucleons behave around a heavier core.

Acknowledgments

This work was supported in part by JSPS KAKENHI Grants No. 25800121 and 15K05072.

References

[1] K. Wildermuth and Y. C. Tang, *A Unified Theory of the Nucleus*. Academic Press, New York (1977).

[2] I. Tanihata, H. Savajols, and R. Kanungo, Recent experimental progress in nuclear halo structure studies, *Prog. Part. Nucl. Phys.* **68**, 215 (2013).

[3] K. Varga, Y. Suzuki, and R. G. Lovas, Microscopic multicluster description of neutron-halo nuclei with a stochastic variational method, *Nucl. Phys. A.* **571**, 447 (1994).

[4] K. Varga, Y. Suzuki, and Y. Ohbayasi, Microscopic multicluster description of the neutron-rich helium isotopes, *Phys. Rev. C.* **50**, 189 (1994).

[5] K. Arai, Y. Suzuki, and K. Varga, Neutron-proton halo structure of the 3.563-MeV 0^+ state in ^{6}Li, *Phys. Rev. C.* **51**, 2488 (1995).

[6] K. Varga, Y. Suzuki, and I. Tanihata, Microscopic four-cluster description of the mirror nuclei ^{9}Li and ^{9}C, *Phys. Rev. C.* **52**, 3013 (1995).

[7] K. Arai, Y. Ogawa, Y. Suzuki, and K. Varga, Structure of the mirror nuclei ^{9}Be and ^{9}B in a microscopic cluster model, *Phys. Rev. C.* **54**, 132 (1996).

[8] K. Arai, Y. Suzuki, and R. G. Lovas, Structure of ^{6}He with an extended three-cluster model, *Phys. Rev. C.* **59**, 1432 (1999).

[9] Y. Ogawa, K. Arai, Y. Suzuki, and K. Varga, Microscopic four-cluster description of ^{10}Be and ^{10}C with the stochastic variational method, *Nucl. Phys. A.* **673**, 122 (2000).

[10] K. Arai, Y. Ogawa, Y. Suzuki, and K. Varga, Microscopic multicluster description of light exotic nuclei with stochastic variational method on correlated Gaussians, *Prog. Theor. Phys. Suppl.* **142**, 97 (2001).

[11] Y. Suzuki, R. G. Lovas, K. Yabana, and K. Varga, *Structure and Reactions of Exotic Nuclei*. Taylor & Francis, London and New York (2003).

[12] H. Horiuchi, K. Ikeda, and K. Katō, Recent developments in nuclear cluster physics, *Prog. Theor. Phys. Suppl.* **192**, 1 (2012).

[13] K. Arai, S. Aoyama, Y. Suzuki, P. Descouvemont, and D. Baye, Tensor force manifestations in *ab initio* study of the ^{2}H$(d,\gamma)^4$He, ^{2}H$(d,p)^3$H, and ^{2}H$(d,n)^3$He reactions, *Phys. Rev. Lett.* **107**, 132502 (2011).

[14] S. Aoyama, K. Arai, Y. Suzuki, P. Descouvemont, and D. Baye, Four-nucleon scattering with a correlated Gaussian basis method, *Few-Body Syst.* **52**, 97 (2012).

[15] P. Descouvemont, D. Baye, Y. Suzuki, S. Aoyama, and K. Arai, Few-body models for nuclear astrophysics, *AIP Adv.* **4**, 041011 (2014).

[16] K. Varga and Y. Suzuki, Precise solution of few-body problems with the stochastic variational method on a correlated Gaussian basis, *Phys. Rev. C.* **52**, 2885 (1995).

[17] K. Varga, Y. Ohbayasi, and Y. Suzuki, Stochastic variational method with noncentral forces, *Phys. Lett. B.* **396**, 1 (1997).

[18] Y. Suzuki and K. Varga, *Stochastic Variational Approach to Quantum-Mechanical Few-Body Problems*. In *Lecture Notes in Physics Monographs*, vol. 54. Springer, Berlin Heidelberg (1998).

[19] Y. Suzuki, W. Horiuchi, M. Orabi, and K. Arai, Global-vector representation of the angular motion of few-body systems II, *Few-Body Syst.* **42**, 33 (2008).

[20] J. Mitroy, S. Bubin, W. Horiuchi, Y. Suzuki, L. Adamowicz, W. Cencek, K. Szalewicz, J. Komasa, D. Blume, and K. Varga, Theory and application of explicitly correlated Gaussians, *Rev. Mod. Phys.* **85**, 693 (2013).
[21] K. Varga, Y. Suzuki, and R. G. Lovas, Microscopic multicluster model of ^{9,10,11}Li, *Phys. Rev. C.* **66**, 041302(R) (2002).
[22] Y. Suzuki, R. G. Lovas, and K. Varga, Study of light exotic nuclei with a stochastic variational method — application to lithium isotopes, *Prog. Theor. Phys. Suppl.* **146**, 413 (2002).
[23] D. M. Brink. In C. Bloch, (ed.), *Proc. Int. School of Physics, "Enrico Fermi", Course XXXVI*, Academic, New York (1966), p. 474.
[24] T. Neff and H. Feldmeier, Clustering and other exotic phenomena in nuclei, *Eur. Phys. J. Special Topics.* **156**, 69 (2008).
[25] Y. Kanada-En'yo, M. Kimura, and A. Ono, Antisymmetrized molecular dynamics and its applications to cluster phenomena, *Prog. Theor. Exp. Phys.* **2012**, 01A202 (2012).
[26] J. P. Elliott, Collective motion in the nuclear shell model. I. Classification schemes for states of mixed configurations, *Proc. R. Soc. A.* **245**, 128 (1958).
[27] J. P. Elliott, Collective motion in the nuclear shell model. II. Classification schemes for states of mixed configurations, *Proc. R. Soc. A.* **245**, 562 (1958).
[28] H. Horiuchi, Kernels of GCM, RGM and OCM and their calculational methods, *Prog. Theor. Phys. Suppl.* **62**, 90 (1977).
[29] K. T. Hecht, E. J. Reske, T. H. Seligman, and W. Zahn, Spectroscopic amplitudes for complex cluster systems, *Nucl. Phys. A.* **356**, 146 (1981).
[30] Y. Suzuki, Structure study of $T = 0$ states in ^{16}O by ^{12}C+α cluster-coupling model. I, *Prog. Theor. Phys.* **55**, 1751 (1976).
[31] Y. Suzuki, Structure study of $T = 0$ states in ^{16}O by ^{12}C+α cluster-coupling model. II, *Prog. Theor. Phys.* **56**, 111 (1976).
[32] G. Rosensteel and D. J. Rowe, Nuclear Sp(3, R) model, *Phys. Rev. Lett.* **38**, 10 (1977).
[33] G. Rosensteel and D. J. Rowe, On the algebraic formulation of collective models III. The symplectic shell model of collective motion, *Ann. Phys.* **126**, 343 (1980).
[34] D. J. Rowe, Microscopic theory of the nuclear collective model, *Rep. Prog. Phys.* **48**, 1419 (1985).
[35] K. T. Hecht, *The Vector Coherent State Method and Its Application to Problems of Higher Symmetries.* In *Lecture Notes in Physics.* Vol. 290. Springer, Berlin Heidelberg (1987).
[36] A. C. Dreyfuss, K. D. Launey, T. Dytrych, J. P. Draayer, and C. Bahri, Hoyle state and rotational features in Carbon-12 within a no-core shell-model framework, *Phys. Lett. B.* **727**, 511 (2013).
[37] K. T. Hecht, Alpha-clustering vs giant $E2$ excitations in ^{16}O, *J. Phys. Soc. Jpn. Suppl.* **44**, 232 (1978).
[38] K. T. Hecht and D. Braunschweig, Sp(6, R) symmetry and α-breakup amplitudes of giant $E2$ excitations in light nuclei, *Nucl. Phys. A.* **295**, 34 (1978).

[39] Y. Suzuki, Cluster and symplectic excitations in nuclei, *Nucl. Phys. A.* **448**, 395 (1986).
[40] Y. Suzuki and K. T. Hecht, Symplectic and cluster excitations in nuclei: Evaluation of interaction matrix elements, *Nucl. Phys. A.* **455**, 315 (1986).
[41] Y. Suzuki and K. T. Hecht, Spin-orbit and tensor interactions in Sp(6, R)-model calculations, *Prog. Theor. Phys.* **77**, 190 (1987).
[42] Y. Suzuki, The nuclear symplectic model for isoscalar giant resonances and its coupling with a cluster basis in ^{20}Ne, *Nucl. Phys. A.* **470**, 119 (1987).
[43] Y. Suzuki and S. Hara, Isoscalar monopole and quadrupole strength of ^{16}O in an α+^{12}C cluster and symplectic mixed basis, *Phys. Rev. C.* **39**, 658 (1989).
[44] Y. Suzuki, Microscopic cluster models versus symplectic and shell models, *J. Phys. Soc. Japan. Suppl.* **58**, 129 (1989).
[45] T. Yamada, Y. Funaki, T. Myo, H. Horiuchi, K. Ikeda, G. Röpke, P. Schuck, and A. Tohsaki, Isoscalar monopole excitations in ^{16}O: α-cluster states at low energy and mean-field-type states at higher energy, *Phys. Rev. C.* **85**, 034315 (2012).
[46] H. Horiuchi and Y. Suzuki, Analysis of alpha-decay widths for molecular viewpoints in light nuclei, *Prog. Theor. Phys.* **49**, 1974 (1973).
[47] R. G. Lovas, R. J. Liotta, A. Insolia, K. Varga, and D. S. Delion, Microscopic theory of cluster radioactivity, *Phys. Rep.* **294**, 265 (1998).
[48] W. Horiuchi and Y. Suzuki, Momentum distribution and correlation of two-nucleon relative motion in ^{6}He and ^{6}Li, *Phys. Rev. C.* **76**, 024311 (2007).
[49] W. Horiuchi and Y. Suzuki, Structure of and $E2$ transition in ^{16}C in a ^{14}C+$n+n$ model, *Phys. Rev. C.* **73**, 037304 (2006).
[50] W. Horiuchi and Y. Suzuki, ^{22}C: An s-wave two-neutron halo nucleus, *Phys. Rev. C.* **74**, 034311 (2006).
[51] W. Horiuchi, Y. Suzuki, P. Capel, and D. Baye, Probing the weakly-bound neutron orbit of ^{31}Ne with total reaction and one-neutron removal cross sections, *Phys. Rev. C.* **81**, 024606 (2010).
[52] S. Saito, Interaction between clusters and Pauli principle, *Prog. Theor. Phys.* **41**, 705 (1969).
[53] V. I. Kukulin and V. N. Pomerantsev, The orthogonal projection method in scattering theory, *Ann. Phys.* **111**, 330 (1978).
[54] W. Horiuchi and Y. Suzuki, Correlated-basis description of α-cluster and delocalized 0^+ states in ^{16}O, *Phys. Rev. C.* **89**, 011601(R) (2014).
[55] M. Wloch, D. J. Dean, J. R. Gour, M. Hjorth-Jensen, K. Kowalski, T. Papenbrock, and P. Piecuch, *Ab initio* coupled-cluster study of ^{16}O, *Phys. Rev. Lett.* **94**, 212501 (2005).
[56] P. Maris, J. P. Vary, and A. M. Shirokov, *Ab initio* no-core full configuration calculations of light nuclei, *Phys. Rev. C.* **79**, 014308 (2009).
[57] E. Epelbaum, H. Krebs, T. A. Lähde, D. Lee, U.-G. Meißner, and G. Rupak, *Ab initio* calculation of the spectrum and structure of ^{16}O, *Phys. Rev. Lett.* **112**, 102501 (2014).

[58] A. Volya and Y. M. Tchuvil'sky, Nuclear clustering using a modern shell model approach *Phys. Rev. C.* 91, 044319 (2015).
[59] H. Horiuchi and K. Ikeda, A molecule-like structure in atomic nuclei of ^{16}O and ^{20}Ne, *Prog. Theor. Phys.* **40**, 277 (1968).
[60] D. R. Thompson, M. LeMere, and Y. C. Tang, Systematic investigation of scattering problems with the resonating-group method, *Nucl. Phys. A.* **286**, 53 (1977).
[61] Y. Suzuki, K. Arai, Y. Ogawa, and K. Varga, Occupation probability of harmonic-oscillator quanta for microscopic cluster-model wave functions, *Phys. Rev. C.* **54**, 2073 (1996).
[62] W. Horiuchi and Y. Suzuki, Tests of a deformable core plus few-nucleon model, *Few-Body Syst.* **55**, 122 (2014).
[63] D. R. Tilley, H. R. Weller, and G. M. Hale, Energy levels of light nuclei $A = 4$, *Nucl. Phys. A.* **541**, 1 (1992).
[64] W. Horiuchi and Y. Suzuki, Inversion doublets of $3N + N$ cluster structure in excited states of ^{4}He, *Phys. Rev. C.* **78**, 034305 (2008).
[65] W. Horiuchi and Y. Suzuki, Excited states and strength functions of ^{4}He in correlated Gaussians, *Few-Body Syst.* **54**, 2407 (2013).
[66] R. Tamagaki, Potential models of nuclear forces at small distances, *Prog. Theor. Phys.* **39**, 91 (1967).
[67] W. Horiuchi and Y. Suzuki, Harmonic-oscillator excitations of precise few-body wave functions, *Phys. Rev. C.* **90**, 034001 (2014).
[68] B. S. Pudliner, V. R. Pandharipande, J. Carlson, S. C. Pieper, and R. B. Wiringa, Quantum Monte Carlo calculations of nuclei with $A \leq 7$, *Phys. Rev. C.* **56**, 1720 (1997).
[69] E. Hiyama, B. F. Gibson, and M. Kamimura, Four-body calculation of the first excited state of ^{4}He using a realistic NN interaction: $^4\mathrm{He}(e,e')^4\mathrm{He}(0_2^+)$ and the monopole sum rule, *Phys. Rev. C.* **70**, 031001(R) (2004).
[70] Y. Suzuki and W. Horiuchi, Significance and properties of internucleon correlation functions, *Nucl. Phys. A.* **818**, 188 (2009).
[71] H. Feldmeier, W. Horiuchi, T. Neff, and Y. Suzuki, Universality of short-range nucleon–nucleon correlations, *Phys. Rev. C.* **84**, 054003 (2011).
[72] T. Neff, H. Feldmeier, and W. Horiuchi, Short-range correlations in nuclei with similarity renormalization group transformations, *Phys. Rev. C.* **92**, 024003 (2015).
[73] S. Bacca, N. Barnea, W. Leidemann, and G. Orlandini, Isoscalar monopole resonance of the alpha particle: A prism to nuclear hamiltonians, *Phys. Rev. Lett.* **110**, 042503 (2013).
[74] W. Horiuchi, Y. Suzuki, and K. Arai, *Ab initio* study of the photoabsorption of ^{4}He, *Phys. Rev. C.* **85**, 054002 (2012).
[75] W. Horiuchi and Y. Suzuki, Spin-dipole strength functions of ^{4}He with realistic nuclear forces, *Phys. Rev. C.* **87**, 034001 (2013).
[76] S. E. Woosley, D. H. Hartmann, R. D. Hoffman, W. C. Haxton, The ν-process, *Astrophys. J.* **356**, 272 (1990).
[77] N. Moiseyev, Quantum theory of resonances: Calculating energies, widths and cross-sections by complex scaling, *Phys. Rep.* **302**, 212 (1998).

[78] S. Aoyama, T. Myo, K. Kato, and K. Ikeda, The complex scaling method for many-body resonances and its applications to three-body resonances, *Prog. Theor. Phys.* **116**, 1 (2006).

[79] P. G. Hansen and B. Jonson, The neutron halo of extremely neutron-rich nuclei, *Europhys. Lett.* **4**, 409 (1987).

[80] K. Ikeda, INS Report No. JHP-7, Japan (1988).

[81] Y. Suzuki, K. Ikeda, and H. Sato, New type of dipole vibration in nuclei, *Prog. Theor. Phys.* **83**, 180 (1990).

[82] D. Mikami, W. Horiuchi, and Y. Suzuki, Electric dipole response of ^{6}He: Halo-neutron and core excitations, *Phys. Rev. C.* **89**, 064303 (2014).

[83] T. Myo, K. Katō, S. Aoyama, and K. Ikeda, Analysis of ^{6}He Coulomb breakup in the complex scaling method, *Phys. Rev. C.* **63**, 054313 (2001).

[84] D. Baye, P. Capel, P. Descouvemont, and Y. Suzuki, Four-body calculation of ^{6}He breakup with the Coulomb corrected eikonal method, *Phys. Rev. C.* **79**, 024607 (2009).

[85] E.C. Pinilla, D. Baye, P. Descouvemont, W. Horiuchi, and Y. Suzuki, Tests of the discretized-continuum method in three-body dipole strengths, *Nucl. Phys. A.* **865**, 43 (2011).

Chapter 8

Symmetry-adapted No-core Shell Model — the *Ab Initio* Picture of Nuclear Collectivity

Jerry P. Draayer[*], Tomas Dytrych[*,†], and Kristina D. Launey[*]
[*]*Department of Physics and Astronomy, Louisiana State University Baton Rouge, LA 70803, USA*
[†]*Nuclear Physics Institute, 250 68 Řež, Czech Republic*

We review exact and partial symmetries of strongly interacting particles in atomic nuclei. These symmetries serve a key role in facilitating studies of the structure and reactions of isotopes across the nuclear chart, using QCD-inspired interactions and high performance computing (HPC) resources. Namely, they provide a relevant organization of the nuclear model space, while providing an efficient way to reduce the ultra-large model space size through a very structured selection of the basis states to physically relevant subspaces. This enables a deeper understanding of and advancing *ab initio* studies for determining the microscopic structure of nuclear systems. This, in turn, can guide explorations of simple patterns in nuclei and how they emerge from first principles, as well as extensions of the theory beyond current limitations toward heavier nuclei and larger model spaces.

Contents

1. Introduction

Major progress in the understanding of the inter-nucleon interaction [1–4] along with the utilization of massively parallel computing resources (e.g., see [5]) have placed *ab initio* (or "from first principles") large-scale simulations at the frontier of nuclear structure explorations. Several *ab initio* nuclear-theory approaches have been recently advanced, including Green's function Monte Carlo (GFMC) [6, 7], no-core shell model (NCSM) [8–12] together with NCSM with a core [13] and importance truncation NCSM [14], coupled-cluster method (CC) [15, 16], lattice effective field theory [17], in-medium SRG [18, 19], symmetry-adapted no-core shell model (SA-NCSM) [20], Monte Carlo NCSM [21], and self-consistent Green's functions [22]. *Ab initio* approaches build upon a "first principles" foundation. They provide a long-missing link that bridges from the many-particle nucleus down to the fundamental blocks, namely, the properties of only two or three nucleons often tied to chiral symmetry-breaking patterns dictated by the underlying quantum chromodynamics (QCD). And while this bridge transforms the nuclear problem into a computational- and data-intensive challenge, it empowers *ab initio* models with two invaluable features: (1) a universal character essential for modeling the coexistence of diverse nuclear substructures and (2) predictive capabilities vital for descriptions of experimentally inaccessible nuclear species far off the valley of stability. As such nuclei are often found key to understanding processes in extreme environments, from stellar explosions to the interior of nuclear reactors or fusion capsules, first-principle nuclear models have been and will be demonstrating a tremendous impact for advancing the frontiers in multiple branches of physics such as astrophysics, neutrino physics, and applied physics [22–32].

In this Chapter, we present a symmetry-guided strategy that utilizes exact as well as partial symmetries for enabling a deeper understanding of and advancing *ab initio* studies for determining the microscopic structure of atomic nuclei. These symmetries naturally provide a physically relevant basis that, for large-scale calculations, allows the model space size to be reduced through a very structured selection of the basis states to physically relevant subspaces. This, in turn, can guide extensions of the theory beyond current limitations. This is crucial, as model space dimensionality and associated computational resource demands grow combinatorially with the number of particles and the spaces in which they reside (so-called, "scale explosion"), thereby limiting the number of active particles or precluding microscopic descriptions of largely deformed spatial structures (Fig. 1).

Symmetries underpin orderly patterns in nuclear dynamics. Experimental evidence supports formation of deformation and rotational patterns, including the dominance of large deformation in low-lying nuclear states, as suggested by enhanced $E2$ transitions and large quadrupole moments. The dominance of large deformation and, by duality, low spin has been demonstrated by symmetry-guided theoretical studies, with further recognition of a new simple structure in nuclei associated with $\mathrm{Sp}(3,\mathbb{R})$ symplectic symmetry, detailed in the Refs. [33, 34] (Secs. 3.2 and 3.4). In this chapter,

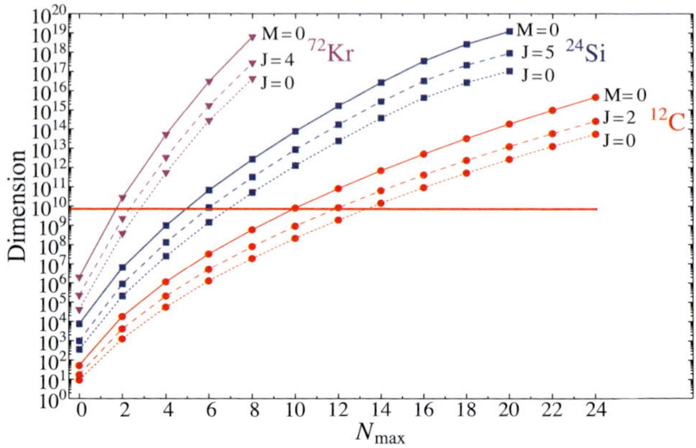

(a)

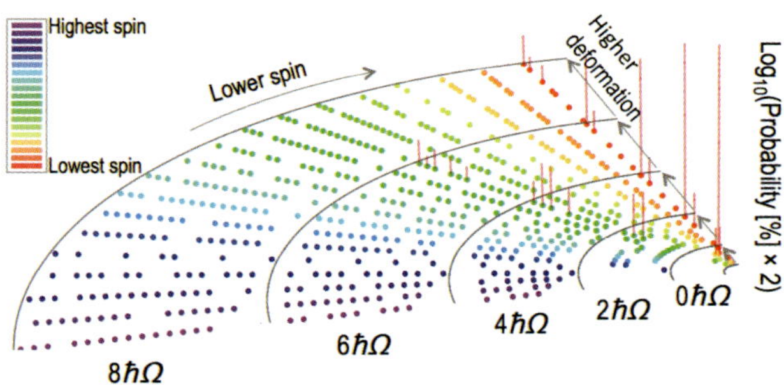

(b)

Fig. 1. (a) Explosive growth of the nuclear model space with the size of the "box" in which particles reside (given by the $N_{\max}$ cutoff) for selected nuclei in the conventional M and J schemes (Sec. 3.1). The largest dimension currently attainable on the best of modern-day computers is shown by the horizontal line (for two-body interactions). (b) Highly structured pattern, revealed in *ab initio* calculations with chiral interactions of the 0^+ ground state (gs) of ^{8}Be. This enables a dramatic reduction of the nuclear model space to physically relevant subspaces and can guide extensions of the *ab initio* theory to heavier nuclei.

we also discuss how such highly structured orderly patterns emerge from first-principle investigations [20, 35, 36] starting with bare nucleon–nucleon (NN) interactions (Secs. 3.3 and 3.5). We discuss the implications of these symmetries to understanding the nature of nuclear dynamics and their key role in expanding the scope of *ab initio* theory (Sec. 4).

2. *Ab initio* Symmetry-adapted No-core Shell Model

The SA-NCSM [20, 37] is a nuclear shell model (SM) [9, 38–40], a many-body "configuration interaction" (CI) method. It solves the Schrödinger equation for A particles, $H\Psi(\vec{r}_1, \vec{r}_2, \ldots, \vec{r}_A) = E\Psi(\vec{r}_1, \vec{r}_2, \ldots, \vec{r}_A)$, which for a complete A-nucleon orthonormal basis becomes a matrix eigenvalue equation (see also Chapter 2). The interaction and basis configurations in the SA-NCSM are adopted as follows.

Interaction — The intrinsic non-relativistic nuclear plus Coulomb interaction Hamiltonian is defined as:

$$H = T_{\text{rel}} + V_{NN} + V_{3N} + \cdots + V_{\text{Coulomb}}, \tag{1}$$

where T_{rel} is the relative kinetic energy $T_{\text{rel}} = \frac{1}{A}\sum_{i<j} \frac{(\vec{p}_i - \vec{p}_j)^2}{2m}$ (m is the nucleon mass), the V_{NN} is the nucleon–nucleon interaction, $V_{NN} = \sum_{i<j}^{A} (V_{NN})_{ij}$, included along with the Coulomb interaction between the protons and possibly, $V_{3N} = \sum_{i<j<k}^{A} (V_{NNN})_{ijk}$, $V_{4N}, \ldots$ interactions. The Hamiltonian may include additional terms, e.g., higher-order electromagnetic interactions such as magnetic dipole–dipole terms.

Basis configurations — The many-nucleon basis is a finite set of antisymmetrized products of single-particle states (Slater determinants), referred to as a "model space". The single-particle states of a three-dimensional spherical harmonic oscillator (HO) are used, $\phi_{\eta l m_l s m_s t_z}(\vec{r}; b)$, where $\eta = 2n_r + l$, l is the particle orbital momentum and $s = \frac{1}{2}$ is its spin with z-projections m_l and m_s, respectively, while t_z distinguishes between protons and neutrons ($b = \sqrt{\hbar/m\Omega}$ is the characteristic length for HO frequency Ω). This basis allows for preservation of translational invariance of the nuclear self-bound system and provides solutions in terms of single-particle wave functions that are analytically known. With larger model spaces, the eigensolutions converge to the exact values. In contrast to the conventional SM theory, the SA-NCSM basis is further reorganized into states of given deformation, referred to as the SU(3) scheme (see Sec. 3.2).

In general, the *ab initio* SM uses high-precision NN ($3N$) potentials fitted to two-nucleon (three-nucleon) data, in particular, to scattering phase shifts and properties of the deuteron (and triton) (see, e.g., [2, 41]). The NN potentials include AV18 [1], CD-Bonn [2], and chiral interactions, such as N^3LO [3, 4]. Another high-precision NN interaction is JISP16 [42] based on J-matrix version of inverse scattering theory that is adjusted, in addition, to binding energies up to $A = 16$; it typically leads to rapid convergence in large-scale SM evaluations and minimizes the contribution of the three-nucleon forces. Various renormalization techniques, such as Okubo–Lee–Suzuki (OLS) [43], similarity renormalization group (SRG) [44], and unitary correlation operator method (UCOM) [45], aim to achieve a softer (renormalized or effective) interaction that enables the use of smaller manageable model spaces.

In the conventional nuclear SM, two principal limitations are encountered: (i) the number of configurations necessary to describe a nuclear state is typically huge and grows combinatorially with the number of particles and the size of the space they occupy (see Fig. 1), and (ii) phenomenological interactions typically yield predictions of nuclear properties that highly diverge outside the nuclear region they were fitted; present high-precision potentials hold predictive power, however, they generate strong — to some degree or another — short-range correlations (coupling to high momenta), and these together with long-range correlations responsible for enhanced collectivity, large spatial deformation and α-cluster substructures (wave function tail spreading to large distances) require ultra-large SM spaces, often inaccessible on the best of modern-day supercomputers. The SA-NCSM approach addresses these limitations by exploiting symmetries and the fact that the wave functions of a quantum mechanical system can be characterized by their invariance properties under certain group transformations. In addition, it recognizes partial symmetries that survive within the nuclear dynamics and that can be used to help reduce the dimensionality of a model space to tractable sizes.

3. Exact and Partial Symmetries of Strongly Interacting Particles: Organization and Selection of the Nuclear Model Space

In the 1950s, two simple models of nuclear structure, complementary in nature, were advanced, and eventually merited a Nobel prize. These are the independent-particle model of Mayer and Jensen [46–48], and the collective model of Bohr and Mottelson [49, 50]. The first of these, which is microscopic in nature, recognizes that nuclei can be described by particles independently moving in a mean field, with the HO potential being a very good first approximation to the average potential experienced by each nucleon in a nucleus. Augmented by a spin-orbit $\mathbf{l} \cdot \mathbf{s}$ term and an orbit–orbit $\mathbf{l}^2$ force to successfully reproduce the "magic numbers" pattern, the model yields single-particle orbits that are energetically separated for the same l but different total angular momentum j, favoring a $j - j$ coupling scheme. The second of these models, the collective model of Bohr and Mottelson recognizes that deformed shapes dominate the nuclear dynamics. While enhanced deformation has been evident in heavy nuclei and those away from closed shells, deformed configurations are found to be important even in a nucleus such as ^{16}O, which is commonly treated as spherical in its ground state, but about 40% of the latter is governed by deformed shapes [51]; in addition, the lowest-lying excited 0^+ states in ^{16}O and their rotational bands are dominated by large deformation (see, e.g., [52]). Bohr and Mottelson offered a simple but important description of nuclei in terms of the deformation of the nuclear surface and associated vibrations and rotations. While this model was not microscopic, it discussed spatial degrees of the combined many-particle system (spatial deformation and rotations of "shapes"), which suggested a relevant $L-S$ coupling scheme.

3.1. *Conventional Schemes and the Shell Model*

Following the success of the independent-particle model of Mayer and Jensen, many valence and NCSMs utilize the $j - j$ coupling scheme.

The NCSM [8] treats all A particles as active (e.g., [9,10,53]), while the valence SM [40, 54] assumes a core of inactive particles and a subset of valence particles within a partially filled valence space that then bears the burden of accounting for all particle correlations. For a major class of valence SMs and their significant role for heavy nuclei, see the Refs. [40,54–56] and Chapter 9. These models also exploit an important exact symmetry of the nuclear Hamiltonian, that is, it is invariant under rotation or a scalar with respect to the SO(3) group. Hence, the nuclear states have definite total angular momentum J and its z-projection M (see Chapter 2). This leads to the J scheme (all particle j's coupled to a good J) and the M scheme (simply constructing Slater determinants with fixed total M), illustrated in Table 1 (see also Fig. 1).

The many-particle SM space consists of nuclear configurations of fixed parity, consistent with the Pauli principle, and truncated by the total HO single-particle energy, $\sum_{i=1}^{A}(\eta_i + 3/2)\hbar\Omega$, typically quoted as the relative energy $N_{\max}\hbar\Omega$ or simply, the $N_{\max}$ cutoff. The $N_{\max}$ cutoff is defined as the maximum number of HO quanta allowed in a many-particle state above the minimum for a given nucleus.

3.2. SU(3) *Scheme and the Elliott model*

The microscopic Elliott model [57–59] and the symplectic SM [33,60] that provides a microscopic formulation of the Bohr–Mottelson collective model have confirmed the relevance of the $L - S$ coupling scheme, while providing a unique and physically relevant organization of the SM space, as discussed next.

In place of the spherical quantum numbers $|\eta l m_l\rangle$, the single-particle HO basis can be specified by $|\eta_z \eta_x \eta_y\rangle$, the HO quanta in the three Cartesian directions, z, x, and y, with $\eta_x + \eta_y + \eta_z = \eta$ ($\eta = 0, 1, 2, \ldots$ for s, p, $sd, \ldots$ shells). For a given HO major shell, the complete SM space is then specified by all distinguishable distributions of η_z, η_x and η_y. E.g., for $\eta = 2$, there are 6 different distributions, $(\eta_z, \eta_x, \eta_y) = (2, 0, 0), (1, 1, 0), (1, 0, 1), (0, 2, 0), (0, 1, 1)$ and $(0, 0, 2)$. The number of these configurations is $\Omega_\eta = (\eta + 1)(\eta + 2)/2$

Table 1. Examples for a two-particle basis state in various model space schemes with single-particle states labeled by $\eta(ls)jmt_z$ (or simply by $\eta ljmt_z$ for $s=\frac{1}{2}$) with $|\eta ljmt_z\rangle = a^\dagger_{\eta ljmt_z}|0\rangle$ ($j-j$ coupling scheme), or with single-particle states labeled by $\eta lm_l sm_s t_z$ with $|\eta lm_l sm_s t_z\rangle = a^\dagger_{\eta lm_l sm_s t_z}|0\rangle$, where $a^\dagger$ is the usual particle creation operator. Incidentally, $a^\dagger_{\eta lm_l sm_s t_z} \equiv a^\dagger_{(\eta\, 0)lm_l sm_s t_z}$ is also an SU(3) tensor of rank $(\lambda\,\mu)=(\eta\,0)$, while the particle annihilation SU(3) tensor of rank $(\lambda\,\mu)=(0\,\eta)$ is given as $\tilde{a}_{(0\,\eta)l-m_l s-m_s t_z} = (-1)^{\eta+l-m_l+s-m_s} a_{\eta lm_l sm_s t_z}$. Underlined labels indicate labels involved in coupling rules; SU(3) multiplicities that are trivial in the 2-particle case have been omitted (see text for details).

Model space scheme	Two-particle basis state
$L-S$ scheme	$\{a^\dagger_{\eta \underline{l}\, \underline{s} t_z} \times a^\dagger_{\eta' \underline{l'}\, \underline{s'} t'_z}\}^{(LS)JM}\|0\rangle$
M scheme	$a^\dagger_{\eta ljmt_z} a^\dagger_{\eta' l' j' m' t'_z}\|0\rangle$, with $m+m'=M$
J scheme	$\{a^\dagger_{\eta l \underline{j} t_z} \times a^\dagger_{\eta' l' \underline{j'} t'_z}\}^{JM}\|0\rangle$
SU(3) scheme	$\{a^\dagger_{\underline{(\eta\,0)}\, \underline{s} t_z} \times a^\dagger_{\underline{(\eta'0)}\, \underline{s'} t'_z}\}^{(\lambda\,\mu)\kappa(LS)JM}\|0\rangle$
Sp(3, ℝ) scheme*	$\{\{\hat{A}^{(2\,0)} \times \hat{A}^{(2\,0)} \cdots \times \hat{A}^{(2\,0)}\}^{(\lambda_n\,\mu_n)} \times$ $\times \underbrace{[a^\dagger_{(\eta_1\,0)} \times a^\dagger_{(\eta_2\,0)}]^{(\lambda_\sigma\,\mu_\sigma)}}_{A=2 \text{ bandhead}}\}^{\rho(\lambda_\omega\,\mu_\omega)}_{\kappa L M_L}\|0\rangle$

Note: *For notational simplicity, the two-particle basis state is given for spatial degrees of freedom only (the associated spin and isospin labels are suppressed) and a $0\hbar\Omega$ bandhead, where $\hat{A}^{(2\,0)}_{\mathfrak{L}M} = \sum_\eta \sqrt{\frac{(\eta+1)(\eta+2)(\eta+3)(\eta+4)}{12}}\{a^\dagger_{\eta+2} \times \tilde{a}_\eta\}^{(20)}_{\mathfrak{L}M} + \hat{A}^{(2\,0)}_{\mathrm{CM};\mathfrak{L}M}$ generates $2\hbar\Omega$ 1p-1h monopole ($\mathfrak{L}=0$) or quadrupole ($\mathfrak{L}=2$) excitations together with a smaller $2\hbar\Omega$ 2p-2h correction for eliminating the spurious center-of-mass (CM) motion (cf. Chapter 4).

(spatial degeneracy) and the associated symmetry is described by the $\mathrm{U}(\Omega_\eta)$ unitary group of the many-particle three-dimensional HO [61]. Each of these (η_z, η_x, η_y) configurations can be unoccupied or can have a maximum of two protons (or neutrons) with up and down spins. An important feature of the SU(3) scheme (Table 1) is that all possible configurations within a major HO shell η (for a fixed number of protons or neutrons) are readily available based on the reduction $\mathrm{U}(\Omega_\eta) \supset \mathrm{SU}(3)$, as detailed in Ref. [62] (an illustrative example for 4 particles in the pf shell, $\eta=3$, is shown in Table 1 of Ref. [37]).

As a simple example for an SU(3)-scheme basis state, consider $A = 2$ protons in the sd shell ($\eta = 2$), or $(sd)^2$, with a particle in the $(2,0,0)$ level with up spin and another in the $(1,1,0)$ level with up spin. The total number of quanta in each direction is $(\eta_z^{\rm tot}, \eta_x^{\rm tot}, \eta_y^{\rm tot}) = (3,1,0)$, or equivalently, $\eta^{\rm tot}(\lambda\,\mu) = 4(2\,1)$, where $\eta^{\rm tot} = \eta_x^{\rm tot} + \eta_y^{\rm tot} + \eta_z^{\rm tot}$, together with $\lambda = \eta_z^{\rm tot} - \eta_x^{\rm tot}$ and $\mu = \eta_x^{\rm tot} - \eta_y^{\rm tot}$ labeling an SU(3) irreducible representation (irrep), in addition to the total intrinsic spin and its projection SM_S. For given $(\lambda\,\mu)$, the quantum numbers κ, L and M_L are given by Elliott [57, 58], according to $\mathrm{SU}(3) \overset{\kappa}{\supset} \mathrm{SO}(3)_L \supset \mathrm{SO}(2)_{M_L}$, where the label κ distinguishes multiple occurrences of the same orbital momentum L in the parent irrep $(\lambda\,\mu)$. For $(\lambda\,\mu) = (2\,1)$ with $\kappa = 1$, $L = 1, 2, 3$, and $M_L = -L, -L+1, \ldots, L$. Hence, the set $\{\eta^A(\lambda\,\mu)\kappa(LS)JM\}$ completely labels a 2-proton SU(3)-scheme basis state — we emphasize that, in general, there is a multiplicity label α_η associated with the reduction $\mathrm{U}(\Omega_\eta) \overset{\alpha_\eta}{\supset} \mathrm{SU}(3)$ that distinguishes multiple occurrences of a $(\lambda\,\mu)$ irrep [37, 62]. Note that there is no dependence on the single-particle orbital momenta, $l_1, l_2, \ldots, l_A$, in the SU(3)-scheme basis states.

We note that the SU(3) scheme provides a classification of the complete SM space (in a single shell as illustrated above and in multiple shells as described in Sec. 3.3) and is related to the $L-S$ and $j-j$ coupling schemes via a unitary transformation. It divides the space into basis states of definite $(\lambda\,\mu)$ quantum numbers of SU(3) that are linked to the intrinsic quadrupole deformation according to the established mapping [63–65]. For example, the simplest cases, $(0\,0)$, $(\lambda\,0)$, and $(0\,\mu)$, describe spherical, prolate, and oblate deformation, respectively, while a general nuclear state is typically a superposition of several hundred various deformed configurations.[a] That is, it is expressed in terms of a reasonable number of "collective"

[a]Following this mapping, quadrupole moments of $(0\,0)$, $(\lambda\,0)$, and $(0\,\mu)$ configurations — in a simple classical analogy to rotating spherical, prolate, and oblate spheroids in the lab frame — are zero, negative, and positive, respectively [66]. Note also that basis states can have little to no deformation, and, e.g., about 60% of the ground state of the closed shell nucleus ^{16}O is described by a single SU(3) basis state, the spherical $(0\,0)$.

basis states, labeled by the intrinsic quadrupole deformation of the many-particle system, $(\lambda\,\mu)$, its rotation, L, and orientation in coordinate space (in addition to the spin degrees of freedom).

Elliott model and selected applications. The seminal work of Elliott [57–59] focused on the key role of SU(3), the exact symmetry of the three-dimensional spherical HO (see also Refs. [68–70]). Elliott's model is a SM that utilizes an SU(3)-scheme basis. For SU(3)-symmetric interactions, the model can be solved analytically. But regardless whether a simple SU(3)-preserving interaction is used (see Figs. 5.1–5.6 of Ref. [71]), or an SU(3)-symmetry breaking

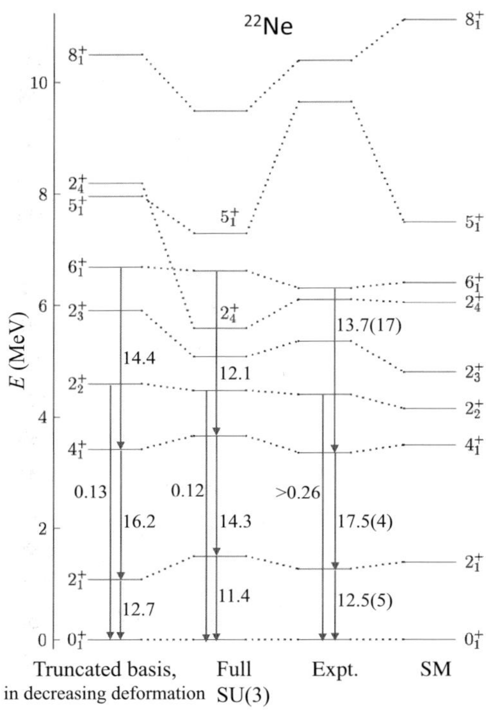

Fig. 2. Elliott model with an SU(3)-preserving interaction plus pairing in the *sd* valence shell: energy spectrum of ^{22}Ne, calculated in selected and full SU(3) model spaces, as compared to SM calculations with an empirical *sd*-shell interaction. $B(E2)$ transition strengths in W.u. are calculated for proton effective charge $e_{\rm eff} = 1.3$. Figure adapted from Ref. [67].

interaction (Fig. 2), the results have revealed a striking feature, namely, the dominance of a few most deformed configurations. This has been shown for *sd*-shell nuclei, such as ^{18,20,22}Ne, ^{22,24}Mg, and ^{28}Si, that have been known to possess a clear collective rotational structure in their low-lying states [57–59]. It has been also observed in heavier nuclei, where pseudo-spin symmetry [72, 73] and its pseudo-SU(3) complement [74] have been shown to play a similar role in accounting for deformation in the upper *pf* and lower *sdg* shells, and in particular, in strongly deformed nuclei of the rare-earth and actinide regions [75, 76]. In this mass region, an approximate "quasi-SU(3)" symmetry has been also suggested [77]. Furthermore, the pairing interaction has been microscopically incorporated into the Elliott model where it breaks the SU(3) symmetry and mixes different $(\lambda\,\mu)$ configurations [78]. It has been shown in Ref. [67] that using an SU(3)-symmetric interaction-plus-pairing yields results close to experiment and to the energies obtained using SM calculations in the full *sd* shell [79] (Fig. 2). It is remarkable that, even in the presence of pairing, comparable results have been obtained in a truncated model space that includes only about 10 most deformed configurations.

3.3. *Multi-shell* SU(3) *Scheme and Symmetry-adapted No-Core Shell Model*

The SA-NCSM [20], reviewed in Ref. [37], is a multi-shell generalization of the SU(3) scheme used in the Elliott model. It adopts the first-principle concept and is a no-core shell model in an SU(3) scheme. The many-nucleon basis states of the SA-NCSM are constructed using efficient group-theoretical algorithms and each state is specified by a particle distribution across the HO shells, many-particle quadrupole deformation $(\lambda\,\mu)$ with its rotation and orientation in space, in addition to intermediate labels and intrinsic spin [20, 80]. Computations of the Hamiltonian matrix elements are based on an SU(3)-type Wigner-Eckart theorem, which factorizes these matrix elements into a product of SU(3) reduced matrix elements and the associated SU(3) coupling coefficients. The

SA-NCSM computational realization, dubbed `LSU3shell` [81, 82], adopts state-of-the-art group-theoretical methods [62] and optimized numerical subroutines [83] for computing SU(3) coupling/recoupling coefficients. There are two major advantages of the SA-NCSM that follow from the use of the SU(3)-scheme basis:

(1) The SU(3)-based organization of the model space allows the complete $N_{\max}$ space to be down-selected to the physically relevant subspace. This makes calculations in larger $N_{\max}$ spaces feasible with current computational resources.
(2) Within the space down-selected to a subset of $(\lambda\,\mu)$ irreps and intrinsic spins S, the spurious CM motion can be factored out exactly [84–86]. This ensures the translational invariance of the SA-NCSM wave functions.

Rotational structure and bands. Within the *ab initio* SA-NCSM framework, it is possible to explore the microscopic nature of important collective configurations. In Refs. [20, 35, 87], we have examined wave functions calculated from first principles: the four lowest-lying isospin-zero $1^+_{\rm gs}$, 3^+_1, 2^+_1, and 1^+_2 states of ^{6}Li; 0^+_{gs}, 2^+_1, and 4^+_1 of ^{8}Be, ^{6}He, and ^{12}C; the lowest 1^+, 3^+, and 0^+ excited states of ^{8}B; and the ground state of ^{16}O (see also Sec. 3.5). The results show that the mixing of $(\lambda\,\mu)$ quantum numbers, induced by the SU(3)-symmetry breaking terms of realistic interactions, exhibits a remarkably simple pattern. For instance, the low-lying states of ^{6}Li, $1^+_{\rm gs}$, 3^+_1, 2^+_1, and 1^+_2, are dominated by the $0\hbar\Omega$ SU(3) irrep $(2\,0)$, as illustrated in Fig. 3 for $1^+_{\rm gs}$. The dominance of this deformation within each of these states clearly identifies them as members of a rotational band, that is, different rotations of the same deformation (consistent with enhanced $E2$ transitions within the band). Indeed, according to Elliott's rule, for $(2\,0)$, $L = 0$ and 2, which couples with spin-1 to yield $J = 1^2, 2, 3$.

Large deformation. In addition to ^{6}Li, Table 2 reveals the preponderance of a single $0\hbar\Omega$ SU(3) irrep in the ground state of ^{8}B, ^{8}Be, ^{12}C, and ^{16}O, associated with triaxial, prolate, oblate, and spherical shapes, respectively. This so-called leading irrep is

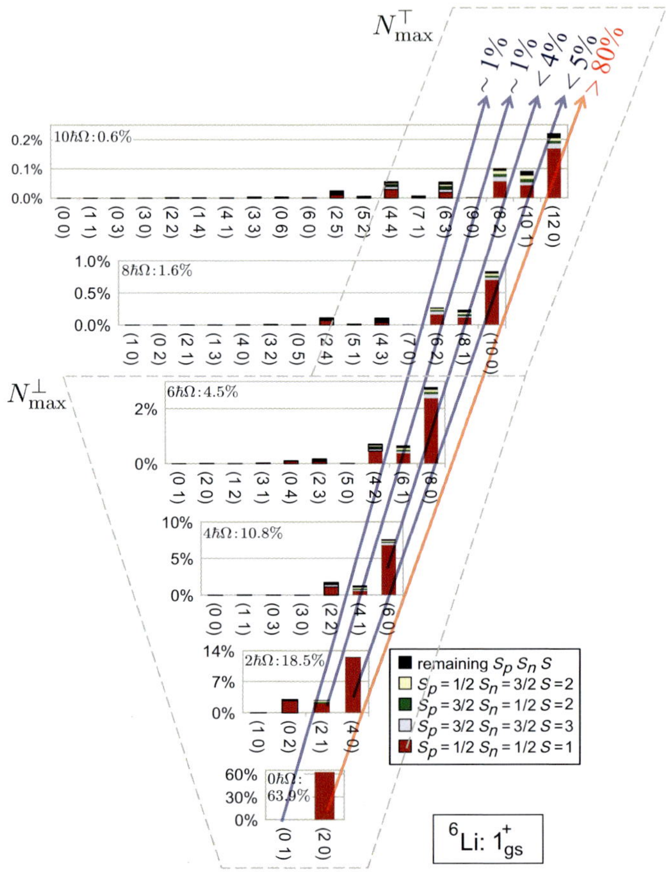

Fig. 3. Probability distributions for proton, neutron, and total intrinsic spins (S_pS_nS) across the Pauli-allowed $(\lambda\,\mu)$ values (horizontal axis, in order of increasing deformation) for the ^{6}Li 1^+ ground state, calculated for $N_{\max} = 10$, $\hbar\Omega = 20$ MeV, and with the JISP16 bare interaction (similar results obtained with the chiral N^{3}LO interaction). The total probability for each $N\hbar\Omega$ subspace is given in the upper left-hand corner of each histogram. The concentration of strengths to the far right demonstrates the dominance of collectivity, which supports a *symmetry-guided* model space selection. The projection onto symplectic vertical slices (with probability $\geq 1\%$) is schematically illustrated for ^{6}Li by arrows and clearly reveals the preponderance of a single symplectic irrep (vertical cone). Figure adapted from Ref. [20].

Table 2. Probability amplitudes of the dominant $(S_p\,S_n\,S)$ spin configurations and the corresponding $0\hbar\Omega$ dominant nuclear deformation $(\lambda_0\,\mu_0)$ for the ground state of selected nuclei calculated in the *ab initio* $N_{\rm max} = 8-14$ SA-NCSM with the bare JISP16 interaction ($\mathrm{NNLO_{opt}}$ [89] is used for ^{20}Ne) for $\hbar\Omega = 20$ MeV. The second most important contribution is shown for ^{12}C — note that the combined $(0\,0\,0), (0\,1\,1), (1\,0\,1)$ contribution for ^{12}C yields 84.79% for the third column and 62.73% for the last column.

Nucleus	$(S_p\,S_n\,S)$	Probability, %	$(\lambda_0\,\mu_0)$	Probability, % $(\lambda_0\,\mu_0)$
^{6}Li	$(\frac{1}{2}\,\frac{1}{2}\,1)$	93.24	$(2\,0)$	57.36
^{8}B	$(\frac{1}{2}\,\frac{1}{2}\,1)$	85.58	$(2\,1)$	56.50
^{8}Be	$(0\,0\,0)$	85.21	$(4\,0)$	55.92
^{12}C	$(0\,0\,0)$	55.60	$(0\,4)$	44.10
	$[(0\,1\,1), (1\,0\,1)]$	[29.19]	$[(1\,2)]$	[18.63]
^{16}O	$(0\,0\,0)$	78.42	$(0\,0)$	60.59
^{20}Ne	$(0\,0\,0)$	79.73	$(8\,0)$	43.93

characterized by the largest value of the intrinsic quadrupole deformation, according to the established geometrical interpretation of the SU(3) labels $(\lambda\,\mu)$ [63–65]. The clear dominance of the single, most deformed $0\hbar\Omega$ configuration within low-lying states of light and intermediate-mass nuclei indicates that the quadrupole–quadrupole interaction of the Elliott SU(3) model of nuclear deformation and rotations [57,58] is realized naturally within an *ab initio* framework. These results corroborate earlier observations based on a simple but useful guide, which involves counting of the number of interacting pairs in different spin–isospin states for a given spatial symmetry and provides an estimate for the overall binding due to one-pion exchange [88].

Low intrinsic spin. The orderly patterns that favor largely deformed spatial configurations, illustrated in Fig. 3 and Table 2, suggest a complementary dominance of low intrinsic proton and neutron spin values. Indeed, the lowest-lying eigenstates in ^{6}Li (likewise for ^{8}B) [20] are almost entirely realized in terms of only four proton/neutron spin-$\frac{1}{2}$ and spin-$\frac{3}{2}$ configurations (Fig. 3), with $(S_pS_nS) = (\frac{1}{2}\frac{1}{2}1)$ carrying about 90% of each eigenstate (Table 2).

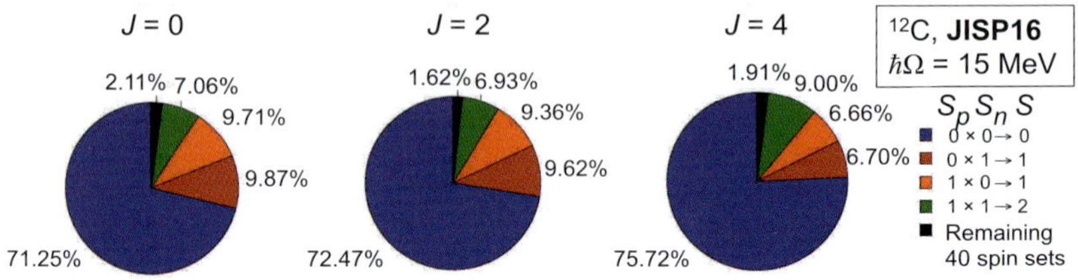

(a)

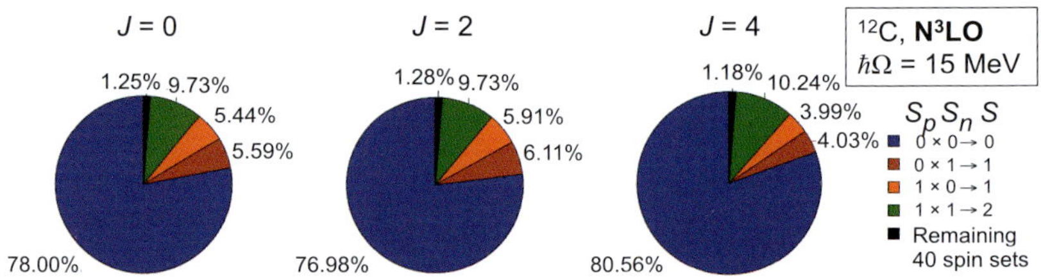

(b)

Fig. 4. Intrinsic spin structure of the $J = 0^+_{\rm gs}, 2^+_1$, and 4^+_1 NCSM states in ^{12}C obtained with (a) JISP16 and (b) N^{3}LO effective interactions (using the OLS procedure) in the $N_{\rm max} = 6$ model space with $\hbar\Omega = 15$ MeV. Similarly, for the 1^+ state (not shown in the figure), only 4 proton/neutron spin-zero and spin-one configurations (with total $S = 1, 2$) contribute at the 94.4% level for JISP16 and 95.3% for N^{3}LO.

Similar results are observed for $S_{p,n} = 0, 1$ configurations in ^{6}He, ^{8}Be, ^{12}C, ^{16}O, and ^{20}Ne [20, 87], and if effective interactions were used (Fig. 4), as shown by the spin decomposition of well-converged NCSM eigenstates in ^{12}C and ^{16}O [35, 51]. A more recent study [53] using SRG effective interactions in the NCSM has arrived at the same conclusions, while also pointing to a similar spin pattern in odd-A p-shell nuclei (cf. Chapter 2).

Model space selection. These patterns suggest a symmetry-guided basis selection philosophy that allows the SA-NCSM to utilize much smaller spaces and to obtain results that are nearly indistinguishable from those obtained in the complete model space [20]. Specifically, the relevant model space can be systematically selected, using a quantified cutoff, starting from the largest deformation/lowest spin (right sector of Fig. 3), and including ever smaller deformation until convergence of results is achieved. One can take advantage of dominant symmetries to refine the SA-NCSM model space, which for the NCSM is fixed by simply specifying the $N_{\max}$ cutoff (or total HO quanta). In the SA-NCSM, the HO quanta distribution in the three spatial directions can guide selections of the model space in high-N $\hbar\Omega$ subspaces, where oscillator quanta are found to be predominantly distributed in z and x directions, but less in y direction, yielding non-negligible SU(3) configurations with enhanced intrinsic quadrupole deformation. Model spaces can then be characterized by a pair of numbers, $\langle N_{\max}^{\perp}\rangle N_{\max}^{\top}$, which implies inclusion of all basis states up through $N_{\max}^{\perp}$ and a subset of basis states beyond this, up through $N_{\max}^{\top}$ (schematically illustrated in Fig. 3). Indeed, SA-NCSM results indicate that observables obtained in the much smaller symmetry-guided selected spaces are practically identical to the corresponding complete-space NCSM counterparts (see Fig. 2 of Ref. [20] and Sec. 4). A crucial advantage of this symmetry-guided concept is that the SA-NCSM with *selected physically relevant model spaces* needs to handle many-body matrices of dimensionality (and number of non-zero matrix elements) much smaller than an equivalent $N_{\max}$ NCSM calculation and hence, can carry out investigations beyond the current $N_{\max}$ NCSM limits (Fig. 5) [82]. Within this context, it is important to emphasize again that for SU(3)-based selected model spaces, the spurious center-of-mass motion is factored out exactly [84–86].

3.4. *Symplectic* Sp(3, ℝ) *Scheme and the Symplectic Model*

For A particles in three-dimensional space, the complete basis for the SM is described by $\mathrm{Sp}(3A,\mathbb{R})\times\mathrm{U}(4)$ [34], where $\mathrm{Sp}(3A,\mathbb{R})$ is the group of all linear canonical transformations of the $3A$-particle phase space and Wigner's supermultiplet group U(4) describes the complementary spin-isospin space. A complete translationally invariant SM basis is classified according to (see, e.g., Refs. [33, 34]),

$$\begin{array}{ccc} \mathrm{Sp}(3(A-1),\mathbb{R}) & \times & \mathrm{U}(4) \\ \cup & & \cup \\ \mathrm{Sp}(3,\mathbb{R})\times\mathrm{O}(A-1) & & \mathrm{SU}(2)_S\times\mathrm{SU}(2)_T \end{array} \qquad (2)$$

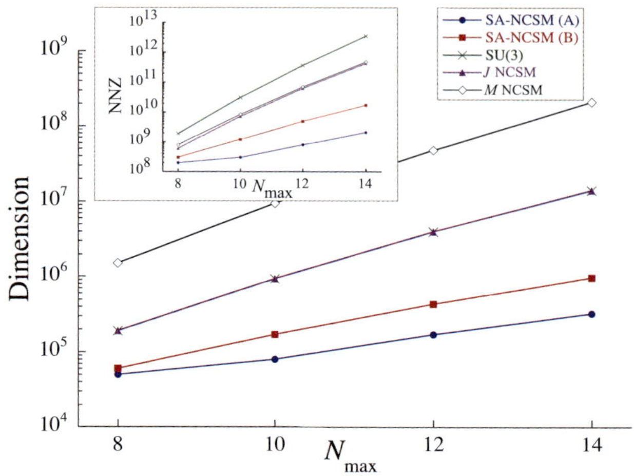

Fig. 5. Matrix dimensions and (inset) number of non-zero (nnz) matrix elements for the typical case of ^{6}Li and $J = 1$ vs. the $N_{\max}$ cutoff for two selected $\langle 6\rangle N_{\max}$ SU(3) model spaces ("A" & "B") and for the complete $N_{\max}$ space for the SU(3), J, and M schemes (NCSM M-scheme dimension for $M = 1$, or all $J \geq 1$). Quoted for an input NN interaction. Figure adapted from Ref. [82].

The Sp(3, $\mathbb{R}$) scheme (Table 1) utilizes the symplectic group Sp(3, $\mathbb{R}$), which consists of all particle-independent linear canonical transformations of the single-particle phase-space observables, $x_{i\alpha} \to \sum_\beta a_{\alpha\beta}x_{i\beta} + b_{\alpha\beta}p_{i\beta}$ and $p_{i\alpha} \to \sum_\beta c_{\alpha\beta}x_{i\beta} + d_{\alpha\beta}p_{i\beta}$ ($i = 1, \ldots, A$ and $\alpha, \beta = x, y, z$) that preserve the commutation relations $[x_{i\alpha}, p_{j\beta}] = i\hbar\delta_{ij}\delta_{\alpha\beta}$ [90] (see Chapter 3).[b] The Sp(3, $\mathbb{R}$) scheme further utilizes the group reduction to classify many-particle basis states $|\sigma n\rho\omega\kappa LM\rangle$ of a symplectic irrep,

$$\begin{array}{ccccccc} \mathrm{Sp}(3,\mathbb{R}) & \supset & \mathrm{U}(3) & \supset & \mathrm{SO}(3) & \supset & \mathrm{SO}(2) \\ \sigma & n\rho & \omega & \kappa & L & & M \end{array}, \tag{3}$$

[b]While the Sp(3, $\mathbb{R}$) scheme utilizes the Sp(3, $\mathbb{R}$) group (transformations acting on x, y, and z components), one can organize the A-particle model space according to the complementary group O($A-1$), with $O(A) \supset$ O($A-1$). The O(A) is the group of orthogonal transformations that act on the "particle-index" space [transformations of nucleon coordinates, $x_{i\alpha} \to \sum_{j=1}^{A} x_{j\alpha}\omega_{ji}$, that leave the O(A) scalars $x_\alpha \cdot x_\beta = \sum_{i=1}^{A} x_{i\alpha}x_{i\beta}$ invariant for $\alpha, \beta = x, y, z$]. This scheme is reviewed in detail in Refs. [33, 34]. The classification of basis states based on the dual O($A-1$) of the reduction (3) underpins the Hyperspherical Harmonics (HH) method [91–93] reviewed in Ref. [94].

where $\sigma \equiv N_\sigma(\lambda_\sigma\, \mu_\sigma)$ labels the $\mathrm{Sp}(3,\mathbb{R})$ irrep, $n \equiv N_n(\lambda_n\, \mu_n)$, $\omega \equiv N(\lambda_\omega\, \mu_\omega)$, and $N = N_\sigma + N_n$ is the total number of HO quanta (ρ and κ are multiplicity labels) [33]. The relation of these symplectic basis states to M-scheme states of the NCSM is provided in Ref. [95]. The key importance of the symplectic symmetry emerges from the physical relevance of its 21 generators, which are directly related to the particle momentum and position coordinates and realize important observables. Namely, the many-particle kinetic energy, the HO potential (or equivalently, the monopole operator), the mass quadrupole moment, and angular momentum operators are all generators of $\mathrm{Sp}(3,\mathbb{R})$ and preserve the symplectic symmetry. In addition, the model includes multi-shell collective vibrations and vorticity degrees of freedom for a description from irrotational to rigid-rotor flows (for symplectic generators and irreps, see Chapter 4).

Symplectic model and selected applications. A significant breakthrough for the nuclear modeling is the microscopic symplectic model, developed by Rosensteel and Rowe [33, 60]. It provides a microscopic framework for understanding deformation-dominated collective phenomena in atomic nuclei and accommodates particle-hole excitations across multiple shells, describing rotations and vibrations of equilibrium deformation [33]. Indeed, the symplectic $\mathrm{Sp}(3,\mathbb{R})$ symmetry underpins the symplectic SM that provides a microscopic formulation of the Bohr–Mottelson collective model and is a multiple-shell generalization of the successful Elliott SU(3) model. The classical realization of this symmetry underpins the dynamics of rotating bodies and has been used, for example, to describe the rotation of deformed stars and galaxies [96].

The symplectic model with $\mathrm{Sp}(3,\mathbb{R})$-preserving interactions have achieved a remarkable reproduction (e.g., see [33,97–99]) of rotational bands, transition rates, and electron scattering form factors for *sd*-shell and rare-earth nuclei, without the need for introducing effective charges (for details of these and other relevant studies, see Chapters 3 and 4). In addition, a SM study that allows for mixing of $\mathrm{Sp}(3,\mathbb{R})$ irreps due to pairing and non-degenerate single-particle energies above a ^{16}O core [100] has found that using only seven Sp(3, $\mathbb{R}$) irreps, which extend up through 15 HO shells, is sufficient to achieve a remarkable reproduction of the ^{20}Ne energy spectrum and $E2$ transition rates (Fig. 6(a)). Recently, alpha-clustering and shape coexistence have been further examined in an Sp(3, $\mathbb{R}$)-based study using self-consistent arguments for ^{16}O [52], as well as in a no-core symplectic shell model (NCSpM) with a symmetry-breaking schematic interaction for ^{12}C [101]. In particular, the $N_{\max} = 20$ NCSpM results for the elusive second 0_2^+ state of ^{12}C (the so-called "Hoyle state") and its 2^+ and 4^+ excitations, which have fostered long-lasting debate, are found to reproduce experimental data, while pointing to a prolate intrinsic deformation for these states [101,102], in agreement with recent *ab initio* calculations using lattice effective

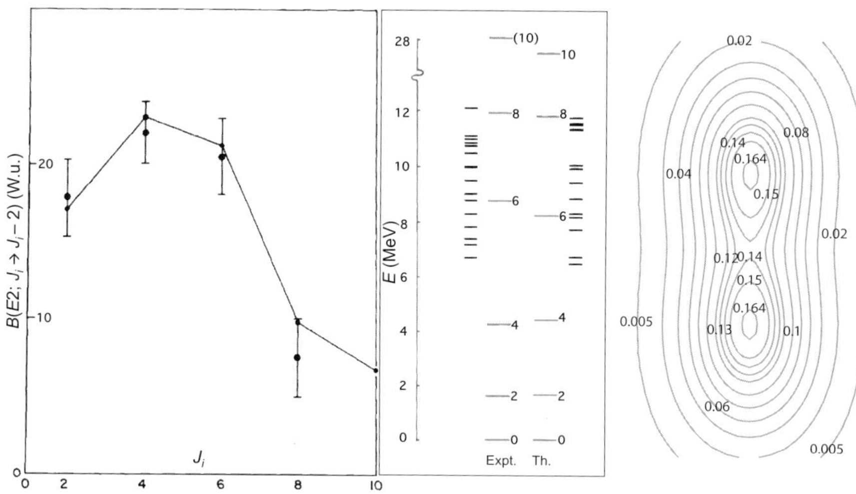

(a) (b)

Fig. 6. (a) Microscopic symplectic model with a set of effective single-particle energies with a $Q \cdot Q$-type interaction + pairing for ^{20}Ne. Left: Calculated $B(E2\downarrow)$ transition strengths *without effective charges* fall within the uncertainties of the corresponding experimental measurements. Right: Calculated energy spectrum for ^{20}Ne as compared to the experiment. Figure adapted from Ref. [100]. (b) One-body density profile *in the intrinsic frame* for the 0^+ ground state of ^{8}Be, as described by a single (4 0) symplectic irrep in the $N_{\max} = 24$ NCSpM [104] — note the formation of overlapping clusters.

field theory (EFT) [17,103] (cf. Chapter 6). In addition, the NCSpM has offered descriptions of phenomena tied to deformation (up to $A = 24$) [105], giant monopole and quadrupole resonances [106], as well as alpha-clustering in ^{8}Be (Fig. 6(b)) [104].

3.5. *Symplectic Symmetry from First Principles*

The existence of the symplectic Sp(3, $\mathbb{R}$) symmetry as approximate symmetry in light nuclei, and hence the propensity of nuclear dynamics toward development of collectivity, has been recently confirmed from first principles with no *a priori* symmetry assumptions [20,35]: the analysis of *ab initio* eigenstates reveals the major dominance of only a few symplectic Sp(3, $\mathbb{R}$) irreps (see arrows in Fig. 3 for ^{6}Li that schematically represent symplectic irreps; similarly, for ^{8}Be and ^{6}He). A remarkable result is that a *single* (2 0) symplectic irrep in ^{6}Li comprises more than 80% of the ground-state wave function, with a total of only 5 Sp(3, $\mathbb{R}$) irreps realizing more than 90% of the state. This clearly reflects the presence of an underlying symplectic Sp(3, $\mathbb{R}$) symmetry of microscopic nuclear collective motion [33,60].

The symplectic symmetry structure is further identified in well-converged *ab initio* wave functions for ^{12}C and ^{16}O, using the bare JISP16 and N^{3}LO NN interactions, as well as their effective counterparts (Fig. 7). Namely, *ab initio* eigensolutions for ^{12}C and ^{16}O have been found to typically project at the 85–90% level onto a few symplectic irreps, starting at the most deformed 0p-0h and $2\hbar\Omega$ 2p-2h configurations [35,107]. The results are nearly independent of whether the bare or effective interactions are used in the analysis and reveal a clear dominance, at the 80–85% level, of only one (in ^{16}O) and three (in ^{12}C) 0p-0h Sp(3, $\mathbb{R}$) irreps across the $\hbar\Omega$ values [51]. In addition, the largest contribution comes from the Sp(3, $\mathbb{R}$) irrep built on the most deformed $0\hbar\Omega$ configuration (Fig. 7, black diamonds). Another striking property of the low-lying eigenstates is revealed when their spin-0 and spin-1 components are examined [51], namely, the Sp(3, $\mathbb{R}$) symmetry within each spin component and hence the geometry of the nuclear system being described is

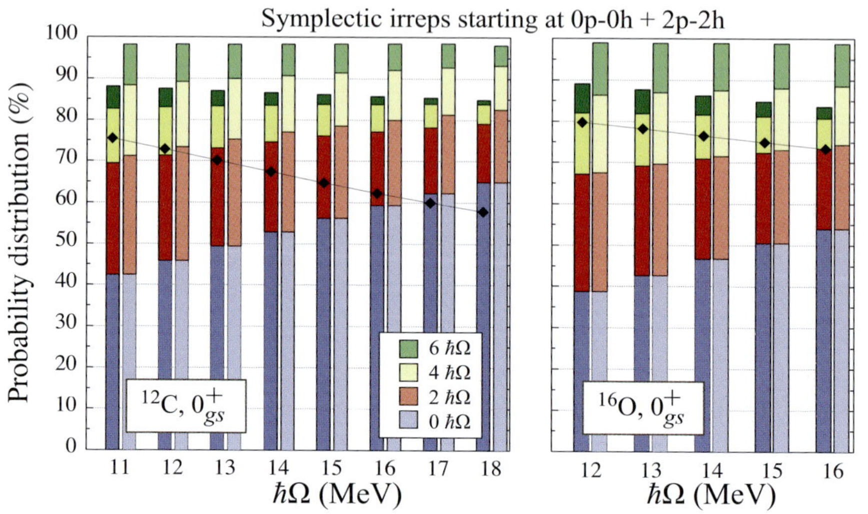

(a) (b)

Fig. 7. Probability distribution over $0\hbar\Omega$ (blue, lowest) to $6\hbar\Omega$ subspaces for the $N_{\max} = 6$ NCSM ground state (right bars) of (a) ^{12}C and (b) ^{16}O, and for the case of the most dominant $\mathrm{Sp}(3,\mathbb{R})$ irreps (left bars) together with the leading $\mathrm{Sp}(3,\mathbb{R})$ irrep contribution (black diamonds), (0 4) for ^{12}C and (0 0) for ^{16}O, across $\hbar\Omega$ values (similarly, for the 2^+_1 and 4^+_1 states [51]). Left bars include irreps of symplectic excitations built on a $2\hbar\Omega$ 2p-2h bandhead (2 particles up a shell). Figure adapted from Ref. [35].

very similar for all members of the ground-state rotational band, as well as nearly independent of the $\hbar\Omega$ oscillator strength and regardless of whether the bare or the effective interactions are used (Fig. 8).

The SA-NCSM in the symplectic scheme provides for further reduction of the model space as compared to the SU(3) scheme, especially for ultra-large $N_{\max}$ model spaces. As $N_{\max}$ increases, the dimension of the $J = 0, 2$, and 4 $\mathrm{Sp}(3,\mathbb{R})$-scheme model space grows very slowly and remains only a small fraction of the complete model space (Fig. 9(a)). The space reduction is even more dramatic in the case of ^{16}O (Fig. 9(b)). This means that a space spanned by a set of symplectic basis states is computationally manageable even when they extend to ultra-large model spaces.

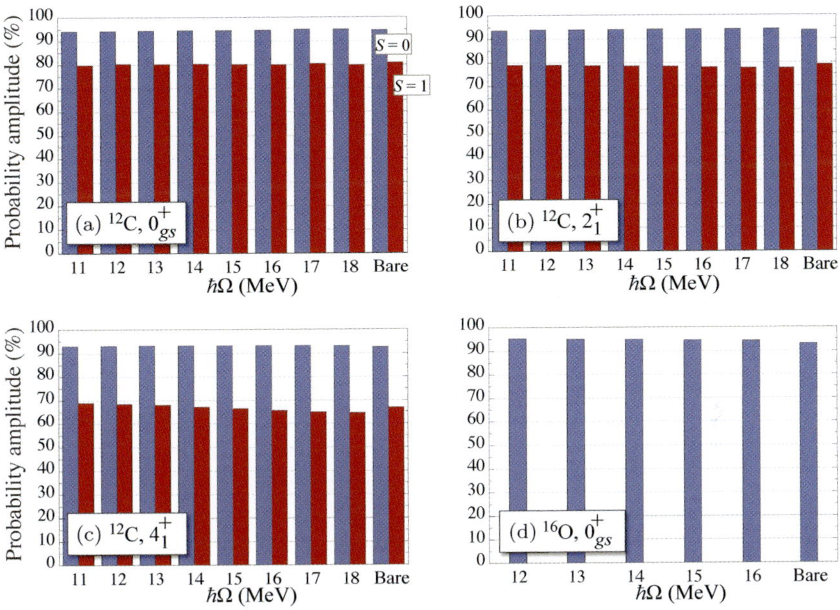

Fig. 8. Projection of the most dominant 0p-0h+2p-2h $S = 0$ (blue, left bars) and $S = 1$ (red, right bars) Sp$(3, \mathbb{R})$ irreps onto the corresponding spin components of the NSCM wave functions for (a) 0^+_{gs}, (b) 2^+_1, and (c) 4^+_1 in ^{12}C, and (d) 0^+_{gs} in ^{16}O, using the effective JISP16 interaction (based on the Lee–Suzuki method) for different $\hbar\Omega$ values and the bare JISP16 interaction (for $\hbar\Omega = 15$ MeV). Figure adapted from Ref. [35].

4. *Ab initio* Symmetry-adapted No-core Shell Model and Selected Applications

4.1. *Spectral and Scattering Properties of Light Nuclei*

Ab initio SA-NCSM calculations for spectral and scattering properties of p-shell nuclei have been found to yield results that are in reasonably close agreement with known experimental data (Figs. 10 and 11) [20, 80, 82]. Calculations are performed in the SU(3) scheme for selected model spaces that expand up to $N_{\max} = 12$, using the JISP16 [42] or NNLO$_{\text{opt}}$ [89] NN interactions, plus the Coulomb potential along with the Gloeckner–Lawson prescription [108] for eliminating spurious center-of-mass excitations.

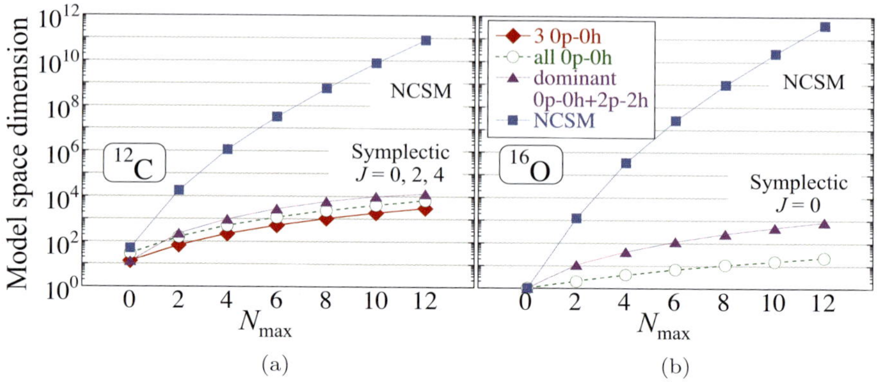

Fig. 9. Model space dimension as a function of the $N_{\max}$ cutoff, as compared to that of the $\mathrm{Sp}(3,\mathbb{R})$ subspace considered in the calculations of Fig. 7: (a) $J = 0, 2,$ and 4 for ^{12}C, and (b) $J = 0$ for ^{16}O. Figure adapted from Ref. [35].

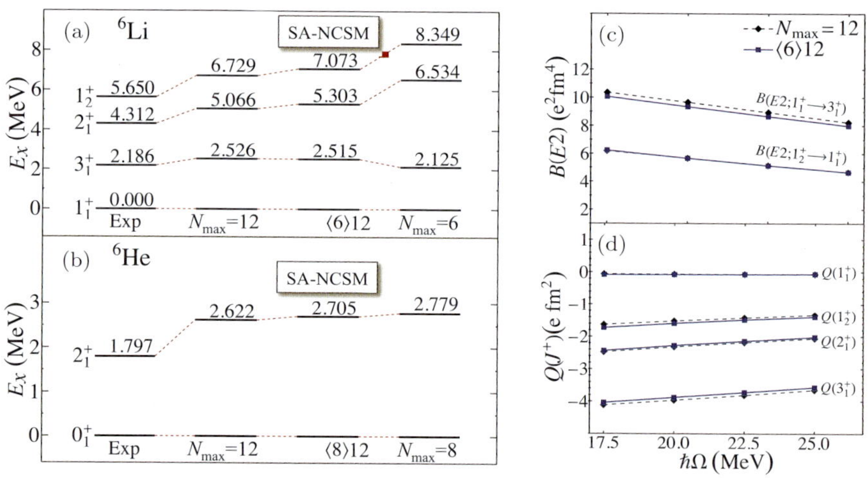

Fig. 10. Experimental (from Ref. [109]) and theoretical excitation energies for (a) the odd-odd ^{6}Li and (b) the halo ^{6}He, calculated for JISP16 and $\hbar\Omega = 20$ MeV in the complete $N_{\max} = 6$ and 12 spaces (or NCSM) and in SU(3)-based selected model spaces. (c) $B(E2)$ transition strengths and (d) quadrupole moments for ^{6}Li without effective charges vs. $\hbar\Omega$ for the complete $N_{\max} = 12$ space (dashed black lines) and $\langle 6\rangle 12$ selected space (solid blue lines). Note that the $B(E2)$ strengths almost double upon increasing the model space from $N_{\max} = 6$ to $N_{\max} = 12$, while the smaller $\langle 6\rangle 12$ SA-NCSM space is sufficient to reproduce the complete-space value. Figure adapted from Ref. [20].

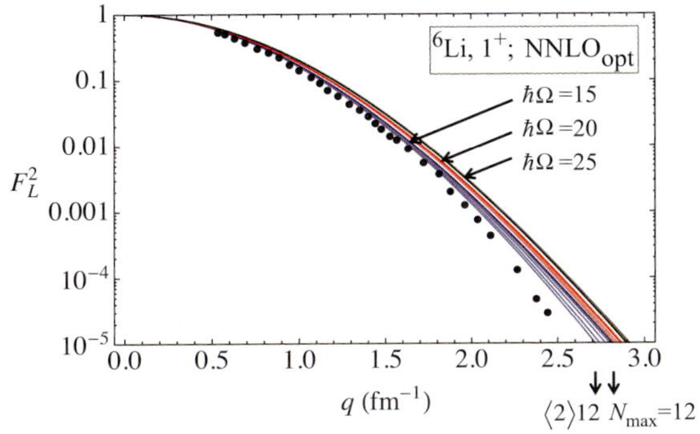

Fig. 11. Longitudinal $C0$ electron scattering form factor F_L^2 vs. momentum transfer q for the ^{6}Li ground state, calculated in the SA-NCSM with a chiral interaction and compared to experiment (dots) [112]. Curve thickness indicates deviations due to the model space selection. Form factors are corrected for the CM motion and the nucleon finite size. Figure from Ref. [80].

The results demonstrate that the SA-NCSM framework can achieve significantly reduced dimensions for equivalent large SM spaces without compromising the accuracy for various observables, including energies, point-particle proton and matter rms radii, electric quadrupole and magnetic dipole moments, reduced electromagnetic $B(E2)$ transition strengths, and form factors. For example, for ^{6}Li and ^{6}He [20, 37], the SA-NCSM excitation energies are found to differ only by 11 keV to a few hundred keV from the corresponding NCSM results (Figs. 10(a) and 10(b)). Similarly, the SA-NCSM eigensolutions yield binding energies that represent 98–99% of the NCSM results, as well as $B(E2)$ strengths and quadrupole moments that track very closely with their complete $N_{\rm max} = 12$ space counterparts (Figs. 10(c) and 10(d), for ^{6}Li). The differences between selected- and complete-space results are found insensitive to the choice of $\hbar\Omega$ and appear sufficiently small as to be inconsequential relative to the residual dependencies on $\hbar\Omega$ and on $N_{\rm max}$. Indeed, the *ab initio* SA-NCSM results agree with those of other *ab initio* models, such as the NCSM (or complete-space SA-NCSM), the variational Monte Carlo (VMC) [6] with the AV18 [1] two-nucleon and Urbana IX [110] three-nucleon interactions and the GFMC with AV18/UIX (Fig. 10 and Table 3). The results

Table 3. ^{6}Li binding energy (E_B), excitation energies (E), electric quadrupole (Q) and magnetic dipole (μ) moments, along with point-nucleon proton (r_p) and matter (r_m) rms radii, as calculated in the $\langle 6\rangle 12$ SA-NCSM with the JISP16 NN interaction (for $\hbar\Omega = 20$ MeV) and compared to other *ab initio* approaches: the complete $N_{\max} = 12$ model space (or NCSM), as well as VMC and GFMC using AV18/UIX interaction (energies taken from Ref. [6]; radii and electromagnetic moments taken from Ref. [111], without contributions from two-body currents). Experimental results (Expt.) taken from Ref. [109] unless otherwise specified.

State		SA-NCSM	NCSM	VMC	GFMC	Expt.
$1^+_{\rm gs}$	E_B [MeV]	30.445	30.951	27.0(1)	31.2(1)	31.99
	rms r_p [fm]	2.112	2.125	2.46(2)		2.43[a]
	rms r_m [fm]	2.106	2.119			2.35(3)[b]
	Q [e fm^2]	−0.080	−0.064	−0.33(18)		−0.0818(17)
	μ [μ_N]	0.839	0.838	0.828(1)		0.822
3^+	E [MeV]	2.515	2.526	3.0(1)	2.7(3)	2.186
	rms r_m [fm]	2.044	2.063			
	Q [e fm^2]	−3.880	−3.965			
	μ [μ_N]	1.866	1.866			
2^+	E [MeV]	5.303	5.066	4.4(1)	4.4(4)	4.312
	rms r_m [fm]	2.180	2.204			
	Q [e fm^2]	−2.279	−2.318			
	μ [μ_N]	1.014	0.970			

Notes: [a]Deduced from the ^{6}Li charge radius of 2.56(5) fm [112];
[b]From Ref. [113].

show good correlations among the different models with, perhaps, the exception of the smaller rms radii obtained with JISP16 (SA-NCSM and NCSM) and the larger magnitude of the $Q_{1^+_{\rm gs}}$ moment in ^{6}Li obtained with the VMC.

Of a special note is the quadrupole moment of the ^{6}Li ground state. In addition to being in excellent agreement to experiment (Fig. 10(d) and Table 3), the SA-NCSM result for $Q_{1^+_{\rm gs}}$ reproduces its challenging sign and nearly vanishing magnitude. In fact, the *ab initio* SA-NCSM has unveiled that even the lightest of nuclei in their ground state exhibit considerable collectivity, as exemplified by the dominance of prolate deformation in the ^{6}Li wave function (Fig. 3) — nonetheless, the considerable contribution of $L = 0$ configurations

($\sim 87\%$) to $1^+_{\rm gs}$ is responsible for the small expectation value of the quadrupole moment, an $L = 2$ operator [20] (cf. Chapter 1).

We have also demonstrated that the SA-NCSM space selection exhibits a relatively small effect on form factors and charge density [80]. In particular, the longitudinal charge electron scattering form factor of the ground state of ^{6}Li (Fig. 11) has been calculated in the *ab initio* SA-NCSM for several selected model spaces, $\langle 2\rangle 12$, $\langle 4\rangle 12, \ldots \langle 10\rangle 12$, together with the complete $N_{\rm max} = 12$ space, using the NNLO$_{\rm opt}$ NN interaction [89] (similarly, for the bare JISP16 interaction). A remarkable result is that the $\langle 6\rangle 12$ selected space yields results that are almost identical to the $N_{\rm max} = 12$ complete-space counterparts for any momentum transfer q. We note that for low q, $q \lesssim 2$ fm^{-1}, the SA-NCSM form factor lies reasonably close to experiment and agrees with the one of the VMC using AV18/UIX without two-body current contributions (compare Fig. 1 of Ref. [6] and Fig. 11). Beyond this, two-body currents become significant for $C0$ [6], leading to deviations from the experimental data for high momentum transfers (Fig. 11).

Finally, to estimate the degree of SU(3)-symmetry mixing in nuclear states, reviewed in Ref. [114], we adopt the Von Neumann information entropy (see, e.g., Ref. [115] and references therein). We consider only the model subspace, up to a given $N_{\rm max}$, within the leading symplectic irrep and estimate the extent to which all possible basis states of given $(\lambda\,\mu)$ mix (for a fixed total M). In particular, for a calculated wave function with probability amplitudes $|w_u|^2$ (e.g., shown in Fig. 3 for ^{6}Li), the Von Neumann information entropy can be defined as,

$$S_I = -\sum_{u=1}^{D} |w_u|^2 \log_D(|w_u|^2), \tag{4}$$

where D is the total number of basis states within the model subspace. Clearly, for no mixing — a nuclear state composed of single $(\lambda\,\mu)$ deformation with $w = 1$ – the entropy is zero; in an extreme scenario, all SU(3)-coupled basis states can be highly mixed with almost equal contributions ($w_u \simeq 1/\sqrt{D}$, for all u), the entropy then becomes one. This entropy has been used, e.g., to study chaotic behavior for pairing-governed nuclei [116] and can be also linked to the Shannon entropy if one-body densities are derived from the wave functions in position and momentum space, similarly to the case of quantum information entropy for a single particle in a given potential (see, e.g., [117]).

The analysis of results reveals that S_I is indeed comparatively small for Be, C, Mg, and Ne isotopes: e.g., for the *ab initio* SA-NCSM ground state of ^{12}C in $N_{\rm max} = 6$ (with the bare JISP16 NN and for $\hbar\Omega = 18$ MeV), the entropy is $S_I = 0.41$, which is projected to decrease to ~ 0.2 in $N_{\rm max} = 20$. This agrees with results in larger SM spaces only possible in the NCSpM with a schematic interaction [101], where, in addition, a slight increase in entropy is noted for cluster-like states, such as the ground state of ^{8}Be and the Hoyle state in ^{12}C (see Table 4 of Ref. [114]).

4.2. *Reaching Intermediate-mass Nuclei and Beyond*

Ab initio investigations of open-shell nuclei in the *sd*-shell region are now made possible by the SA-NCSM [118]. First SA-NCSM calculations for ^{24}Si, important for modeling and understanding X-ray burst nucleosynthesis, and for its mirror nucleus ^{24}Ne (Fig. 12(a)) have been achieved with SRG-evolved chiral N^{3}LO interactions [3] in an $N_{\rm max} = \langle 2\rangle 6$ symmetry-selected model space of 3.5×10^6 basis states [32, 119]. Note that the corresponding $N_{\rm max} = 6$ NCSM M-scheme calculations require a model space of $\sim 10^{11}$ dimensionality, which is currently inaccessible (see Fig. 1(a)). While it is difficult to experimentally study the short-lived ^{24}Si, $E2$ transitions are currently available for its mirror ^{24}Ne, with the SA-NCSM $B(E2; 2_1^+ \rightarrow 0_1^+)$ strength found to be in close agreement without using effective charges (Fig. 12(a)). We emphasize that conclusive results will require the exploration of the bare $3N$ chiral interaction, along with the SRG-induced many-body components of the interaction and observables.

Furthermore, the description of the ^{20}Ne ground-state rotational band is available now from first principles and reveals an interesting

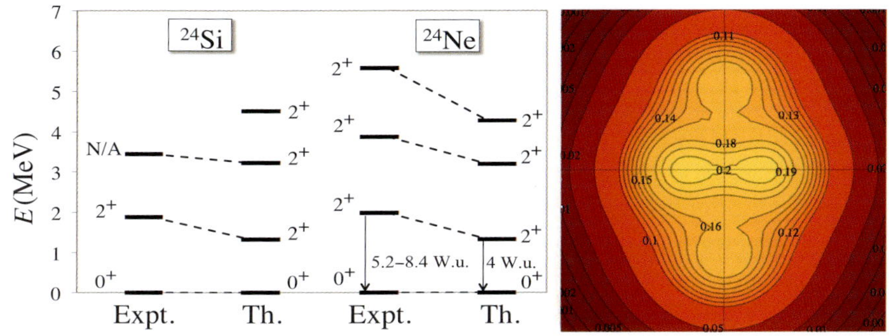

(a) (b)

Fig. 12. (a) First SA-NCSM calculations with chiral interactions for open-shell intermediate-mass nuclei, ^{24}Si and ^{24}Ne, performed in a model space of 9 HO shells and using the SRG-N^{3}LO NN interaction for an SRG cutoff $\lambda_c = 2$ fm^{-1} (for $\hbar\Omega = 15$ MeV). Figure adapted from [119]. (b) *Ab initio* SA-NCSM one-body density profile for the ^{20}Ne ground state in the challenging model space of 13 HO shells for $\hbar\Omega = 15$ MeV and with the NNLO$_{\rm opt}$ NN interaction; simulations performed on the Blue Waters system.

substructure formation (Fig. 12(b)). *Ab initio* SA-NCSM calculations are performed in the challenging $N_{\rm max} = \langle 2\rangle 10$ model space. For example, for the 6^+ state there are 51×10^6 basis states in the SA-NCSM, which can be compared to the inaccessible $N_{\rm max} = 10$ NCSM model space of 10^{12} (or 4×10^{11}) dimensionality required for the corresponding M-scheme (or J-scheme) calculations.

Clearly, exploiting the symmetry-guided concept allows us, within the framework of an HPC setting, to reduce the demand for computational resources and reach heavier nuclei. While medium-mass nuclei are expected to exhibit a stronger SU(3) mixing in low-N $\hbar\Omega$ subspaces as a result of richer subspaces — e.g., the dimension of the $0\hbar\Omega$ subspace is 4 for the ^{6}Li ground state and 4×10^4 for the ^{48}Cr ground state, — the relevant $(\lambda\,\mu)S$ configurations typically remain a fraction of the complete subspaces. It is important to emphasize that such an SU(3) mixing is readily accounted for in the SA-NCSM, by the complete horizontal span of the SA-NCSM model space in low-N $\hbar\Omega$ subspaces, while the vertical selection that expands to high $N_{\rm max}$ values is guided by symplectic excitations built on the most dominant SU(3) configurations. And while it is clear that selected model spaces, without compromising the accuracy of results, can reduce the memory demand in computations, utilizing symmetries leads to time-consuming evaluations of the many-body Hamiltonian. Nonetheless, it transforms a memory-bound unfeasible problem in a complete model space into a more CPU-intensive problem in a much smaller selected model space, which is attainable on today's petascale architecture. This allows a theory of the symmetry-guided large-scale shell-model type to extend the reach of the conventional schemes by exploiting symmetry-guided principles.

5. Summary and Outlook

We have reviewed exact and partial symmetries of atomic nuclei that have been long recognized, and only recently exposed from first principles and exploited in *ab initio* theories. Such partial symmetries include the deformation-related SU(3) and the symplectic $\mathrm{Sp}(3,\mathbb{R})$ for descriptions of rotations and vibrations of equilibrium deformation. We have focused on an open-core shell-model theory and the use of symmetries in large-scale nuclear simulations that permit symmetry mixing. In such calculations, symmetries are not used to restrict the nuclear dynamics, but rather to expose appropriate degrees of freedom to organize and down-select physically relevant spaces from what would otherwise be unwieldy at best or even prohibitively large model spaces. We have discussed an important new development, the SA-NCSM, that has demonstrated, for the first time, that observed collective phenomena in light and intermediate-mass nuclei emerge naturally from *ab initio* considerations. The results, supporting experimental evidence, underscore the strong dominance of configurations with large quadrupole deformation and low intrinsic spins, and symplectic excitations thereof.

The findings reviewed here start to unveil new physics, namely, understanding the mechanism on how simple patterns in nuclei and a diversity of nuclear properties emerge from a fundamental level.

Acknowledgements

We thank D. Rowe, G. Rosensteel, and C. Bahri, together with the PetaApps Collaboration, J. P. Vary, P. Maris, U. Catalyurek, and M. Sosonkina, for useful discussions. This work was supported by the U.S. NSF (OCI-0904874, ACI -1516338), the U.S. DOE (DE-SC0005248), Southeastern Universities Research Association (SURA), Louisiana State University (through its direct support of J.P.D.'s work with SURA), and the Czech Science Foundation under Grant No. 16-16772S. This work also benefitted from computing resources provided by Blue Waters, as well as the Louisiana Optical Network Initiative and HPC resources provided by LSU (`www.hpc.lsu.edu`). A portion of the computational resources were provided by the National Energy Research Scientific Computing Center, which is supported by the Office of Science of the U.S. DOE under Contract No. DE-AC02-05CH11231, and by an INCITE award, "Nuclear Structure and Nuclear Reactions", from the DOE Office of Advanced Scientific Computing. T. D. acknowledges support from Michal Pajr and CQK Holding.

References

[1] R. B. Wiringa, V. G. J. Stoks, and R. Schiavilla, *Phys. Rev. C.* **51**, 38 (1995).
[2] R. Machleidt, *Phys. Rev. C.* **63**, 024001 (2001).
[3] D. R. Entem and R. Machleidt, *Phys. Rev. C.* **68**, 041001 (2003).
[4] E. Epelbaum, *Prog. Part. Nucl. Phys.* **57**, 654 (2006).
[5] D. Langr, I. Simecek, P. Tvrdik, T. Dytrych, and J. P. Draayer. Adaptive-Blocking Hierarchical Storage Format for Sparse Matrices. In *Proc. Federated Conf. Computer Science and Information Systems (FEDCSIS), 2012*, (2012), p. 545.
[6] R. B. Wiringa and R. Schiavilla, *Phys. Rev. Lett.* **81**, 4317 (1998).
[7] S. C. Pieper, K. Varga, and R. B. Wiringa, *Phys. Rev. C.* **66**, 044310 (2002).
[8] P. Navrátil, J. P. Vary, and B. R. Barrett, *Phys. Rev. Lett.* **84**, 5728 (2000).
[9] B. Barrett, P. Navrátil, and J. Vary, *Prog. Part. Nucl. Phys.* **69**, 131 (2013).
[10] P. Maris, J. P. Vary, and P. Navrátil, *Phys. Rev. C.* **87**, 014327 (2013).
[11] P. Navrátil, S. Quaglioni, I. Stetcu, and B. R. Barrett, *J. Phys. G.* **36**, 083101 (2009).
[12] S. Baroni, P. Navrátil, and S. Quaglioni, *Phys. Rev. Lett.* **110**, 022505 (2013).

[13] A. F. Lisetskiy, B. R. Barrett, M. K. G. Kruse, P. Navratil, I. Stetcu, and J. P. Vary, *Phys. Rev. C.* **78**, 044302 (2008).
[14] R. Roth and P. Navrátil, *Phys. Rev. Lett.* **99**, 092501 (2007).
[15] M. Wloch, D. J. Dean, J. R. Gour, M. Hjorth-Jensen, K. Kowalski, T. Papenbrock, and P. Piecuch, *Phys. Rev. Lett.* **94**, 212501 (2005).
[16] G. Hagen, T. Papenbrock, D. J. Dean, and M. Hjorth-Jensen, *Phys. Rev. Lett.* **101**, 092502 (2008).
[17] E. Epelbaum, H. Krebs, D. Lee, and U.-G. Meissner, *Phys. Rev. Lett.* **106**, 192501 (2011).
[18] K. Tsukiyama, S. K. Bogner, and A. Schwenk, *Phys. Rev. Lett.* **106**, 222502 (2011).
[19] S. K. Bogner, H. Hergert, J. D. Holt, A. Schwenk, S. Binder, A. Calci, J. Langhammer, and R. Roth, *Phys. Rev. Lett.* **113**, 142501 (2014).
[20] T. Dytrych, K. D. Launey, J. P. Draayer, P. Maris, J. P. Vary, E. Saule, U. Catalyurek, M. Sosonkina, D. Langr, and M. A. Caprio, *Phys. Rev. Lett.* **111**, 252501 (2013).
[21] T. Abe, P. Maris, T. Otsuka, N. Shimizu, Y. Utsuno, and J. Vary, *Phys. Rev. C.* **86**, 054301 (2012).
[22] A. Cipollone, C. Barbieri, and P. Navratil, *Phys. Rev. Lett.* **111**, 062501 (2013).
[23] R. B. Wiringa and S. C. Pieper, *Phys. Rev. Lett.* **89**, 182501 (2002).
[24] A. C. Hayes, P. Navratil, and J. P. Vary, *Phys. Rev. Lett.* **91**, 012502 (2003).
[25] P. Maris, A. M. Shirokov, and J. P. Vary, *Phys. Rev. C.* **81**, 021301(R) (2010).
[26] P. Navrátil and S. Quaglioni, *Phys. Rev. Lett.* **108**, 042503 (2012).
[27] G. Hagen, M. Hjorth-Jensen, G. R. Jansen, R. Machleidt, and T. Papenbrock, *Phys. Rev. Lett.* **109**, 032502 (2012).
[28] S. Bacca, N. Barnea, W. Leidemann, and G. Orlandini, *Phys. Rev. Lett.* **110**, 042503 (2013).
[29] G. Hupin, S. Quaglioni, and P. Navrátil, *Phys. Rev. C.* **90**, 061601 (2014).
[30] E. Epelbaum, H. Krebs, T. A. Lähde, D. Lee, U.-G. Meißner, and G. Rupak, *Phys. Rev. Lett.* **112** (10), 102501 (2014).
[31] C. Romero-Redondo, S. Quaglioni, P. Navrátil, and G. Hupin, *Phys. Rev. Lett.* **113**, 032503 (2014).
[32] T. Dytrych, K. D. Launey, and J. P. Draayer, *Symmetry-adapted No-core Shell Model*, In *McGraw-Hill Yearbook of Science and Technology*. McGraw-Hill (2014).
[33] D. J. Rowe, *Reports Progr. Phys.* **48**, 1419 (1985).
[34] D. J. Rowe, *Prog. Part. Nucl. Phys.* **37**, 265 (1996).
[35] T. Dytrych, K. D. Sviratcheva, C. Bahri, J. P. Draayer, and J. P. Vary, *Phys. Rev. Lett.* **98**, 162503 (2007).
[36] T. Dytrych, J. P. Draayer, K. D. Sviratcheva, C. Bahri, and J. P. Vary. Dominance of low spin and high deformation in *Ab Initio* approaches to the structure of light nuclei. In M. Milin, T. Niksic, S. Szilner, and D. Vretenar, (eds.), "*Nuclear Structure And Dynamics '09*": *Proc. Int.*

Conf., Dubrovnik, Croatia, 2009. Vol. 1165, AIP Conference Proceedings (2009).

[37] K. D. Launey, T. Dytrych, and J. P. Draayer, *Prog. Part. Nucl. Phys.* **89**, 101 (review) (2016).

[38] P. Brussard and P. Glaudemans, *Shell-Model Applications in Nuclear Spectroscopy*. North-Holland Publishing Company, Amsterdam (1977).

[39] I. Shavitt, *Molecular Physics*. **94**, 3 (1998).

[40] E. Caurier, G. Martinez-Pinedo, F. Nowacki, A. Poves, and A. P. Zuker, *Rev. Mod. Phys.* **77**, 427 (2005).

[41] D. Gazit, S. Quaglioni, and P. Navratil, *Phys. Rev. Lett.* **103**, 102502 (2009).

[42] A. Shirokov, J. Vary, A. Mazur, and T. Weber, *Phys. Lett. B.* **644**, 33 (2007).

[43] K. Suzuki and S. Y. Lee, *Prog. Theor. Phys.* **64**, 2091 (1980).

[44] S. Bogner, R. Furnstahl, and R. Perry, *Phys. Rev. C.* **75**, 061001(R) (2007).

[45] R. Roth, S. Reinhardt, and H. Hergert, *Phys. Rev. C.* **77**, 064003 (2008).

[46] M. G. Mayer, *Phys. Rev.* **75**, 1969 (1949).

[47] O. Haxel, J. H. D. Jensen, and H. E. Suess, *Phys. Rev.* **75**, 1766 (1949).

[48] M. G. Mayer, *Phys. Rev.* **78**, 16 (1950).

[49] A. Bohr and B. R. Mottelson, *Nuclear Structure*. Vol. 1, Benjamin, New York (1969).

[50] B. R. Mottelson, Nobel lectures, *Physics 1971-1980* (1992).

[51] T. Dytrych, K. D. Sviratcheva, C. Bahri, J. P. Draayer, and J. P. Vary, *Phys. Rev. C.* **76**, 014315 (2007).

[52] D. J. Rowe, G. Thiamova, and J. L. Wood, *Phys. Rev. Lett.* **97**, 202501 (2006).

[53] C. W. Johnson, *Phys. Rev. C.* **91**, 034313 (2015).

[54] B. Brown, *Prog. Part. Nucl. Phys.* **47**, 517 (2001).

[55] S. Koonin, D. Dean, and K. Langanke, *Phys. Rept.* **278**, 1 (1997).

[56] Y. Alhassid, A. Mukherjee, H. Nakada, and C. Ozen, *J. Phys.: Conf. Ser.* **403**, 012012 (2012).

[57] J. P. Elliott, *Proc. Roy. Soc. A.* **245**, 128 (1958).

[58] J. P. Elliott, *Proc. Roy. Soc. A.* **245**, 562 (1958).

[59] J. P. Elliott and M. Harvey, *Proc. Roy. Soc. A.* **272**, 557 (1962).

[60] G. Rosensteel and D. J. Rowe, *Phys. Rev. Lett.* **38**, 10 (1977).

[61] V. Bargmann and M. Moshinsky, *Nucl. Phys.* **18**, 697 (1960).

[62] J. P. Draayer, Y. Leschber, S. C. Park, and R. Lopez, *Comput. Phys. Commun.* **56**, 279 (1989).

[63] G. Rosensteel and D. J. Rowe, *Ann. Phys. N.Y.* **104**, 134 (1977).

[64] Y. Leschber and J. P. Draayer, *Phys. Letts. B.* **190**, 1 (1987).

[65] O. Castaños, J. P. Draayer, and Y. Leschber, *Z. Phys. A.* **329**, 33 (1988).

[66] D. J. Rowe and J. L. Wood, *Fundamentals of Nuclear Models: Foundational Models*. World Scientific, Singapore (2010).

[67] C. Vargas, J. Hirsch, and J. Draayer, *Nucl. Phys. A.* **690**, 409 (2001).

[68] M. Moshinsky, *Rev. Mod. Phys.* **34**, 813 (1962).

[69] M. Moshinsky, J. Patera, R. T. Sharp, and P. Winternitz, *Ann. Phys. (N.Y.)*. **95**, 139 (1975).
[70] D. J. Millener, *J. Math. Phys.* **19**, 1513 (1978).
[71] M. Harvey, *Adv. Nucl. Phys.* **1**, 67 (1968).
[72] K. T. Hecht and A. Adler, *Nucl. Phys. A.* **137**, 129 (1969).
[73] A. Arima, M. Harvey, and M. K. Shimizu, *Phys. Lett.* **30B**, 517 (1969).
[74] R. R. Raju, J. Draayer, and K. Hecht, *Nucl. Phys. A.* **202**, 433 (1973).
[75] J. Draayer. In A. Frank and K. Wolf, (eds.), *Proc. Symmetries in Physics*, Cocoyoc, Mexico, 1991, Springer-Verlag, Berlin (1991).
[76] C. Bahri, J. Draayer, and S. Moszkowski, *Phys. Rev. Lett.* **68**, 2133 (1992).
[77] A. Zuker, J. Retamosa, A. Poves, and E. Caurier, *Phys. Rev. C.* **52**, R1741 (1995).
[78] C. Bahri, J. Escher, and J. P. Draayer, *Nucl. Phys. A.* **592**, 171 (1995).
[79] B. Preedom and B. Wildenthal, *Phys. Rev. C.* **6**, 1633 (1972).
[80] T. Dytrych, A. C. Hayes, K. D. Launey, J. P. Draayer, P. Maris, J. P. Vary, D. Langr, and T. Oberhuber, *Phys. Rev. C.* **91**, 024326 (2015).
[81] T. Dytrych, *LSU3shell code, Louisiana State University, available under the GNU General Public License at the git repository.* http://sourceforge.net/projects/lsu3shell (2013).
[82] T. Dytrych, P. Maris, K. D. Launey, J. P. Draayer, J. P. Vary, M. Caprio, D. Langr, U. Catalyurek, and M. Sosonkina, *Comput. Phys. Commun.* **207**, 202 (2016).
[83] Y. Akiyama and J. P. Draayer, *Comput. Phys. Commun.* **5**, 405 (1973).
[84] B. J. Verhaar, *Nucl. Phys.* **21**, 508 (1960).
[85] K. T. Hecht, *Nucl. Phys. A.* **170**, 34 (1971).
[86] D. Millener. In J. Draayer and J. Janecke, (eds.), *Group Theory and Special Symmetries in Nuclear Physics*. World Scientific, Singapore (1992), p. 276.
[87] T. Dytrych, J. P. Draayer, K. D. Launey, M. A. Caprio, and D. Langr, *J. of Phys.: Conf. Ser.* **403**, 012015 (2012).
[88] R. B. Wiringa, *Phys. Rev. C.* **73**, 034317 (2006).
[89] A. Ekström, G. Baardsen, C. Forssén, G. Hagen, M. Hjorth-Jensen, G. R. Jansen, R. Machleidt, W. Nazarewicz, *et al.*, *Phys. Rev. Lett.* **110**, 192502 (2013).
[90] D. Rowe, *AIP Conf. Proc.* **1541**, 104 (2013).
[91] E. Borel, *Introduction Geometrique a Quelques Theories Physiques.* Gauthier-Villars, Paris (1914).
[92] N. Barnea, W. Leidemann, and G. Orlandini, *Nucl. Phys. A.* **650**, 427 (1999).
[93] N. K. Timofeyuk, *Phys. Rev. C.* **69**, 034336 (2004).
[94] W. Leidemann and G. Orlandini, *Prog. Part. Nucl. Phys.* **68**, 158 (2013).
[95] T. Dytrych, K. D. Sviratcheva, J. P. Draayer, C. Bahri, and J. P. Vary, *J. Phys. G.* **35**, 123101 (2008).
[96] G. Rosensteel, *Astrophys. J.* **416**, 291 (1993).
[97] C. Bahri and D. J. Rowe, *Nucl. Phys. A.* **662**, 125 (2000).
[98] D. Peterson and K. Hecht, *Nucl. Phys. A.* **344**, 361 (1980).

[99] J. Escher and J. P. Draayer, *Phys. Rev. Lett.* **82**, 5221 (1999).

[100] J. Draayer, K. Weeks, and G. Rosensteel, *Nucl. Phys. A.* **419**, 1 (1984).

[101] A. C. Dreyfuss, K. D. Launey, T. Dytrych, J. P. Draayer, and C. Bahri, *Phys. Lett. B.* **727**, 511 (2013).

[102] K. D. Launey, A. C. Dreyfuss, J. P. Draayer, T. Dytrych, and R. B. Baker, *J. Phys. Ser.* **569**, 012061 (2014).

[103] E. Epelbaum, H. Krebs, T. Lähde, D. Lee, and U.-G. Meißner, *Phys. Rev. Lett.* **109**, 252501 (2012).

[104] K. D. Launey, T. Dytrych, J. P. Draayer, G. K. Tobin, M. C. Ferriss, D. Langr, A. C. Dreyfuss, P. Maris, J. P. Vary, and C. Bahri. Symmetry-adapted No-core Shell Model for Light Nuclei. In J. H. Hamilton and A. V. Ramayya, (eds.), *Proc. 5th Int. Conf. Fission and properties of neutron-rich nuclei, ICFN5, 2012*, Sanibel Island, Florida, World Scientific, Singapore (2013), p. 29.

[105] G. K. Tobin, M. C. Ferriss, K. D. Launey, T. Dytrych, J. P. Draayer, and C. Bahri, *Phys. Rev. C.* **89**, 034312 (2014).

[106] A. C. Dreyfuss, K. D. Launey, T. Dytrych, J. P. Draayer, R. B. Baker, C. M. Deibel, and C. Bahri, *submitted to Phys. Rev. C* (2016).

[107] T. Dytrych, K. D. Sviratcheva, C. Bahri, J. P. Draayer, and J. Vary, *J. Phys. G.* **35**, 095101 (2008).

[108] D. H. Gloeckner and R. D. Lawson, *Phys. Lett. B.* **53**, 313 (1974).

[109] D. R. Tilley *et al.*, *Nucl. Phys. A.* **708**, 3 (2002).

[110] B. S. Pudliner, V. R. Pandharipande, J. Carlson, and R. B. Wiringa, *Phys. Rev. Lett.* **74**, 4396 (1995).

[111] B. S. Pudliner, V. R. Pandharipande, J. Carlson, S. C. Pieper, and R. B. Wiringa, *Phys. Rev. C.* **56**, 1720 (1997).

[112] G. C. Li, I. Sick, R. R. Whitney, and M. R. Yearian, *Nucl. Phys. A.* **162**, 583 (1971).

[113] I. Tanihata, *Phys. Lett. B.* **206**, 592 (1988).

[114] K. D. Launey, T. Dytrych, J. P. Draayer, G.-H. Sun, and S.-H. Dong, *Int. J. Mod. Phys. E.* **24**, 1530005 (2015).

[115] H. A. Weidenmüller and G. E. Mitchell, *Rev. Mod. Phys.* **81**, 539 (2009).

[116] X. Guan, K. D. Launey, J. Gu, F. Pan, and J. P. Draayer, *Phys. Rev. C.* **88**, 044325 (2013).

[117] G. Yañez-Navarro, G.-H. Sun, T. Dytrych, K. Launey, S.-H. Dong, and J. Draayer, *Ann. Phys.* **348**, 153 (2014).

[118] T. Dytrych, K. D. Launey, J. P. Draayer, *et al.*, *to be submitted* (2016).

[119] J. P. Draayer, T. Dytrych, K. D. Launey, A. C. Dreyfuss, and D. Langr, *J. of Phys.: Conf. Ser.* **578**, 012010 (2015).

PART 4

Symmetries and Emergent Features in Medium-mass and Heavy Nuclei

Chapter 9

Auxiliary-field Quantum Monte Carlo Methods in Nuclei

Yoram Alhassid
Center for Theoretical Physics, Sloane Physics Laboratory
Yale University, New Haven, Connecticut 06520, USA
yoram.alhassid@yale.edu

Auxiliary-field quantum Monte Carlo methods enable the calculation of thermal and ground state properties of correlated quantum many-body systems in model spaces that are many orders of magnitude larger than those that can be treated by conventional diagonalization methods. We review recent developments and applications of these methods in nuclei using the framework of the configuration-interaction shell model.

Contents

1. Introduction

A major challenge in nuclear theory is understanding the properties of nuclei from the underlying interactions between their constituents. There has been much progress in applying *ab initio* methods such as the Green's function Monte Carlo [1, 2], the no-core shell model [3, 4], and the symmetry-adapted no-core shell model [5] to calculate nuclear properties, but these approaches are limited to light nuclei. The coupled-cluster method [6] has been applied to light nuclei and mid-mass nuclei near shell closure. Density functional theory [7,8] is applicable across the table of nuclei, but as a mean-field approximation it can miss important correlations and the description of excited states requires extensions of the theory.

The configuration-interaction (CI) shell-model approach [9] is a suitable framework to include correlations beyond the mean-field approximation. The CI shell model is widely used in nuclear, atomic and molecular physics. However, the dimensionality of the many-particle model space scales combinatorially with the number of valence single-particle orbitals and the number of valence nucleons, hindering its applications in mid-mass and heavy nuclei.

The auxiliary-field Monte Carlo (AFMC) approach [10–13], also known in nuclear physics as the shell model Monte Carlo (SMMC) method [14–17], enables calculations in model spaces that are many orders of magnitude larger than those that can be treated by conventional diagonalization methods. AFMC is a powerful technique for calculating thermal and ground-state properties. As a finite-temperature method, it has been applied mainly to the calculation of statistical and collective properties of nuclei. In particular, AFMC is the state-of-the-art method for the microscopic calculation of nuclear level densities in the presence of correlations. (See Chapter 6 for the use of AFMC in *ab initio* applications to light nuclei.)

While fermionic quantum Monte Carlo methods are often limited by the so-called sign problem that leads to large statistical errors, the dominant components [18] of shell-model nuclear interactions have a good sign in AFMC and are often sufficient for realistic calculation of statistical and collective properties of nuclei. The smaller bad-sign components can be treated by the extrapolation method of Ref. [15].

AFMC methods have been applied to other correlated quantum many-body systems. In condensed matter physics, AFMC has been used to study strongly correlated electron systems [12, 19, 20]. In quantum chemistry, it was applied to study the electronic structure of molecules, such as the recent study of the chromium dimer [21]. In cold atom physics, AFMC methods were used to study the thermodynamics of the two-species Fermi gas with contact interaction for both the uniform gas [22, 23] and the harmonically trapped gas [24], and the ground state of the Fermi gas in its unitary limit [25]. AFMC simulations were recently carried out in studies of the neutron matter equation of state [26].

Here we review the finite-temperature AFMC method in the context of the CI nuclear shell model, and in particular its recent developments and applications to mid-mass and heavy nuclei. Earlier applications were discussed in Refs. [16, 17] and references therein. In Sec. 2 we discuss the finite-temperature formalism of the AFMC method in the framework of both the grand canonical and canonical ensembles. The canonical ensemble, in which the number of particles is fixed, is particularly useful in applications to nuclei. In Sec. 3 we discuss various projection methods in AFMC, and in Sec. 4 we describe the calculation of state densities. In Sec. 5 we discuss recent applications of AFMC to mid-mass nuclei, and in particular in the calculation of level densities, spin distributions and pairing gaps. In Sec. 6 we describe recent applications to heavy lanthanide nuclei, including the emergence of collectivity in the framework of the CI shell model, the calculation of state densities from the underlying Hamiltonian and the description of nuclear deformation in a rotationally invariant framework. We conclude in Sec. 7 with a summary and outlook.

2. Auxiliary-field Monte Carlo Method

The AFMC method is based on the Hubbard–Stratonovich (HS) [27, 28] representation of the Gibbs ensemble, as we discuss in Sec. 2.1 The proper ensemble to describe nuclei is the canonical ensemble with fixed numbers of protons and neutrons, and in Sec. 2.2 we describe how this is accomplished using an exact particle-number projection [29]. The Monte Carlo sampling method is briefly discussed in Sec. 2.3 A rule to determine good-sign interactions in the grand canonical ensemble is discussed in Sec. 2.4 For a good-sign interaction, the Monte Carlo sign remains good when projecting on the canonical ensemble with even number of particles, but a sign problem emerges when projecting on an odd number of particles at low temperatures. In Sec. 2.5 we describe a recent method we introduced to carry out accurate calculation of the ground-state energy of an odd-particle system despite the odd-particle sign problem [30].

2.1. *The Hubbard-Stratonovich Transformation*

The CI shell-model Hamiltonian $\hat{H}$ contains a one-body part described by single-particle orbitals i and single-particle energies ϵ_i and a residual two-body interaction characterized by its two-body matrix elements $v_{ij,kl}$. The two-body interaction term can be brought to a diagonal form

$$\hat{H} = \sum_i \epsilon_i \hat{n}_i + \frac{1}{2}\sum_\alpha v_\alpha \hat{\rho}_\alpha^2, \tag{1}$$

where $\hat{\rho}_\alpha$ are linear combinations of one-body densities $\hat{\rho}_{ij} = a_i^\dagger a_j$, and v_α are the interaction "eigenvalues".

The Gibbs density operator $e^{-\beta H}$ at inverse temperature $\beta = 1/kT$ can be viewed as the many-body evolution operator in imaginary time β. The HS transformation [27, 28] expresses this propagator as a functional integral over one-body propagators describing non-interacting nucleons in time-dependent external fields $\sigma(\tau)$ $(0 \le \tau \le \beta)$. The HS transformation is derived by dividing the time interval $(0, \beta)$ into N_t time slices of length $\Delta\beta$ each, and factorizing $e^{-\beta\hat{H}} = (e^{-\Delta\beta\hat{H}})^{N_t}$. For each time slice $\Delta\beta$, we have to order $(\Delta\beta)^2$

$$e^{-\Delta\beta\hat{H}} \approx \prod_i e^{-\Delta\beta\epsilon_i\hat{n}_i} \prod_\alpha e^{-\frac{1}{2}\Delta\beta v_\alpha \hat{\rho}_\alpha^2}. \tag{2}$$

Each factor in the product over α in Eq. (2) can be written as an integral over an auxiliary variable σ_α

$$e^{-\frac{1}{2}\Delta\beta v_\alpha \hat{\rho}_\alpha^2} = \sqrt{\frac{\Delta\beta|v_\alpha|}{2\pi}} \int d\sigma_\alpha e^{-\frac{1}{2}\Delta\beta|v_\alpha|\sigma_\alpha^2} e^{-\Delta\beta|v_\alpha|s_\alpha\sigma_\alpha\hat{\rho}_\alpha}, \tag{3}$$

where $s_\alpha = \pm 1$ for $v_\alpha < 0$ and $s_\alpha = \pm i$ for $v_\alpha > 0$. Using a set of auxiliary fields $\sigma_\alpha(\tau_n)$ at each time slice $\tau_n = n\Delta\beta$ and taking the limit of large N_t, we obtain the HS transformation

$$e^{-\beta \hat{H}} = \int \mathcal{D}[\sigma] G_\sigma \hat{U}_\sigma, \tag{4}$$

where

$$G_\sigma = e^{-\frac{1}{2}\int_0^\beta |v_\alpha| \sigma_\alpha^2(\tau) d\tau} \tag{5}$$

is a Gaussian weight. $\hat{U}_\sigma$ in Eq. (4) is given by

$$\hat{U}_\sigma = \mathcal{T} e^{-\int_0^\beta \hat{h}_\sigma(\tau) d\tau}, \tag{6}$$

where $\mathcal{T}$ denotes time ordering and

$$\hat{h}_\sigma(\tau) = \sum_i \epsilon_i \hat{n}_i + \sum_\alpha s_\alpha |v_\alpha| \sigma_\alpha(\tau) \hat{\rho}_\alpha \tag{7}$$

is a one-body Hamiltonian describing nucleons moving in external time-dependent auxiliary fields $\sigma_\alpha(\tau)$. The measure in the functional integral (4) over the auxiliary fields is

$$\mathcal{D}[\sigma] \equiv \prod_{\alpha,n} \left[d\sigma_\alpha(\tau_n) \sqrt{\Delta\beta |v_\alpha| / 2\pi} \right]. \tag{8}$$

Using the HS transformation (4), the thermal expectation of an observable $\hat{O}$ can be written as

$$\langle \hat{O} \rangle = \frac{\mathrm{Tr}\,(\hat{O} e^{-\beta \hat{H}})}{\mathrm{Tr}\,(e^{-\beta \hat{H}})} = \frac{\int \mathcal{D}[\sigma] G_\sigma \langle \hat{O} \rangle_\sigma \mathrm{Tr}\,\hat{U}_\sigma}{\int \mathcal{D}[\sigma] G_\sigma \mathrm{Tr}\,\hat{U}_\sigma}, \tag{9}$$

where $\langle \hat{O} \rangle_\sigma \equiv \mathrm{Tr}\,(\hat{O}\hat{U}_\sigma)/\mathrm{Tr}\,\hat{U}_\sigma$ is the expectation value of $\hat{O}$ for non-interacting particles in external auxiliary fields $\sigma(\tau)$.

2.1.1. *Algebraic structure of the Hubbard-Stratonovich transformation*

The set of all one-body densities $a_i^\dagger a_j$ ($i, j = 1, \ldots, N_s$, where N_s is the number of single-particle orbitals) forms a Lie algebra, i.e., the commutator of any two such one-body densities is a linear combination of one-body densities. The corresponding operators of the form $\exp(\sum_{ij} c_{ij} a_i^\dagger a_j)$ (where c_{ij} are c-numbers) therefore describe a Lie group. The one-body propagator $\hat{U}_\sigma$ in Eq. (6) is a time-ordered product of such group elements and therefore is also a group element. The operator $\hat{U}_\sigma$ defines a propagator in Fock space for any number of particles. Of particular interest is its representation in the single-particle space, described by an $N_s \times N_s$ matrix $\mathbf{U}_\sigma$. As we discuss below, the quantities appearing in the integrands of Eq. (9) can be expressed in terms of this single-particle representation.

We first discuss the grand canonical ensemble, in which the traces are evaluated over the many-particle Fock space with all possible particle numbers. A chemical potential μ (which determines the average number of particles in the ensemble) is introduced by replacing the single-particle energies ϵ_i with $\epsilon_i - \mu$.

2.1.2. *One-body observables*

The trace of $\hat{U}_\sigma$ over the complete Fock space is given in terms of the matrix $\mathbf{U}_\sigma$ by

$$\mathrm{Tr}\, \hat{U}_\sigma = \det(\mathbf{1} + \mathbf{U}_\sigma). \tag{10}$$

Equation (10) can be thought of as the grand-canonical partition function of non-interacting fermions in a given set of time-dependent external fields $\sigma(\tau)$.

The grand canonical expectation value of a one-body operator $\hat{O} = \sum_{i,j} O_{ij} a_i^\dagger a_j$ can be calculated from

$$\langle a_i^\dagger a_j \rangle_\sigma \equiv \frac{\mathrm{Tr}\, (a_i^\dagger a_j \hat{U}_\sigma)}{\mathrm{Tr}\, \hat{U}_\sigma} = \left[\frac{1}{\mathbf{1} + \mathbf{U}_\sigma^{-1}}\right]_{ji}. \tag{11}$$

2.1.3. *Two-body observables*

Since $\hat{U}_\sigma$ describes an uncorrelated ensemble, we can use Wick's theorem to calculate the grand canonical expectation value of a two-body operator

$$\langle a_i^\dagger a_j^\dagger a_l a_k \rangle_\sigma = \langle a_i^\dagger a_k \rangle_\sigma \langle a_j^\dagger a_l \rangle_\sigma - \langle a_i^\dagger a_l \rangle_\sigma \langle a_j^\dagger a_k \rangle_\sigma, \tag{12}$$

where the expectation values on the right-hand side are given by Eq. (11).

2.2. *Particle-number Projection*

The nucleus is a finite-size system, and it is important to consider the canonical ensemble with fixed numbers of protons and neutrons. We are therefore interested in canonical expectation values. Such quantities can be calculated using an exact particle-number projection. Since the number N_s of single-particle orbitals is finite, we can describe the particle-number projection by a discrete Fourier transform. The canonical partition function of $\hat{U}_\sigma$ for particle number $\mathcal{A}$ is given by Ref. [29]

$$\mathrm{Tr}_{\mathcal{A}} U_\sigma = \frac{e^{-\beta\mu\mathcal{A}}}{N_s} \sum_{m=1}^{N_s} e^{-\mathrm{i}\varphi_m \mathcal{A}} \det\left(\mathbf{1} + e^{\mathrm{i}\varphi_m} e^{\beta\mu} \mathbf{U}_\sigma\right), \qquad (13)$$

where $\varphi_m = 2\pi m/N_s \quad (m = 1, \ldots, N_s)$ are quadrature points and μ is a chemical potential introduced to stabilize the numerical evaluation of the Fourier sum. Similarly for a one-body observable $\hat{O} = \sum_{i,j} O_{i,j} a_i^\dagger a_j$

$$\mathrm{Tr}_{\mathcal{A}}(\hat{O}\hat{U}_\sigma) = \frac{e^{-\beta\mu\mathcal{A}}}{N_s} \sum_{m=1}^{N_s} e^{-\mathrm{i}\varphi_m \mathcal{A}} \det(\mathbf{1} + e^{\mathrm{i}\varphi_m + \beta\mu} \mathbf{U}_\sigma) \times \mathrm{tr}\left(\frac{1}{\mathbf{1} + e^{-\mathrm{i}\varphi_m - \beta\mu} \mathbf{U}_\sigma^{-1}} \mathbf{O}\right), \qquad (14)$$

where $\mathbf{O}$ is the matrix with elements O_{ij}. In the actual nuclear calculations we project on both neutron number N and proton number Z.

2.3. *Monte Carlo Sampling*

The integrands in Eq. (9) are calculated by matrix algebra in the single-particle space [see, e.g., Eqs. (10) and (11)]. However, the number of integration variables $\sigma_\alpha(\tau)$ is very large. For small but finite $\Delta\beta$, this multi-dimensional integral can be evaluated exactly (up to a statistical error) by Monte Carlo methods.

In the applications of AFMC to nuclei we carry out the Monte Carlo sampling in the canonical ensemble. For a nucleus of $\mathcal{A}$ nucleons we define the positive-definite weight function

$$W_\sigma \equiv G_\sigma |\mathrm{Tr}_{\mathcal{A}}\, \hat{U}_\sigma|. \tag{15}$$

Next we define the W-weighted average of a quantity X_σ that depends on the auxiliary field configuration σ by

$$\overline{X_\sigma} \equiv \frac{\int D[\sigma] W_\sigma X_\sigma \Phi_\sigma}{\int D[\sigma] W_\sigma \Phi_\sigma}, \tag{16}$$

where

$$\Phi_\sigma \equiv \mathrm{Tr}_{\mathcal{A}}\, U_\sigma / |\mathrm{Tr}_{\mathcal{A}}\, U_\sigma| \tag{17}$$

is the Monte Carlo sign. With this definition, the canonical thermal expectation of an observable $\hat{O}$ can be written as

$$\langle \hat{O} \rangle = \overline{\mathrm{Tr}_{\mathcal{A}}(\hat{O}\hat{U}_\sigma)/\mathrm{Tr}_{\mathcal{A}}\hat{U}_\sigma}. \tag{18}$$

In AFMC, a random walk is performed in the space of auxiliary fields $\sigma \equiv \{\sigma_\alpha(\tau_m)\}$ that samples the σ-fields according to the positive-definite distribution W_σ. The average $\overline{X_\sigma}$ is then estimated from

$$\overline{X_\sigma} \approx \frac{\sum_k X_{\sigma^{(k)}} \Phi_{\sigma^{(k)}}}{\sum_k \Phi_{\sigma^{(k)}}}, \tag{19}$$

where $\sigma^{(k)}$ are M uncorrelated samples. The statistical error of $\overline{X_\sigma}$ can be estimated from the variance of the "measurements" $X_{\sigma^{(k)}}$. Though a standard random walk can be constructed by the Metropolis algorithm, a modification based on Gaussian quadratures improves its efficiency [31].

2.4. *Sign Problem and Good-sign Interactions*

Assuming the Hamiltonian in Eq. (1) is time-reversal invariant, it can be rewritten in the form

$$\hat{H} = \sum_i \epsilon_i \hat{n}_i + \frac{1}{2} \sum_\alpha \tilde{v}_\alpha (\rho_\alpha \bar{\rho}_\alpha + \bar{\rho}_\alpha \rho_\alpha), \tag{20}$$

where $\bar{\rho}_\alpha$ is the time-reverse density of ρ_α and $\tilde{v}_\alpha$ are real. When all the interaction eigenvalues $\tilde{v}_\alpha$ in the representation (20) are *negative*, the grand canonical one-body partition function $\mathrm{Tr}\,\hat{U}_\sigma$ is positive for any sample σ. Such interactions are known as good-sign interactions in AFMC.

To prove the above sign rule, we consider the one-body Hamiltonian that appears in the HS decomposition for the ensemble described by the Hamiltonian (20)

$$\hat{h}_\sigma = \sum_i \epsilon_i \hat{n}_i + \sum_\alpha (\tilde{v}_\alpha s_\alpha \sigma^*_\alpha \rho_\alpha + \tilde{v}_\alpha s_\alpha \sigma_\alpha \bar{\rho}_\alpha). \tag{21}$$

When all $\tilde{v}_\alpha < 0$, $s_\alpha = 1$ for all α and the one-body Hamiltonian (21) is invariant under time reversal, i.e., $\bar{h}_\sigma = h_\sigma$. Since the spins of the single-particle states are half integers, it follows that the eigenstates of the propagator matrix $\mathbf{U}_\sigma$ appear in time-reversed pairs with complex conjugate eigenvalues $\{\lambda_i, \lambda_i^*\}$. The grand canonical partition of $\hat{U}_\sigma$ in Eq. (10) can then be written as

$$\mathrm{Tr}\,\hat{U}_\sigma = \prod_i |1+\lambda_i|^2 \tag{22}$$

and is positive for any configuration σ of the auxiliary fields.

The dominant collective components of effective nuclear interactions are attractive [18], and in our calculations of statistical and collective properties of nuclei discussed here we used good-sign interactions (see Secs. 5.1 and 6.1). Small bad-sign components of realistic effective nuclear interactions can be treated following the extrapolation method of Ref. [15].

2.5. *Circumventing the Odd Particle-number Sign Problem*

For a good-sign interaction, the particle-number projected partition $\mathrm{Tr}_{\mathcal{A}}\hat{U}_\sigma$ for an even number of particles $\mathcal{A}$ remains almost always positive and the particle-projected Monte Carlo sign is good for an even–even nucleus. However, for an odd number of particles, the projected partition can be negative for certain field configurations ($\mathrm{Tr}\,\hat{U}_\sigma$ is always real for a good-sign interaction). The average sign $\langle\Phi_\sigma\rangle$ is then smaller than 1 and it decreases with increasing values of β. At low temperatures, this so-called odd particle-number sign problem becomes severe. Consequently, it is difficult to determine an accurate ground-state energy for odd-mass and odd–odd nuclei.

We introduced a method [30] to circumvent the odd particle-number sign problem by using the imaginary-time single-particle Green's functions of the even-particle system to determine an accurate ground-state energy of the odd-particle system.

The CI shell-model Hamiltonian is rotationally invariant, and the single-particle orbitals $i = (\nu m)$ with $\nu = (nlj)$ are characterized by a principal quantum number n, orbital angular momentum l, total angular momentum j and its z projection m. For any ν, we define the scalar Green's function

$$G_\nu(\tau) = \frac{\mathrm{Tr}_{\mathcal{A}}\left[e^{-\beta\hat{H}}\mathcal{T}\sum_m a_{\nu m}(\tau)a^\dagger_{\nu m}(0)\right]}{\mathrm{Tr}_{\mathcal{A}}\; e^{-\beta\hat{H}}}, \tag{23}$$

where $\mathcal{T}$ denotes time ordering and $a_{\nu m}(\tau) \equiv e^{\tau\hat{H}}a_{\nu m}e^{-\tau\hat{H}}$ is an annihilation operator of a particle at imaginary time τ $(-\beta \leq \tau \leq \beta)$ in an orbital $i = (\nu m)$. Using the HS representation and the notation of Eq. (16), we obtain

$$G_\nu(\tau) = \begin{cases} \overline{\sum_m [\mathbf{U}_\sigma(\tau)(\mathbf{1} - \langle\hat{\rho}\rangle_\sigma)]_{\nu m,\nu m}} & \text{for } \tau > 0 \\ \\ \overline{\sum_m [\langle\hat{\rho}\rangle_\sigma \mathbf{U}_\sigma^{-1}(|\tau|)]_{\nu m,\nu m}} & \text{for } \tau \leq 0 \end{cases}. \tag{24}$$

Here, $\mathbf{1}$ is the identity matrix in the single-particle space, and $\langle\hat{\rho}\rangle_\sigma$ is a matrix whose $\nu m, \nu' m'$ matrix element is $\langle\hat{\rho}_{\nu m,\nu' m'}\rangle_\sigma$ where $\hat{\rho}_{\nu m,\nu' m'} = a^\dagger_{\nu' m'}a_{\nu m}$.

Consider an even–even nucleus $\mathcal{A} \equiv (Z, N)$. Assuming that the ground state of this nucleus has zero spin, the Green's functions behave asymptotically in τ as $G_\nu(\tau) \sim e^{-\Delta E_{J=j}(\mathcal{A}_\pm)|\tau|}$ [$\mathcal{A}_\pm$ denotes the even–odd nuclei $(Z, N \pm 1)$ when ν is a neutron orbital, and the odd-even nuclei $(Z \pm 1, N)$ when ν is a proton orbital]. The $+$ and $-$ correspond, respectively, to $\tau > 0$ and $\tau \leq 0$, and $\Delta E_{J=j}(\mathcal{A}_\pm)$ is the difference between the energies of the lowest spin J eigenstate of the $\mathcal{A}_\pm$-particle nucleus and the ground state of the $\mathcal{A}$-particle nucleus. In this asymptotic regime we determine $\Delta E_j(\mathcal{A}_\pm)$ from the

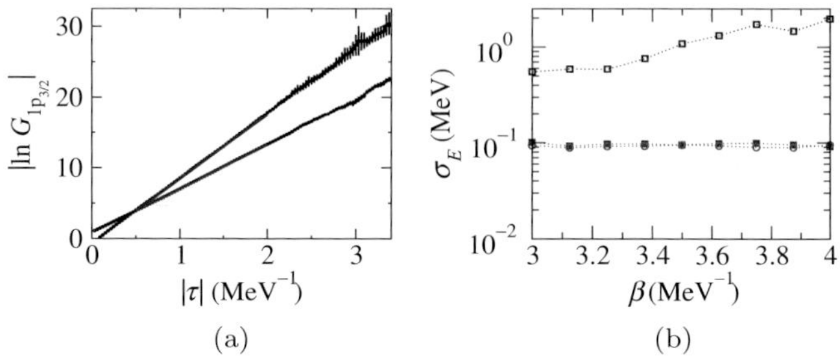

Fig. 1. (a) Extracting the ground-state energy of the odd–even nucleus ^{57}Fe in the imaginary time Green's function method. The absolute value of logarithm of the AFMC Green's function for the neutron orbital $\nu = 1p_{3/2}$ in ^{56}Fe (lower curve) and ^{58}Fe (upper curve) vs. $|\tau|$ at $\beta = 4$ MeV^{-1}. The solid lines are linear fits for 0.5 MeV$^{-1} \leq |\tau| \leq 2$ MeV^{-1}. (b) The statistical errors for the energy of ^{57}Fe in the Green's function method (solid squares) are compared with statistical errors in direct AFMC calculations (open squares). A logarithmic scale is used for the statistical errors. We also show for comparison the statistical errors associated with the energy of the even–even ^{56}Fe nucleus (open circles, almost on top of the solid squares). Adapted from Ref. [30].

slope of $\ln G_\nu(\tau)$. We then minimize $\Delta E_j(\mathcal{A}_\pm)$ over all possible values of j to find the difference between the ground-state energy of the $\mathcal{A}_\pm$ nuclei and the ground-state energy of the $\mathcal{A}$ nucleus, $E_{\rm gs}(\mathcal{A})$. Since $E_{\rm gs}(\mathcal{A})$ and $G_\nu(\tau)$ characterize the even–even nucleus, they can be calculated in AFMC without a sign problem (for a good-sign interaction).

We demonstrate the method for calculating the ground-state energy of ^{57}Fe in Fig. 1 (the model space and interaction used are discussed in Sec. 5). The left panel shows the absolute value of the logarithm of the Green's function for the neutron $1p_{3/2}$ orbital vs. $|\tau|$ for the even–even nuclei ^{56}Fe and ^{58}Fe. The right panel shows that the statistical errors of the ground-state energy of ^{57}Fe extracted from the Green's function method are much smaller than the statistical errors in direct AFMC calculations.

3. Projection Methods

The grand canonical traces we take in, e.g., Eq. (9), are over the complete Fock space, in which case we can use Eqs. (10) and (11).

Canonical traces in the finite nucleus can be calculated via particle-number projection; see Eqs. (13) and (14).

The calculation of thermal observables at given values of good quantum numbers such as spin and parity, requires additional projections. Parity projection in AFMC was discussed in Refs. [32–35]. In Sec. 3.1 we discuss spin projection [36] in AFMC. For isospin projection in AFMC see Ref. [37].

It is also possible to project on observables that do not commute with the Hamiltonian. An example is the projection on the axial quadrupole operator [38] discussed in Sec. 3.2.

3.1. *Spin Projection*

We first discuss the projection on the z component $\hat{J}_z$ of the spin operator [36]. We define the projected partition function for an eigenvalue M of $\hat{J}_z$ to be $Z_M(\beta) = \text{Tr}_M e^{-\beta \hat{H}}$. Using the HS transformation and the notation of Eq. (16), we have

$$\frac{Z_M(\beta)}{Z(\beta)} = \overline{\frac{\text{Tr}_M \hat{U}_\sigma}{\text{Tr} \hat{U}_\sigma}}. \tag{25}$$

The M-projected partition $\text{Tr}_M \hat{U}_\sigma$ can be calculated using the Fourier representation of $\delta(\hat{J}_z - M)$

$$\text{Tr}_M\, \hat{U}_\sigma = \frac{1}{2J_s + 1} \sum_{k=-J_s}^{J_s} e^{-\mathrm{i}\varphi_k M} \text{Tr}\, (e^{\mathrm{i}\varphi_k \hat{J}_z} \hat{U}_\sigma), \tag{26}$$

where $\varphi_k = \pi k/(J_s + 1/2)$ for $k = -J_s, \ldots, J_s$. Since $\hat{J}_z$ is a one-body operator, we can use the group property (see Sec. 2.1.1) to represent $e^{\mathrm{i}\varphi_k \hat{J}_z} \hat{U}_\sigma$ in the single-particle space by the matrix $e^{\mathrm{i}\varphi_k \mathbf{J}_z} \mathbf{U}_\sigma$ ($\mathbf{J}_z$ is a diagonal $N_s \times N_s$ matrix with the magnetic quantum numbers of the corresponding orbitals along its diagonal). The grand canonical trace is then given by

$$\text{Tr}\, (e^{\mathrm{i}\varphi_k \hat{J}_z} \hat{U}_\sigma) = \det(1 + e^{\mathrm{i}\varphi_k \mathbf{J}_z} \mathbf{U}_\sigma). \tag{27}$$

In practice, we also project on fixed numbers of protons and neutrons.

The spin-projected partition function $Z_J(\beta) = \mathrm{Tr}_J e^{-\beta\hat{H}}$, is calculated using the identity $\mathrm{Tr}_J e^{-\beta\hat{H}} = \mathrm{Tr}_{M=J} e^{-\beta\hat{H}} - \mathrm{Tr}_{M=J+1} e^{-\beta\hat{H}}$ (valid since $e^{-\beta\hat{H}}$ is a rotationally invariant operator). Using the HS transformation, we find

$$\frac{Z_J(\beta)}{Z(\beta)} = \overline{\frac{\mathrm{Tr}_{M=J}\hat{U}_\sigma}{\mathrm{Tr}\hat{U}_\sigma} - \frac{\mathrm{Tr}_{M=J+1}\hat{U}_\sigma}{\mathrm{Tr}\hat{U}_\sigma}}. \tag{28}$$

We can similarly calculate the spin-projected energies, $E_M(\beta)$ and $E_J(\beta)$.

3.2. *Axial Quadrupole Projection*

Here the observable of interest, the axial quadrupole operator $\hat{Q}_{20} = \sum_i (2z_i^2 - x_i^2 - y_i^2)$, does not commute with the Hamiltonian, i.e., $[\hat{H}, \hat{Q}_{20}] \neq 0$. Its distribution $P_T(q)$ is given by Ref. [38]

$$P_T(q) = \sum_n \delta(q - q_n) \sum_m \langle q, n|e, m\rangle^2 e^{-\beta e_m}, \tag{29}$$

where $|q, n\rangle$ are eigenstates of $\hat{Q}_{20}$ satisfying $\hat{Q}_{20}|q, n\rangle = q_n|q, n\rangle$ and similarly $|e, m\rangle$ are eigenstates of $\hat{H}$ with $\hat{H}|e, m\rangle = e_m|e, m\rangle$. We note that in a finite model space the spectrum of $\hat{Q}_{20}$ is discrete.

$P_T(q)$ can be calculated in AFMC by a projection on $\hat{Q}_{20}$ using the Fourier transform of the Dirac δ function

$$P_T(q) \equiv \frac{\mathrm{Tr}\,[\delta(\hat{Q}_{20} - q)e^{-\beta\hat{H}}]}{\mathrm{Tr}\,e^{-\beta\hat{H}}} = \frac{\int_{-\infty}^{\infty} \frac{d\varphi}{2\pi} e^{-\mathrm{i}\varphi q}\,\mathrm{Tr}\,(e^{\mathrm{i}\varphi\hat{Q}_{20}} e^{-\beta\hat{H}})}{\mathrm{Tr}\,e^{-\beta\hat{H}}} \tag{30}$$

together with the HS representation (4) of $e^{-\beta\hat{H}}$. In practice, we divide an interval $[-q_{\max}, q_{\max}]$ into $2M+1$ intervals of length $\Delta q = 2q_{\max}/(2M+1)$ and use a discrete Fourier representation for each sample σ

$$\mathrm{Tr}[\delta(\hat{Q}_{20} - q_m)\hat{U}_\sigma] \approx \frac{1}{2q_{\max}} \sum_{k=-M}^{M} e^{-\mathrm{i}\varphi_k q_m} \mathrm{Tr}(e^{\mathrm{i}\varphi_k \hat{Q}_{20}} \hat{U}_\sigma), \tag{31}$$

where $q_m = m\Delta q$ ($m = -M, \ldots, M$) and $\varphi_k = \pi k/q_{\max}$ ($k = -M, \ldots, M$). Since $\hat{Q}_{20}$ is a one-body operator, we have

$$\mathrm{Tr}(e^{\mathrm{i}\varphi_k \hat{Q}_{20}} \hat{U}_\sigma) = \det\left(\mathbf{1} + e^{\mathrm{i}\varphi_k \mathbf{Q}_{20}} \mathbf{U}_\sigma\right), \tag{32}$$

where $\mathbf{Q}_{20}$ is the matrix representing $\hat{Q}_{20}$ in the single-particle space.

4. State Densities

The nuclear state density is among the most important statistical nuclear properties. It is an integral part of the Hauser–Feshbach theory [39] of statistical nuclear reactions and appears in the Fermi golden rule for transition rates. However, the calculation of the state density in the presence of correlations is a challenging many-body problem, and most calculations are based on mean-field approximations such as the Hartree–Fock (HF) and the Hartree–Fock–Bogoliubov (HFB) approximations [40, 41]. AFMC offers a state-of-the-art method for calculating state densities beyond the mean field in very large model spaces that are required at finite excitation energies [32, 42–47].

In AFMC, we calculate the canonical thermal energy as the expectation value of the Hamiltonian, $E(\beta) = \langle H \rangle$. The canonical partition function $Z(\beta)$ can then be calculated by integrating the thermodynamic identity $-\partial \ln Z/\partial\beta = E(\beta)$. We find

$$\ln Z(\beta) = \ln Z(0) - \int_0^\beta E(\beta) d\beta, \tag{33}$$

where $Z(0)$ is the total number of many-particle states with Z protons and N neutrons in the model space. The state density $\rho(E)$ at energy E is related to the partition function by an inverse Laplace transform

$$\rho(E) = \frac{1}{2\pi \mathrm{i}} \int_{-\mathrm{i}\infty}^{\mathrm{i}\infty} d\beta\, e^{\beta E} Z(\beta). \tag{34}$$

The average state density is obtained by evaluating Eq. (34) in the saddle-point approximation [48]

$$\rho(E) \approx (2\pi T^2 C)^{-1/2} e^{S(E)}, \tag{35}$$

where $S(E)$ is the canonical entropy and C is the canonical heat capacity given by

$$S = \ln Z + \beta E; \quad C = \frac{dE}{dT}. \tag{36}$$

The value of β used in Eqs. (35) and (36) is determined by the saddle-point condition $E = -\partial \ln Z/\partial\beta = E(\beta)$.

Similar formulas apply for the calculation of the level density at a given value of a good quantum number by starting from the corresponding projected thermal energy as a function of β. An example is the calculation of the spin-dependent level densities $\rho_J(E)$ by using the spin-projected energies $E_J(\beta)$ discussed in Sec. 3.1

5. Applications to Mid-mass Nuclei

We first discuss applications of AFMC to mid-mass nuclei. In Sec. 5.1 we describe the model space and the interaction used. In Sec. 5.2 we discuss a recent calculation of level densities in a family of nickel isotopes. In Sec. 5.3 we use the spin-projection method of Sec. 3.1 to calculate the spin distributions for several nuclei in the iron region, and in Sec. 5.4 we present results for pairing gaps in families of mid-mass isotopes calculated from odd–even mass differences.

5.1. *Model Space and Interaction*

We carried out AFMC studies of mid-mass nuclei using the $fpg_{9/2}$ shell for both protons and neutrons. The Hamiltonian used is an isoscalar. The single-particle energies are determined from a Woods–Saxon (WS) potential plus spin-orbit interaction with the parameters in Ref. [48]. The interaction is given by [32]

$$-g\hat{P}^{(0,1)\dagger} \cdot \hat{\tilde{P}}^{(0,1)} - \chi \sum_\lambda k_\lambda \hat{O}^{(\lambda,0)} \cdot \hat{O}^{(\lambda,0)}, \tag{37}$$

where

$$\hat{P}^{(\lambda,T)\dagger} = \frac{\sqrt{4\pi}}{2(2\lambda+1)} \sum_{ab} \langle j_a \| Y_\lambda \| j_b \rangle [a^\dagger_{j_a} \times a^\dagger_{j_b}]^{(\lambda,T)},$$

$$\hat{O}^{(\lambda,T)} = \frac{1}{\sqrt{2\lambda+1}} \sum_{ab} \left\langle j_a \left\| \frac{dV}{dr} Y_\lambda \right\| j_b \right\rangle [a^\dagger_{j_a} \times \tilde{a}_{j_b}]^{(\lambda,T)} \tag{38}$$

and $(\cdot)$ denotes a scalar product in both spin (λ) and isospin (T). The modified annihilation operator $\tilde{a}$ is defined by $\tilde{a}_{j,m,m_t} = (-)^{j-m+\frac{1}{2}-m_t} a_{j,-m,-m_t}$, and $\hat{\tilde{P}}^{(\lambda,T)}$ is similarly defined. V in Eq. (38) is the central part of the single-particle potential, and the three multipole interaction terms ($\lambda = 2, 3, 4$) are obtained by expanding the separable surface-peaked interaction $v(\mathbf{r}, \mathbf{r}') = -\chi(dV/dr)(dV/dr')\delta(\hat{\mathbf{r}} - \hat{\mathbf{r}}')$. The interaction strength χ is determined by a self-consistency condition [49]. Core polarization effects are taken into account by renormalizing χ with the factors k_λ. We use $k_2 = 2$, $k_3 = 1.5$, and $k_4 = 1$, in overall agreement with realistic effective nuclear interactions in this shell. The pairing strength g is determined so as to reproduce in particle-number projected BCS the pairing gaps determined from experimental odd–even mass differences for spherical nuclei in the mass region $A = 40 - 80$ with $Z = 20$, $N = 28$, $Z = 28$, or $N = 40$. We determined a constant mean value of $g = 0.212$ MeV.

5.2. *Level Densities in Nickel Isotopes*

The density ρ discussed in Sec. 4 is the *state* density where the magnetic degeneracy $2J + 1$ of levels with spin J is included in the counting of states. However, the density measured in the experiments is often the *level* density $\tilde{\rho}$, where each level is counted just once, irrespective of its magnetic degeneracy. In Ref. [50] we showed that it is possible to calculate directly the level density in AFMC by using

$$\tilde{\rho} = \begin{cases} \rho_{M=0} & \text{even-mass nucleus} \\ \rho_{M=1/2} & \text{odd-mass nucleus} \end{cases}, \tag{39}$$

where ρ_M is the $\hat{J}_z$-projected density (see Sec. 3.1). This density can be calculated as in Sec. 4 by replacing $E(\beta)$ with $E_M(\beta)$.

In Fig. 2 we show the AFMC level densities (solid circles) for a family of nickel isotopes $^{59-64}$Ni as a function of excitation energy [51] calculated using Eq. (39). Accurate ground-state energies for the odd-mass nickel isotopes were determined by the Green's function method discussed in Sec. 2.5. The calculated level densities are in good agreement with experimental data: proton evaporation spectra

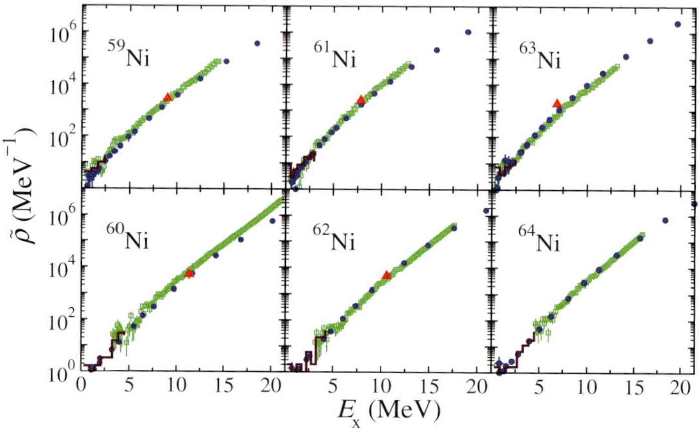

Fig. 2. Level densities $\tilde{\rho}$ in a family of nickel isotopes $^{59-64}$Ni: theory vs. experiment. The AFMC level densities (solid circles) are compared with level counting at low excitation energies (histograms), neutron resonance data (triangles) [53] when available, and level densities extracted from proton evaporation spectra (squares forming quasi-continuous lines) [52]. Taken from Ref. [51].

(squares forming quasi-continuous lines) [52], level counting data (histograms) at low excitation energies and neutron resonance data (triangles) [53] when available.

5.3. *Spin Distributions*

We used the spin projection method of Sec. 3.1 to calculate the spin-projected energies $E_J(\beta)$ as a function of β. We can then use the saddle-point approximation of Sec. 4 to calculate the spin-dependent level densities $\rho_J(E_x)$ as a function of excitation energy E_x.

In a statistical model in which the spins of individual nucleons are coupled randomly to total spin $\mathbf{J}$ the spin distribution ρ_J/ρ follows the spin cutoff model [54]

$$\frac{\rho_J(E_x)}{\rho(E_x)} = \frac{(2J+1)}{2\sqrt{2\pi}\sigma^3} e^{-\frac{J(J+1)}{2\sigma^2}}, \tag{40}$$

where the parameter $\sigma = \sigma(E_x)$ is known as the spin cutoff parameter. The spin-projected density $\rho_J(E_x)$ in Eq. (40) does not include the $2J+1$ magnetic degeneracy and is normalized by

$\sum_J (2J+1)\rho_J(E_x) \approx \rho(E_x)$. The spin cutoff parameter is related to the thermal moment of inertia $\mathcal{I}$ (at temperature T) by

$$\sigma^2 = \frac{\mathcal{I}T}{\hbar^2}. \tag{41}$$

In our AFMC studies of mid-mass nuclei we found that the spin cutoff formula (40) works well at higher excitation energies with a rigid-body moment of inertia. However, in even–even nuclei we found [36] at low excitations an odd–even staggering effect in J. Also, we observed in such even–even nuclei a strong suppression of the moment of inertia at low excitations, an effect associated with pairing correlations.

In Fig. 3 we show the spin distribution ρ_J/ρ for the odd–even nucleus ^{55}Fe, the even–even nucleus ^{56}Fe and the odd–odd nucleus ^{60}Co at the excitation energies specified in the figure. The solid squares with statistical errors are the AFMC results of Ref. [36]. The solid lines describe empirical distributions determined from experimentally known low-lying levels [55, 56]. The dashed lines describe similar empirical curves but with larger values of the spin

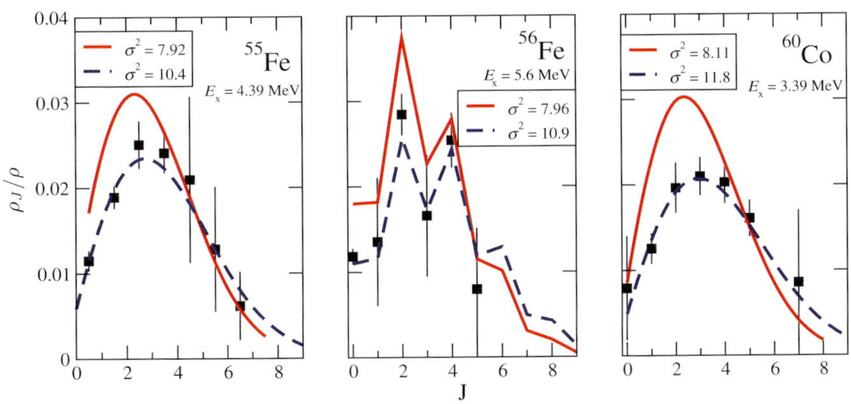

Fig. 3. Spin distributions ρ_J/ρ as a function of spin J for ^{55}Fe (left panel), ^{56}Fe (middle panel) and ^{60}Co (right panel) at excitation energies E_x. The AFMC distributions, taken from Ref. [36], are shown by solid squares with statistical errors. The solid lines are empirical results [55] based on systematic studies of low-lying experimental levels and the dashed lines describe similar distributions but with larger values of the spin cutoff parameter σ. Taken from Ref. [55].

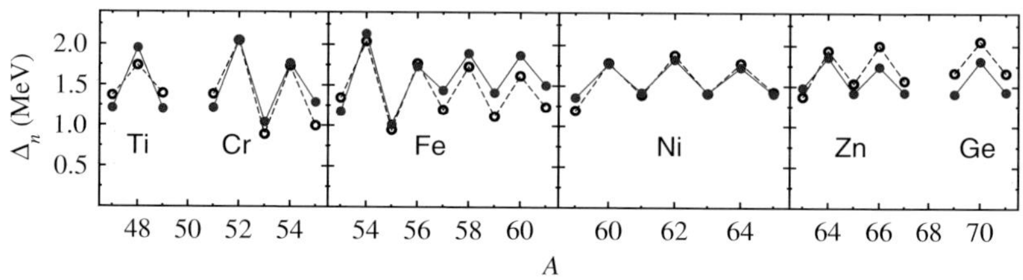

Fig. 4. Neutron pairing gaps Δ_n from odd–even mass differences in families of isotopes in the iron region. The AFMC results (solid circles) are compared with experimental values (open circles). Adapted from Ref. [30].

cutoff parameter to account for the higher excitation energies used in the calculations. The empirical distributions are given by the spin cutoff model for the odd–even ^{55}Fe and the odd–odd ^{60}Co nuclei. The staggering in the even–even ^{56}Fe nucleus is described empirically by a spin cutoff formula multiplied by a factor $1 + x$ where $x \approx 0.23$ for even spin values, $x \approx -0.23$ for odd spin values, and $x \approx 1.02$ for $J = 0$. This empirical staggering is in good agreement with the AFMC predictions of Ref. [36].

5.4. *Pairing Gaps*

Pairing gaps can be calculated from odd-even mass differences [48]. Using the method of Sec. 2.5 to circumvent the odd-particle sign problem, we calculated ground-state energies of odd-mass nuclei. We can then find accurate neutron pairing gaps from the second-order difference in the ground-state energy as a function of the number of neutrons. In Fig. 4 we show neutron pairing gaps Δ_n in families of isotopes in the iron region. The AFMC gaps are compared with gaps determined from the experiments.

6. Applications to Heavy Nuclei

Here we discuss recent applications of AFMC to heavy rare-earth nuclei. In Sec. 6.1 we present the model space and interaction used. In Sec. 6.2 we demonstrate the emergence of various types of collectivity in heavy nuclei using the framework of the CI shell model. In Sec. 6.3 we present results for the AFMC state densities in families of samarium and neodymium isotopes. Finally, in Sec. 6.4 we discuss the finite-temperature distributions of the quadrupole deformation using the rotationally invariant framework of the CI shell model.

6.1. *Model Space and Interaction*

The AFMC method was extended to heavy nuclei using a proton–neutron formalism, in which protons and neutrons can occupy different shells [46]. In particular, we applied the method to the lanthanides. The single-particle model space we used is composed of the orbitals $0g_{7/2}, 1d_{5/2}, 1d_{3/2}, 2s_{1/2}, 0h_{11/2}, 1f_{7/2}$ for protons and $0h_{11/2}, 0h_{9/2}, 1f_{7/2}, 1f_{5/2}, 2p_{3/2}, 2p_{1/2}, 0i_{13/2}, 1g_{9/2}$ for neutrons. The total number of single-particle states is 40 for protons and 66 for neutrons. The single-particle levels and wave functions are determined from a central WS potential plus spin-orbit interaction. The interaction we used is similar to the one used in Sec. 5.1 and for the proton–neutron formalism it is given by

$$- \sum_{\nu=p,n} g_\nu \hat{P}_\nu^\dagger \hat{P}_\nu - \sum_\lambda \chi_\lambda : (\hat{O}_{\lambda;p} + \hat{O}_{\lambda;n}) \cdot (\hat{O}_{\lambda;p} + \hat{O}_{\lambda;n}) : , \qquad (42)$$

where $\hat{P}_\nu^\dagger = \sum_{nljm} (-)^{j+m+l} a^\dagger_{\alpha jm;\nu} a^\dagger_{\alpha j-m;\nu}$ ($\nu = p, n$ and $\alpha = nl$) is the $J = 0$ pair creation operator, :: denotes normal ordering, and $\hat{O}_{\lambda;\nu} = \frac{1}{\sqrt{2\lambda+1}} \sum_{ab} \langle j_a || \frac{dV_{\rm WS}}{dr} Y_\lambda || j_b \rangle [a^\dagger_{\alpha j_a;\nu} \times \tilde{a}_{\alpha j_b;\nu}]^{(\lambda)}$ is a surface-peaked multipole operator [$\tilde{a}_{jm} = (-1)^{j-m} a_{j-m}$]. We include quadrupole, octupole and hexadecupole terms with corresponding strengths $\chi_\lambda = \chi k_\lambda$. The parameter χ is determined self-consistently [49] and k_λ are renormalization factors that account for core polarization effects. The values used for g_ν and k_λ are given in Refs. [46,47].

The extension to heavy nuclei had required overcoming a number of technical challenges. A typical excitation energy in even–even rare-earth nuclei ($\sim$ 100 keV) is an order of magnitude smaller than a typical excitation energy in even–even mid-mass nuclei. It is then necessary to carry out the calculations to much lower temperatures to reach the ground-state energy. The larger values of β and the larger band width of the single-particle spectrum require more computationally intensive calculations and lead to ill-conditioned matrices $\mathbf{U}_\sigma$. Stabilization methods were introduced in strongly correlated electron systems in the grand canonical ensemble [19, 20] and we extended them to the canonical ensemble [46]. Since there are N_s terms in the Fourier sum of the canonical projection (see Sec. 2.2), the computational cost scales as N_s^4. We recently introduced a novel stabilization method [57] for the canonical ensemble that reduces this scaling to N_s^3. The method was used in AFMC studies of cold atoms [24] but it is also useful in the application of AFMC to nuclei.

6.2. *Emergence of Collectivity in the Configuration-interaction Shell-model Approach*

Heavy nuclei are known to exhibit various types of collectivity that are well described by empirical models. However, a microscopic description in the framework of the CI shell model has been mostly lacking. A particularly important question is whether rotational collectivity, typical of deformed nuclei, can be described within a truncated spherical shell-model approach. Various types of collectivity are usually identified by their energy level schemes. However, while AFMC enables CI shell-model studies of heavy nuclei in very large model spaces, it is difficult to use for extracting detailed spectroscopic information. In Ref. [46] we identified a thermal observable, $\langle \hat{\mathbf{J}}^2 \rangle_T$ ($\hat{\mathbf{J}}$ is the total angular momentum of the nucleus), whose low-temperature behavior distinguishes between different types of collectivity. Assuming an even–even nucleus with either a vibrational or rotational ground-state band, we find

$$\langle \hat{\mathbf{J}}^2 \rangle_T \approx \begin{cases} 30 \dfrac{e^{-E_{2^+}/T}}{\left(1 - e^{-E_{2^+}/T}\right)^2} & \text{vibrational band} \\[2ex] \dfrac{6}{E_{2^+}} T & \text{rotational band,} \end{cases} \tag{43}$$

where E_{2^+} is the excitation energy of the lowest 2^+ level and T is temperature. In Fig. 5 we show the low-temperature behavior of $\langle \hat{\mathbf{J}}^2 \rangle_T$ for ^{162}Dy (left panel) and ^{148}Sm (right panel). The AFMC results (circles) are compared with the best fit (using the 2^+

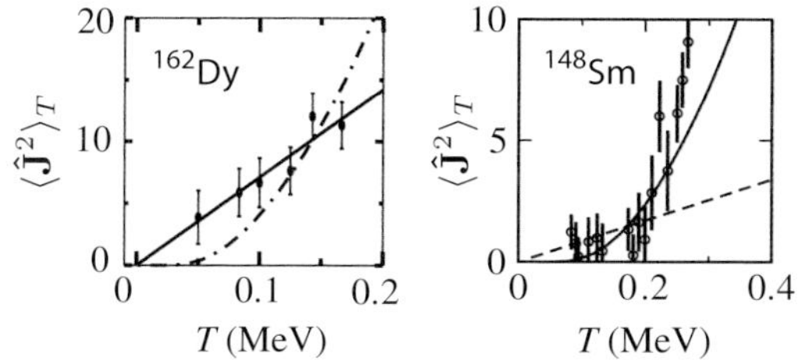

Fig. 5. (Left) $\langle \hat{\mathbf{J}}^2 \rangle_T$ vs. temperature T in ^{162}Dy. The AFMC results (solid circles) are well described by a fit to the rotational band model (solid line). The dashed-dotted line is a fit to the vibrational model. Adapted from Ref. [46]. (Right) $\langle \hat{\mathbf{J}}^2 \rangle_T$ vs. temperature T in ^{148}Sm. The SMMC results (open circles) are in better agreement with the vibrational model (solid line) than with the rotational model (dashed line). Taken from Ref. [58].

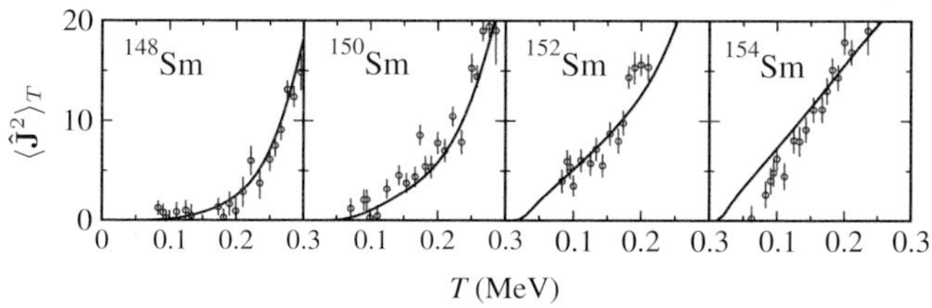

Fig. 6. $\langle \hat{\mathbf{J}}^2 \rangle_T$ vs. temperature T for the even-mass samarium isotopes $^{148-154}$Sm. The AFMC results (open circles with statistical error bars) are compared with curves that are calculate from experimental data as discussed in the text (solid lines). Adapted from Ref. [47].

excitation energy as the fit parameter) to the formulas in Eq. (43) for a nucleus with a rotational or vibrational band structure. We observe that in ^{162}Dy, the AFMC results for $\langle \hat{\mathbf{J}}^2 \rangle_T$ are in agreement with a straight line characterizing the rigid response to temperature in a rotational nucleus, while in ^{148}Sm the AFMC results are in agreement with a softer response to temperature describing a vibrational nucleus.

6.2.1. *Crossover from vibrational to rotational collectivity*

The observable $\langle \hat{\mathbf{J}}^2 \rangle_T$ can also be used to describe the crossover from vibrational to rotational collectivity. This is demonstrated in Fig. 6 for a family of even–even samarium isotopes $^{148-154}$Sm. The AFMC results for $\langle \hat{\mathbf{J}}^2 \rangle_T$ (open circles with statistical errors) are compared with values deduced from the experiments (solid lines). We observe a crossover from a soft response to temperature in the vibrational nucleus ^{148}Sm to a rigid response in the rotational nucleus ^{154}Sm. The AFMC results are in overall agreement with the experimentally deduced curves. The latter are calculated from

$$\langle \hat{\mathbf{J}}^2 \rangle_T = \frac{1}{Z(T)} \left(\sum_i^N J_i(J_i+1)(2J_i+1)e^{-E_i/T} \right. \\ \left. + \int_{E_N}^{\infty} dE_x \, \rho(E_x) \, \langle \hat{\mathbf{J}}^2 \rangle_{E_x} \, e^{-E_x/T} \right), \tag{44}$$

where the partition function $Z(T)$ is

$$Z(T) = \sum_i^N (2J_i+1)e^{-E_i/T} + \int_{E_N}^{\infty} dE_x \rho(E_x) e^{-E_x/T}. \tag{45}$$

The summations in Eqs. (44) and (45) are over a set of experimental levels with excitation energies E_i and spins J_i that is complete up to an energy E_N. Above E_N the summation is replaced by an integral over a back-shifted Fermi gas formula whose parameters are determined empirically from level counting at low excitation energies and the neutron resonance data at the neutron separation energy.

6.3. *State Densities*

We calculated the state densities of families of even–even samarium and neodymium isotopes using the method of Sec. 4. Figure. 7 shows the AFMC state densities (open circles) of even mass $^{148-154}$Sm and $^{144-152}$Nd isotopes. The calculated densities are compared with level counting data at low excitation energies (histograms) and neutron resonance data (triangles).

For odd-mass samarium and neodymium isotopes, AFMC calculations can be carried out in practice up to $\beta \sim 5$ MeV^{-1}. At higher values of β, the statistical errors become too large because of the odd-particle sign problem. The method of Sec. 2.5 is too time consuming for heavy nuclei and requires additional developments.

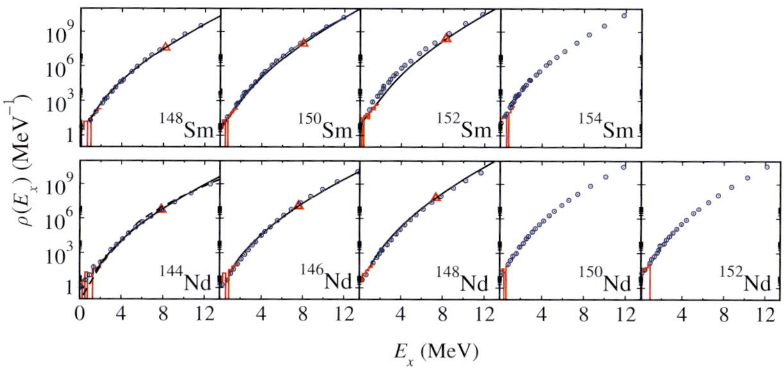

Fig. 7. State densities of even-mass samarium isotopes (top row) and neodymium isotopes (bottom row) vs. excitation energy E_x. The AFMC results (open circles) are overall in good agreement with level counting data at low excitation energies (histograms) and neutron resonance data (triangles) when available. The latter are converted to state densities assuming a spin cutoff formula (40) with a rigid-body moment of inertia in Eq. (41). The solid lines describe empirical back-shifted Fermi gas formula with parameters determined from the experiments. Adapted from Refs. [47, 59].

As an alternative we implemented a method [60] to determine the ground-state energy of a heavy odd-mass nucleus by a one-parameter fit of its AFMC thermal energy (at $T \geq 0.2\,\text{MeV}$ for which AFMC results are available) to the thermal energy that is determined from experimental data. The results are shown in Fig. 8 for odd-mass samarium and neodymium isotopes. We observe good agreement with experimental data.

6.4. *Nuclear Deformation in a Rotationally Invariant Framework*

Nuclear deformation is an important concept in understanding the structure of heavy nuclei. However, it is introduced in the context of a mean-field approximation (e.g., Hartree-Fock, HF, or Hartree-Fock-Bogoliubov HFB), in which the rotational invariance is broken. An important question is whether we can observe model-independent signatures of deformation within a framework that preserves rotational symmetry and still calculate an intrinsic deformation in such a framework.

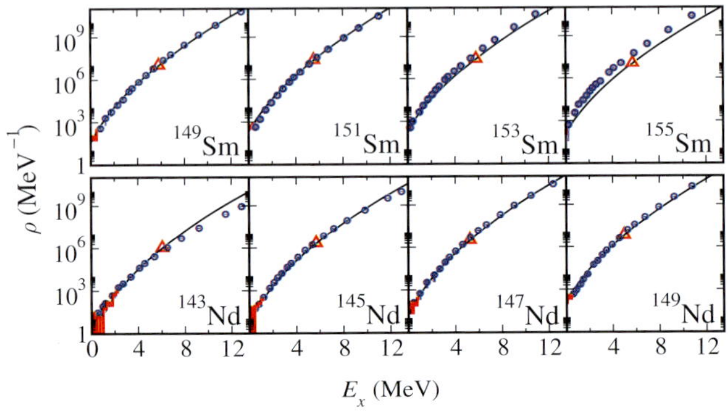

Fig. 8. State densities vs. excitation energy E_x in odd-mass samarium (top row) and neodymium (bottom row) isotopes. The AFMC densities are compared with experimental data. Symbols and lines are as in Fig. 7. Adapted from Ref. [60].

6.4.1. *Quadrupole distributions in the laboratory frame*

We used the quadrupole-projection method of Sec. 3.2 to determine the distribution $P_T(q)$ of the axial quadrupole operator $\hat{Q}_{20}$ (see Eq. (30)) in the laboratory frame for heavy rare-earth nuclei. In Fig. 9 we show the distributions $P_T(q)$ for ^{154}Sm at three temperatures. In a mean-field approximation, such as the HFB approximation, this nucleus is deformed in its ground state (i.e., at $T = 0$) and it undergoes a shape transition to a spherical nucleus at a certain critical temperature. At a low temperature ($T = 0.1$ MeV) we find that $P_T(q)$ is a skewed distribution, and it is in qualitative agreement with the axial quadrupole distribution of a prolate rigid rotor with an intrinsic quadrupole moment that is the same as the ground-state value found in HFB (dashed line in Fig. 9). The distribution $P_T(q)$ remains skewed at the shape transition temperature around $T = 1.2$ MeV, suggesting that deformation effects survive beyond the shape transition temperature. At a high temperature of $T = 4$ MeV, we observe a symmetric distribution $P_T(q)$ that is close to a Gaussian.

In contrast, similar calculations for ^{148}Sm result in Gaussian-like distributions already at low temperatures. This is consistent with ^{148}Sm being spherical in its ground state within the HFB approximation. We conclude that the distribution $P_T(q)$ can be used

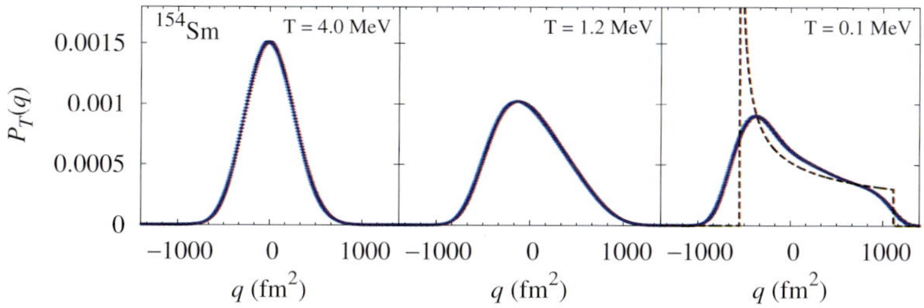

Fig. 9. Axial quadrupole distributions $P_T(q)$ (see text) in ^{154}Sm at temperatures $T = 0.1$ MeV, $T = 1.2$ MeV (close to the HFB shape transition temperature) and $T = 4$ MeV. The dashed line at $T = 0.1$ MeV is the corresponding quadrupole distribution for a prolate rotor with the same intrinsic quadrupole moment as that found in HFB. Adapted from Ref. [38].

as model-independent signature of deformation in a framework that preserves rotational symmetry.

6.4.2. *Quadrupole distributions in the intrinsic frame*

The quadrupole distribution $P_T(q)$ calculated in AFMC describes the laboratory-frame distribution. However, in modeling dynamical nuclear processes, such as fission, it is often of interest to determine the statistical nuclear properties as a function of the intrinsic deformation. The intrinsic frame is a concept introduced in the context of a mean-field approximation and the challenge is to describe intrinsic deformation in the context of the rotationally invariant CI shell-model framework without resorting to a mean-field approximation.

The distribution of the quadrupole tensor components $q_{2\mu}$ ($\mu = -2, \ldots, 2$) is invariant under rotations. We can therefore expand the logarithm of the distribution in the so-called quadrupole invariants [61,62]. There are three such invariants up to fourth order, which in terms of the intrinsic quadrupole deformation parameters,[a] β and γ, are given by β^2, $\beta^3 \cos 3\gamma$ and β^4. Expanding to fourth order,

[a] We use β to denote both the inverse temperature and the deformation parameter. The meaning should be clear from the context.

we have

$$-\ln P_T(\beta,\gamma) = N + A\beta^2 - B\beta^3\cos 3\gamma + C\beta^4 + \cdots\,, \tag{46}$$

where A, B, C are temperature-dependent parameters and N is a normalization constant. Eq. (46) resembles the Landau expansion of the free energy in which the quadrupole tensor is treated as the order parameter of the shape transition [63–66]. We can determine the parameters A, B, C in Eq. (46) from the expectation values of the three lowest-order quadrupole invariants. These expectation values can be calculated as a function of A, B, C using the density distribution P_T in Eq. (46) together with the corresponding volume element

$$\prod_\mu dq_{2\mu} = \frac{1}{2}\beta^4|\sin 3\gamma|\, d\beta\, d\gamma\, d\Omega, \tag{47}$$

where Ω are the Euler angles characterizing the orientation of the intrinsic frame. On the other hand, the expectation values of the three lowest-order quadrupole invariants are related to moments of $\hat{Q}_{20}$ in the laboratory frame [38]

$$\langle \hat{Q}\cdot\hat{Q}\rangle = 5\langle \hat{Q}_{20}^2\rangle;\, \langle(\hat{Q}\times\hat{Q})\cdot\hat{Q}\rangle = -5\sqrt{\frac{7}{2}}\langle \hat{Q}_{20}^3\rangle\,;$$
$$\langle(\hat{Q}\cdot\hat{Q})^2\rangle = \frac{35}{3}\langle \hat{Q}_{20}^4\rangle \tag{48}$$

and can therefore be directly calculated from the AFMC distribution $P_T(q)$.

In Fig. 10 we show the calculated curves, $-\ln P_T(\beta,\gamma=0)$, vs. deformation β for ^{154}Sm at three temperatures. Even though these curves are calculated in the CI shell-model approach, which preserves rotational invariance, they seem to mimic the behavior of the free energy surfaces within a mean-field approximation describing a shape transition from a deformed prolate nucleus to a spherical nucleus.

The distributions $P_T(\beta,\gamma)$ at constant temperature T can be converted to level densities $\rho(E_x,\beta,\gamma)$ as a function of excitation

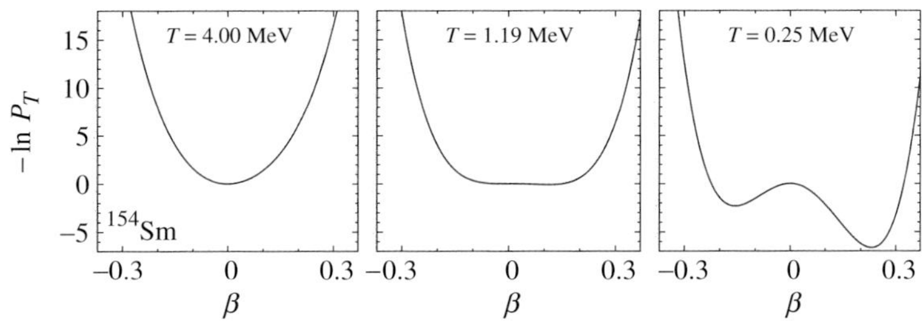

Fig. 10. $-\ln P_T(\beta, \gamma = 0)$ in Eq. (46) vs. quadrupole deformation parameter β in ^{154}Sm at temperatures $T = 0.25, 1.19$, and 4 MeV. Taken from Ref. [67].

energy E_x and intrinsic deformation β, γ using the saddle-point approximation.

7. Conclusion and Outlook

The AFMC method is a powerful technique for calculating thermal and ground-state properties of many-particle fermionic systems in very large model spaces. It has been applied to strongly correlated electron systems, molecules, cold atomic Fermi gases and nuclei. Here, we reviewed recent developments and applications of AFMC to nuclei in the framework of the nuclear CI shell-model approach. Of particular importance in nuclear applications of AFMC is the use of the canonical ensemble with fixed numbers of protons and neutrons. AFMC is a state-of-the-art method for calculating statistical properties of nuclei and, in particular, level densities and their dependence on good quantum numbers such as spin and parity. We also discussed the use of AFMC to calculate collective nuclear properties, e.g., pairing gaps and quadrupole deformation. We presented applications to mid-mass nuclei in the iron region and to heavy lanthanide nuclei.

Interesting future AFMC studies in nuclei include the calculation of statistical nuclear properties as a function of the intrinsic quadrupole deformation, and the extension of the method to other mass regions in the table of nuclei, such as the actinides and unstable nuclei. The CI shell-model Hamiltonians are specific to each mass region. It would be useful to derive effective CI shell-model Hamiltonians from a theory that is valid globally across the table of nuclei, such as density functional theory. First steps in mapping an energy density functional onto a shell-model Hamiltonian were discussed in Refs. [68, 69].

Acknowledgments

This work was supported in part by the Department of Energy Grant No. DE-FG-0291-ER-40608. I would like to thank G.F. Bertsch, M. Bonett-Matiz, L. Fang, C.N. Gilbreth, S. Liu, A. Mukherjee, H. Nakada, and C. Özen for their collaboration on various parts of the work reviewed above. The research presented here used resources of the National Energy Research Scientific Computing Center, which is supported by the Office of Science of the U.S. Department of Energy under Contract No. DE-AC02-05CH11231. It also used resources provided by the facilities of the Yale University Faculty of Arts and Sciences High Performance Computing Center.

References

[1] J. Carlson and R. Schiavilla, *Rev. Mod. Phys.* **70**, 743 (1998).
[2] S. C. Pieper and R. B. Wiringa, *Ann. Rev. Nucl. Part. Sci.* **51**, 53 (2001).
[3] B. R. Barrett, P. Navrtil and J. P. Vary, *Prog. Part. Nucl. Phys.* **69**, 131 (2013).
[4] I. Stetcu and J. Rotureau, *Prog. Part. Nucl. Phys.* **69**, 182 (2013).
[5] J. P. Draayer, T. Dytrych, K. D. Launey and D. Langr, *Prog. Part. Nucl. Phys.* **67**, 516 (2012).
[6] G. Hagen, T. Papenbrock, M. Hjorth-Jensen and D. J. Dean, *Rep. Prog. Phys.* **77**, 096302 (2014).
[7] M. Bender, P. -H. Heenen and P.-G. Reinhard, *Rev. Mod. Phys.* **75**, 121 (2003).
[8] G. F. Bertsch, D. J. Dean and W. Nazarewicz, *SciDAC Review.* **6**, 42 (2007).
[9] E. Caurier, G. Martnez-Pinedo, F. Nowacki, A. Poves and A. P. Zuker, *Rev. Mod. Phys.* **77**, 427 (2005).
[10] D. J. Scalapino and R. L. Sugar, *Phys. Rev. Lett.* **46**, 519 (1980).
[11] R. Blankenbecler, D. J. Scalapino and R. L. Sugar, *Phys. Rev. D.* **24**, 2278 (1981).
[12] J. E. Hirsch. *Phys. Rev. B.* **31**, 4403 (1985).
[13] G. Sugiyama and S.E. Koonin, *Ann. Phys.* **168**, 1 (1986).

[14] G. H. Lang, C. W. Johnson, S. E. Koonin and W. E. Ormand, *Phys. Rev. C.* **48**, 1518 (1993).
[15] Y. Alhassid, D. J. Dean, S. E. Koonin, G. Lang and W. E. Ormand, *Phys. Rev. Lett.* **72**, 613 (1994).
[16] S. E. Koonin, D. J. Dean and K. Langanke, *Phys. Rep.* **278**, 2 (1997).
[17] Y. Alhassid, *Int. J. Mod. Phys. B.* **15**, 1447 (2001).
[18] M. Dufour and A. P. Zuker, *Phys. Rev. C.* **54**, 1641 (1996).
[19] E. Y. Loh, Jr. and J. E. Gubernatis. In W. Hanke and Y. V. Kopaev, (eds.), *Electronic Phase Transitions.* North Holland, Amsterdam (1992).
[20] W. von der Linden, *Phys. Rep.* **220**, 53 (1992).
[21] W. Purwanto, S. Zhang and H. Krakauer, *J. Chem. Phys.* **142**, 064302 (2015).
[22] A. Bulgac, J. E. Drut and P. Magierski, *Phys. Rev. A.* **78**, 023625 (2008).
[23] A. Bulgac, M. N. Forbes and P. Magierski. In W. Zwerger, (ed.), *Lecture Notes Phy.* Vol. 836, Chapter 9, Springer (2012) pp. 305–373.
[24] C. N. Gilbreth and Y. Alhassid, *Phys. Rev. A.* **88**, 063643 (2013).
[25] J. Carlson, S. Gandolfi, K. E. Schmidt and S. Zhang, *Phys. Rev. A.* **84**, 061602 (R) (2011).
[26] G. Wlazlowski, J. W. Holt, S. Moroz, A. Bulgac and K. J. Roche, *Phys. Rev. Lett.* **113**, 182503 (2014).
[27] J. Hubbard, *Phys. Rev. Lett.* **3**, 77 (1959).
[28] R. L. Stratonovich, *Dokl. Akad. Nauk. S. S. S. R.* **115**, 1097 (1957).
[29] W. E. Ormand, D. J. Dean, C. W. Johnson, G. H. Lang and S. E. Koonin, *Phys. Rev. C.* **49**, 1422 (1994).
[30] A. Mukherjee and Y. Alhassid, *Phys. Rev. Lett.* **109**, 032503 (2012).
[31] D. J. Dean, S. E. Koonin, G. H. Lang, P. B. Radha and W. E. Ormand, *Phys. Lett. B.* **317**, 275 (1993).
[32] H. Nakada and Y. Alhassid, *Phys. Rev. Lett.* **79**, 2939 (1997).
[33] H. Nakada and Y. Alhassid, *Phys. Lett. B.* **436**, 231 (1998).
[34] Y. Alhassid, G. F. Bertsch, S. Liu and H. Nakada, *Phys. Rev. Lett.* **84**, 4313 (2000).
[35] C. Özen, K. Langanke, G. Martinez-Pinedo and D. J. Dean, *Phys. Rev. C.* **75**, 064307 (2007).
[36] Y. Alhassid, S. Liu and H. Nakada, *Phys. Rev. Lett.* **99**, 162504 (2007).
[37] H. Nakada and Y. Alhassid, *Phys. Rev. C.* **78**, 051304 (R) (2008).
[38] Y. Alhassid, C. N. Gilbreth and G. F. Bertsch, *Phys. Rev. Lett.* **113**, 262503 (2014).
[39] W. Hauser and H. Feshbach, *Phys. Rev.* **87**, 366 (1952).
[40] S. Hilaire and S. Goriely, *Nucl. Phys. A.* **779**, 63 (2006).
[41] Y. Alhassid, G. F. Bertsch, C. N. Gilbreth and H. Nakada, *Phys. Rev. C.* **93**, 044320 (2016).
[42] W. E. Ormand, *Phys. Rev. C.* **56**, 1678 (R) (1997).
[43] K. Langanke, *Phys. Lett. B.* **438**, 235 (1998).
[44] Y. Alhassid, S. Liu and H. Nakada, *Phys. Rev. Lett.* **83**, 4265 (1999).
[45] Y. Alhassid, G. F. Bertsch and L. Fang, *Phys. Rev. C.* **68**, 044322 (2003).
[46] Y. Alhassid, L. Fang and H. Nakada, *Phys. Rev. Lett.* **101**, 082501 (2008).

[47] C. Özen, Y. Alhassid and H. Nakada, *Phys. Rev. Lett.* **110**, 042502 (2013).
[48] A. Bohr and B. R. Mottelson, *Nuclear Structure* Vol. 1. Benjamin, New York (1969).
[49] Y. Alhassid, G. F. Bertsch, D. J. Dean and S. E. Koonin, *Phys. Rev. Lett.* **77**, 1444 (1996).
[50] Y. Alhassid, M. Bonett-Matiz, S. Liu and H. Nakada, *Phys. Rev. C.* **92**, 024307 (2015).
[51] M. Bonett-Matiz, A. Mukherjee and Y. Alhassid, *Phys. Rev. C.* **88**, 011302 (R) (2013).
[52] A. V. Voinov, S. M. Grimes, C. R. Brune, T. Massey and A. Schiller, *EPJ Web of Conferences* **21**, 05002 (2012); A. V. Voinov (private communication).
[53] R. Capote, M. Herman, P. Oblozinsky, P. G. Young, S. Goriely, T. Belgya, A. V. Ignatyuk, A. J. Koning, S. Hilaire, V. A. Plujko, M. Avrigeanu, O. Bersillon, M. B. Chadwick, T. Fukahori, Zhigang Ge, Yinlu Han, S. Kailas, J. Kopecky, V. M. Maslov, G. Reffo, M. Sin, E. Sh. Soukhovitskii and P. Talou, *Nucl. Data Sheets* **110**, 3107 (2009).
[54] T. Ericson, *Adv. Phys.* **9**, 425 (1960).
[55] T. von Egidy and D. Bucurescu, *Phys. Rev. C.* **78**, 051301 (R) (2008).
[56] T. von Egidy and D. Bucurescu, *Phys. Rev. C.* **80**, 054310 (2009).
[57] C. N. Gilbreth and Y. Alhassid, *Comput. Phys. Commun.* **188**, 1 (2015).
[58] Y. Alhassid, M. Bonett-Matiz, A. Mukherjee, H. Nakada and C. Özen, *J. Phys.: Conf. Ser.* **580**, 012009 (2015).
[59] Y. Alhassid, Y. C. Özen and H. Nakada, *Nuclear Data Sheets* **118**, 233 (2014).
[60] C. Özen, Y. Alhassid and H. Nakada, *Phys. Rev. C.* **91**, 034329 (2015).
[61] K. Kumar, *Phys. Rev. Lett.* **28**, 249 (1972).
[62] D. Cline, *Ann. Rev. Nucl. Part. Sci.* **36**, 683 (1986).
[63] S. Levit and Y. Alhassid, *Nucl. Phys. A.* **413**, 439 (1984).
[64] Y. Alhassid, S. Levit and J. Zingman, *Phys. Rev. Lett.* **57**, 539 (1986).
[65] Y. Alhassid. In Y. Abe, (ed.), *New Trends in Nuclear Collective Dynamics.* Springer Verlag, NewYork (1992).
[66] Y. Alhassid, *Nucl. Phys. A.* **553**, 137c (1993).
[67] Y. Alhassid, G. F. Bertsch, C. N. Gilbreth, H. Nakada and C. Özen. In F. Cerutti, M. Chadwick, A. Ferrari, T. Kawano and P. Schoofs, (eds.), *Proc. Fourteen International Conference on Nuclear Reaction Mechanisms, CERN-proceedings-2015-001*, CERN, Geneva (2015) pp. 41–48.
[68] Y. Alhassid, G. F. Bertsch, L. Fang and B. Sabbey, *Phys. Rev. C.* **74**, 034301 (2006).
[69] R. Rodriguez-Guzman, Y. Alhassid and G. F. Bertsch, *Phys. Rev. C.* **77**, 064308 (2008).

Chapter 10

Lie Density Functional Theory

George Rosensteel
Department of Physics, Tulane University
New Orleans, LA 70118, USA

Density functional theory (DFT) generalizes to Lie DFT, which applies to Lie dynamical groups in many-body quantum mechanics. In Lie DFT, densities are elements of the dual space to a Lie algebra of one-body operators. In this setting, the paper proves an extended Hohenberg–Kohn theorem which guarantees the existence of a dual space energy functional that is minimized by the density of the exact ground state. Lie DFT is relevant to nuclear physics when collective degrees of freedom, say in a deformed isotope, are more significant than an independent fermion picture, like Hartree–Fock or Kohn–Sham mean-field theories. Lie DFT offers a distinct computational advantage: Lie DFT calculations are made in a small finite-dimensional dual space even though the representation space of the dynamical group is large or infinite-dimensional.

Contents

1. Introduction

In its simplest formulation, density functional theory (DFT) is a theory for densities of non-degenerate ground states. The Hohenberg–Kohn theorem proves the existence of a universal energy functional minimized by the density of the exact ground-state wave function [1, 2]. Although DFT is used routinely in quantum chemistry [3], it applies to solid state [4, 5] and nuclear physics [6] too.

This Chapter reviews a DFT generalization in which the densities are elements of the dual space to a Lie algebra of quantum microscopic observables (for definitions of a Lie algebra of observables, a symmetry group, and a dynamical group, see Chapter 3). The Hohenberg–Kohn theorem is proven in this context, and Sec. 3 establishes the existence of the energy functional on the dual space. In the typical case of a dynamical group, a natural mean-field theory exists on the coadjoint orbits in the Lie algebra's dual space.

Lie DFT employs its own distinctive problem-solving methods and mathematical language. The mathematical framework is adopted from the Kirillov-Souriau-Kostant-Vogan orbit method [7–10] and the modern symplectic geometric formulation of classical mechanics [11].

In a typical application of Lie DFT, the energy functional is a low degree polynomial in the enveloping algebra (that is, a polynomial in the algebra's generators). The choice is dictated primarily by practical considerations and is supported usually by a physical argument. *As with DFT, an energy functional can be improved by adding more terms which incorporate more realistic many-body correlations into the theory. Note, however, that Lie DFT chooses an algebra that includes collective and highly correlated degrees of freedom directly into the space of densities.* A Lie DFT density does not correspond to an independent-particle wave function, as in Kohn–Sham calculations.

This paper chooses two paradigm examples to illustrate Lie DFT, the Lipkin–Meshkov–Glick (LMG) model based on $\mathfrak{su}(2)$, and the Elliott $\mathfrak{su}(3)$ model. Energy functions are adopted that are physically plausible, but make no claim to be ideal choices. They do illustrate the mathematical constructions and physics results of Lie DFT.

The significance of Lie DFT to nuclear structure and its relation to a conventional group representation theory (cf. Chapters 2 and 3) can be exemplified by the ground-state rotational band of ^{12}C. Figure 1(a) shows the lowest 0^+ and 2^+ eigenstates of a realistic Hamiltonian in terms of basis states that have good SU(3) $(\lambda\,\mu)$ quantum numbers familiar from the Elliott model [12], that is, they are also eigenvectors of the $\mathfrak{su}(3)$ Casimir operators, $\hat{\mathcal{C}}_2$ and $\hat{\mathcal{C}}_3$ (see Sec. 4.2). In these calculations, group representation theory is used to organize the model space into basis states that respect the SU(3) symmetry group. The results reveal the dominance of a single $\mathfrak{su}(3)$ irreducible representation (irrep) [see $(\lambda\,\mu) = (0\,4)$ in Fig. 1(a)]. This corroborates early findings of the Elliott model, which has provided a reasonable description of states of a rotational band based on a single irrep [12]. In addition, the band structure is indicative of the presence of a quasi-dynamical symmetry [13]. This implies that the coefficients $a^J_{(\lambda\,\mu)}$ in the expansion $|J\rangle = \sum_{(\lambda\,\mu)} a^J_{(\lambda\,\mu)}|(\lambda\,\mu)J\rangle$ for the 0^+ and 2^+ states are (approximately) independent of J. In the case of no J-dependence $a_{(\lambda\,\mu)}$, for each Casimir operator, the expectation value $\langle\hat{\mathcal{C}}_k\rangle = \langle J|\hat{\mathcal{C}}_k|J\rangle = \sum_{(\lambda\,\mu)} |a_{(\lambda\,\mu)}|^2 \mathcal{C}_k(\lambda,\mu)$ is constant for all J. Their

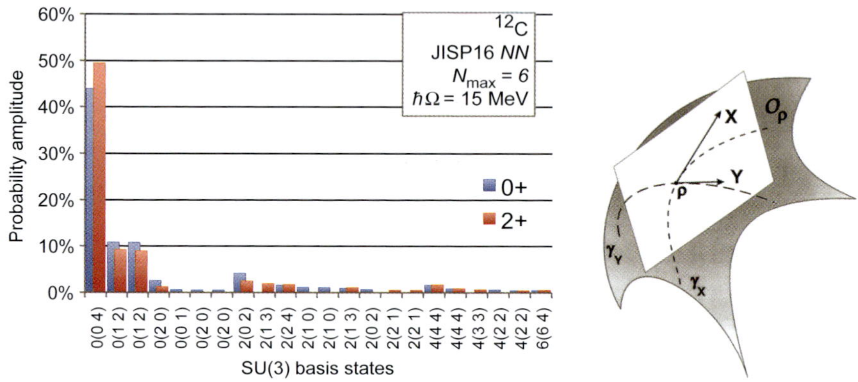

(a) (b)

Fig. 1. (a) Probability distribution for the *ab initio* 0^+ and 2^+ states of the ^{12}C ground-state rotational band over the $\mathfrak{su}(3)$ basis states labeled as $n(\lambda\,\mu)$, where $n = 0, \ldots, 6$ denotes the total harmonic-oscillator excitations, and $(\lambda\,\mu)$ label an $\mathfrak{su}(3)$ irreducible representation; the expectation values of $\hat{\mathcal{C}}_2$ and $\hat{\mathcal{C}}_3$ are ~ 20 and ~ -11, respectively, for both 0^+ and 2^+ states and hence, the densities of these states lie on a single coadjoint orbit. (b) Schematic illustration of a coadjoint orbit surface $\mathcal{O}_\rho$ through the density ρ; the Lie algebra elements X and Y are geometrically viewed as tangent vectors to the curves γ_X and γ_Y in the coadjoint orbit surface $\mathcal{O}_\rho$ (see Sec. 4.3).

constant values determine a coadjoint orbit (Fig. 1(b)). For ^{12}C, the expectation values of both $\mathfrak{su}(3)$ Casimir operators are practically the same for both 0^+ and 2^+ states, and hence, the densities of the band states lie on a common level surface of the algebra's Casimirs, which is a single coadjoint orbit (see Table A.1 for comparisons of representation and coadjoint orbit models and a summary of main notations).

2. Definitions

Let $\mathcal{H}$ denote the Hilbert space of N-fermion antisymmetrized state vectors, $\mathcal{P}(\mathcal{H})$ its complex projective space, and H the self-adjoint Hamiltonian operator on $\mathcal{H}$. In Lie DFT, $\mathfrak{g}$ is a real linear vector space of hermitian operators defined on $\mathcal{H}$, and $\mathfrak{g}^*$ is its dual space, i.e., the vector space of real-valued linear functionals on $\mathfrak{g}$. Denote the pairing between a dual element $\rho \in \mathfrak{g}^*$ and an operator $X \in \mathfrak{g}$

by $\langle \rho, X\rangle \equiv \rho(X)$. Define the moment map $M : \mathcal{P}(\mathcal{H}) \to \mathfrak{g}^*$ from the projective space of states into the dual space by

$$\langle M(\phi), X\rangle := \frac{\langle \phi | X\phi\rangle}{\langle \phi|\phi\rangle} \quad \text{for all } X \in \mathfrak{g}, \tag{1}$$

where $\phi \in \mathcal{H}$, $M(\phi) \in \mathfrak{g}^*$. Marsden and Ratiu [14] provides a brief history of the moment map in classical dynamics.

Conventional DFT is a special case of Lie DFT. In DFT, $\mathfrak{g}^*$ is the space of measurable $\mathcal{L}^1$ functions

$$\mathfrak{g}^* = \left\{\rho : \mathbf{R}^3 \to \mathbf{R} \;\middle|\; \int |\rho(\vec{r})|\, d^3r < \infty\right\}, \tag{2}$$

$\mathfrak{g}$ is a subspace $\mathcal{L}^\infty$ of real-valued measurable functions $v(\vec{r})$ on $\mathbf{R}^3$, which are identified with one-body multiplication operators $\sum_{n=1}^{N} v(\vec{r}_n)$ on $\mathcal{H}$. The pairing is given by $\langle \rho, v\rangle = \int \rho(\vec{r})v(\vec{r})d^3r$. If ϕ is normalized, then the moment map $M(\phi) = \rho$ is

$$\rho(\vec{r}) = N \int ||\phi(\vec{r}, \vec{r}_2, \ldots, \vec{r}_N)||^2\, d^3r_2 \ldots d^3r_N. \tag{3}$$

A density ρ in the image of the moment map is called N-representable. Densities in the dual space that are not N-representable are unphysical and must be excluded from DFT theory because these dual elements do not correspond to any quantum state vector.

Define the energy mapping $F : \mathcal{P}(\mathcal{H}) \to \mathbf{R}$ from the projective space to the real numbers by

$$F(\phi) := \frac{\langle \phi|H\phi\rangle}{\langle \phi|\phi\rangle}, \tag{4}$$

and, for each fixed $X \in \mathfrak{g}$, define $F_X : \mathcal{P}(\mathcal{H}) \to \mathbf{R}$ by

$$F_X(\phi) := F(\phi) + \langle M(\phi), X\rangle. \tag{5}$$

3. Generalized Hohenberg–Kohn Theorem

Define the set of states $\mathcal{S}$ consisting of every $\phi \in \mathcal{P}(\mathcal{H})$ that is an absolute nondegenerate minimum of F_X for some $X \in \mathfrak{g}$. Set $\mathcal{C}$ equal to the image of $\mathcal{S}$ under the moment map, $\mathcal{C} = M(\mathcal{S}) \subset \mathfrak{g}^*$.

Theorem 1. *The moment map $M : \mathcal{S} \to \mathcal{C}$ is an injection (that is, for two distinct states in $\mathcal{S}$ the moment map yields two distinct densities in $\mathcal{C}$).*

Proof. The proof of this theorem is similar to the original Hohenberg-Kohn argument [1]. Suppose, to the contrary, that there are two distinct states $\phi_1, \phi_2 \in \mathcal{S}$, $\phi_1 \neq \phi_2$, yet their densities are equal, $M(\phi_1) = M(\phi_2)$. There are operators $X_1, X_2 \in \mathfrak{g}$ such that ϕ_1 is an absolute non-degenerate minimum for F_{X_1} and ϕ_2 is an absolute non-degenerate minimum for F_{X_2}. Since $\phi_1 \neq \phi_2$, the non-degeneracy of the minima implies

$$\begin{aligned} E_1 &\equiv F_{X_1}(\phi_1) < F_{X_1}(\phi_2), \\ E_2 &\equiv F_{X_2}(\phi_2) < F_{X_2}(\phi_1). \end{aligned}$$

$\mathfrak{g}$ is a vector space, so $X_1 - X_2 \in \mathfrak{g}$ and

$$\begin{aligned} E_1 &< F_{X_1}(\phi_2) = F(\phi_2) + \langle M(\phi_2), (X_1 - X_2) + X_2 \rangle \\ &= F_{X_2}(\phi_2) + \langle M(\phi_2), (X_1 - X_2) \rangle \text{ or} \\ E_1 &< E_2 + \langle M(\phi_2), (X_1 - X_2) \rangle. \end{aligned}$$

By repeating the argument for interchanged 1 and 2,

$$E_2 < E_1 + \langle M(\phi_1), (X_2 - X_1) \rangle.$$

Adding the two equations, and recalling that $M(\phi_1) = M(\phi_2)$, leads to the contradiction

$$E_1 + E_2 < E_2 + E_1.$$

Hence, for two distinct states in $\mathcal{S}$ the moment map yields two distinct densities in $\mathcal{C}$. □

Hence, for each $X \in \mathfrak{g}$, the mapping $\tilde{F}_X = F_X \circ M^{-1} : \mathcal{C} \to \mathbf{R}$ is well-defined. The state ϕ minimizes F_X on S if and only if $\rho = M(\phi)$ minimizes $\tilde{F}_X$ on $\mathcal{C}$.

A density ρ is said to be v- or, in the general setting, X-representable if $\rho \in \mathcal{C} = M(\mathcal{S})$. The energy functional must be minimized on the set of X-representable densities. If the X-representable densities can only be characterized by their defining property as an image of a set of N-fermion wave functions, this minimization is equivalent to the original many-body eigenvalue problem on Fock space. Another problem is that the existence theorem's proof does not provide a construction of the energy functional.

These two problems are circumvented by the constrained search formulation of Levy [15,16]. Consider the image of the moment map $\mathcal{C}_N = M(\mathcal{P}(\mathcal{H}))$, for which $\mathfrak{g}^* \supset \mathcal{C}_N \supset \mathcal{C}$. A density ρ is said to be N-representable if $\rho \in \mathcal{C}_N$. Define the energy functional

$$E : \mathcal{C}_N \to \mathbf{R}$$
$$E[\rho] = \inf_{\phi \in M^{-1}(\rho)} F(\phi), \tag{6}$$

and also $E_X[\rho] = E[\rho] + \langle \rho, X \rangle$ for $\rho \in \mathcal{C}_N$.

Theorem 2. *The ground-state energy E_0 satisfies the minimization condition*

$$E_0 = \inf_{\rho \in \mathcal{C}_N} E_X[\rho],$$

and the ground-state density ρ_0 attains this minimized energy. If ρ is X-representable, then $E_X[\rho] = \tilde{F}_X(\rho)$.

Proof. The ground-state energy minimizes the expectation of the total Hamiltonian:

$$\begin{aligned} E_0 &= \min_{\phi \in \mathcal{H}} F_X(\phi) \\ &= \min_{\rho \in \mathcal{C}_N} \left\{ \min_{\phi \in M^{-1}(\rho)} (F(\phi) + \langle M(\phi), X \rangle) \right\} \\ &= \min_{\rho \in \mathcal{C}_N} (E[\rho] + \langle \rho, X \rangle). \end{aligned}$$

If $\phi_0 \in \mathcal{S}$ is a non-degenerate ground state, $\rho_0 \in \mathcal{C}$ its density, and $E_0 = E_X[\rho_0]$ the ground-state energy, then

$$E_0 = F_X(\phi_0) = \tilde{F}_X(\rho_0).$$

□

Thus, according to the generalized Levy's theorem, it is sufficient to minimize the energy functional on the space of all N-representable densities instead of all X-representable densities.

To determine the universal energy functional $E[\rho]$, the Lagrange multiplier theorem, suitably extended to Banach spaces [14], may be applied to eliminate the constraint to $\phi \in M^{-1}(\rho)$ from the functional's defining Eq. (6). This is achieved by regarding the pairing $\langle \rho, X \rangle$ as a Lagrange multiplier term while the operator X is the Lagrange multiplier itself. Suppose the vector space of operators is self-dual, i.e., the dual space to $\mathfrak{g}^*$ is $\mathfrak{g}$, and it is a Banach space, i.e., a norm $\|X\|$ is defined on $\mathfrak{g}$. Both conditions are satisfied, for example, if the space is finite-dimensional. Suppose that $M^{-1}(\rho)$ is a submanifold of $\mathcal{P}(\mathcal{H})$.

Theorem 3. *The following are equivalent conditions on* $\phi_0 \in M^{-1}(\rho)$:

(i) ϕ_0 *is a critical point of* $F(\phi)$ *restricted to* $M^{-1}(\rho)$; *and*
(ii) *there is a* $X_0 \in \mathfrak{g}$ *such that* (X_0, ϕ_0) *is a critical point of* $F_X(\phi)$ *on the space* $\mathfrak{g} \times \mathcal{P}(\mathcal{H})$.

The universal energy functional is given by $E[\rho] = F(\phi_0)$ when the critical point corresponds to an absolute minimum. To test for a minimum, let α be a real positive constant, (X_0, ϕ_0) a critical point of $F_X(\phi)$, and set

$$F_X^{\alpha}(\phi) = F_X(\phi) + \alpha \|X - X_0\|^2, \tag{7}$$

that also has a critical point at (X_0, ϕ_0). If the restriction of $F(\phi)$ to $M^{-1}(\rho)$ has a non-degenerate minimum at ϕ_0, then for sufficiently large positive α, $F_X^{\alpha}(\phi)$ is minimized at (X_0, ϕ_0). This observation may be used to identify minima using the unconstrained second derivative test.

4. Dynamical Lie Algebra

Lie DFT applies to a dynamical Lie algebra $\mathfrak{g}$. In this case, the Hilbert space is an irreducible representation space and is a proper subspace of fermion Fock space. The self-adjoint Hamiltonian operator H is the representation of an element of the universal enveloping algebra $U(\mathfrak{g})$, a polynomial in the algebra's generators.

The energy function is often derived from a Hamiltonian operator by a natural construction. Suppose the Hamiltonian is a polynomial in the basis $Z_i \in \mathfrak{g}$ for $i = 1, 2, \ldots, \dim\, \mathfrak{g}$. The enveloping algebra element maps into a unique element of the symmetric tensor product of the algebra, and it may be expressed as a polynomial $\mathcal{E}(\zeta_1, \zeta_2, \ldots, \zeta_k)$, where $\zeta_i = \langle \rho, Z_i \rangle$, cf. Appendix of Refs. [17, 18]. This simple construction of the energy from an enveloping algebra element is purely algebraic and is adopted in this Chapter.

To improve on it, a correlation energy that ultimately depends on the representation and its ground-state density must be added to the symmetric algebra's energy function. Such an exact energy functional must exist according to the generalized Hohenberg–Kohn theorem. Unfortunately, the energy functional definition is formal and no known method constructs the Hohenberg–Kohn energy directly. This theoretical defect is shared with DFT and is the subject of most current DFT investigations in physics and chemistry [19, 20].

4.1. *Coadjoint Group Action*

Let G denote the simply connected and connected Lie group of unitary operators corresponding to the representation of $\mathfrak{g}$. The group action of G on $\mathcal{H}$ naturally induces the coadjoint action Ad_g^* of G on the dual space of densities. To show this, suppose $\phi \in \mathcal{H}$ and $g \cdot \phi \in \mathcal{H}$ is the result of the unitary representation of $g \in$ G acting on ϕ. If $\rho = M(\phi)$, then $M(g \cdot \phi) = Ad_g^*(\rho)$, since

$$\begin{aligned}\langle M(g \cdot \phi), X \rangle &= \langle g \cdot \phi | X\, g \cdot \phi \rangle / \langle g \cdot \phi | g \cdot \phi \rangle \\ &= \langle \phi | g^{-1} X g \cdot \phi \rangle / \langle \phi | \phi \rangle \\ &= \langle \rho, Ad_{g^{-1}}(X) \rangle \\ &= \langle Ad_g^*(\rho), X \rangle \end{aligned} \tag{8}$$

for all $X \in \mathfrak{g}$. Thus the coadjoint action leaves the space of N-representable densities invariant, $Ad_g^* : \mathcal{C}_N \to \mathcal{C}_N$.

If G is a symmetry group of the Hamiltonian H, then the energy functional transforms according to the adjoint transformation,

$$F_X \circ Ad_g^* = F_{Ad_{g^{-1}}(X)}, \tag{9}$$

for $X \in \mathfrak{g}$ and $g \in \mathrm{G}$. If $\rho \in \mathcal{C}$ is an absolute non-degenerate minimum for F_X, then $Ad_g^*(\rho)$ is an absolute non-degenerate minimum for $F_{Ad_{g^{-1}}(X)}$. Therefore, if G is a symmetry group for the Hamiltonian, the coadjoint G-action leaves the space of X-representable densities $\mathcal{C}$ invariant.

In conventional DFT, the abelian Lie algebra integrates to an abelian group of unitary operators

$$\mathrm{G} = \left\{ g = \exp\left(\mathrm{i} \sum_{n=1}^{N} f(\vec{r}_n) \right), f \in \mathfrak{g} \right\}, \tag{10}$$

and G is the isotropy subgroup at each density, $Ad_g^*(\rho) = \rho$. If $\phi \in \mathcal{H}$ and $M(\phi) = \rho \in \mathcal{C}_N$, then the energy functional satisfies

$$E[\rho] \leq \inf_{f \in \mathfrak{g}} F(g \cdot \phi). \tag{11}$$

This equation provides an upper bound on the exact energy functional and may be used to approximate it.

For a non-abelian Lie algebra $\mathfrak{g}$, the Lie group G is not typically contained in the isotropy subgroup G_ρ at $\rho \in \mathcal{C}_N$. The coadjoint orbit $\mathcal{O}_\rho$ consists of all points $Ad_g^*(\rho) \in \mathcal{C}_N$ as g ranges over the group G. Each orbit space $\mathcal{O}_\rho$ is diffeomorphic to the homogeneous coset space G/G_ρ. Hence, the space of N-representable densities is a union of disjoint coadjoint orbits each diffeomorphic to a homogeneous space.

4.2. *Casimir Elements*

In Lie DFT, the Hilbert space is an irreducible representation of the dynamical Lie algebra. The N-representable densities are characterized conveniently for computational purposes as level surfaces of the algebra's Casimirs. Casimirs are elements in the center of the enveloping algebra, i.e., a Casimir commutes with all elements of the algebra and any polynomial of the algebra's elements. For a semisimple Lie algebra, the dimension of the center, which is the number of independent Casimirs, equals the algebra's rank.

The representation of a Casimir element $\tilde{\mathcal{C}}_k$ is a Casimir operator $\hat{\mathcal{C}}_k$ on the representation space. For an irreducible representation π_λ with highest weight λ of a compact Lie group, each vector in the irreducible representation space is a simultaneous eigenvector for all the Casimir operators. Conversely, a simultaneous eigenspace of all the Casimir operators in a representation space is necessarily an irreducible representation space.

For the dual space of a Lie algebra, a set of Lie DFT concepts parallels the essential ideas of irreducible representation theory. Each Casimir element $\tilde{\mathcal{C}}_k$ in the enveloping algebra defines a smooth real-valued function $\mathcal{C}_k$ on the algebra's dual space. A Casimir function is constant on each coadjoint orbit. Conversely, a common level surface of the Casimir functions for a compact Lie group is necessarily a single coadjoint orbit. Recall that a level surface of a real-valued function F on a space V is the subset of points $p \in V$ for which $F(p) =$ constant; a common level surface of a set of functions is the intersection of level surfaces for all the functions. Therefore, Lie DFT N-representable densities lie on a level surface of the algebra's Casimirs. Because the Casimirs are polynomials, the surface is an algebraic variety.

An integral coadjoint orbit $\mathcal{O}_{\lambda+\delta}$ is a common level surface of the Casimirs evaluated at the sum of the dominant integral weight λ plus δ, where δ is one-half the sum of the positive roots. Each level surface $\mathcal{O}_{\lambda+\delta}$ has a remarkable property: the volume of the orbit equals the dimension of the irreducible representation space,

$$\text{vol}\ \mathcal{O}_{\lambda+\delta} = \dim \pi_\lambda. \tag{12}$$

4.3. *Mean-field Dynamics*

Every Lie algebra element X defines a tangent vector to the Casimir level surface through the N-representable density ρ (Fig. 1). This vector is the tangent to the curve $\gamma_X(t) = Ad_g^*(\rho)$ where $g = \exp(itX)$. Note that the elements of the isotropy subalgebra $\mathfrak{g}_\rho$ correspond to null tangent vectors, and the tangent space is isomorphic to the coset vector space $\mathfrak{g}/\mathfrak{g}_\rho$. For any two tangent vectors $X, Y \in \mathfrak{g}/\mathfrak{g}_\rho$ to the coadjoint orbit through ρ, the bilinear form

$$\omega_\rho(X,Y) = \mathrm{i}\langle \rho, [X,Y]\rangle, \tag{13}$$

is antisymmetric, closed, and non-degenerate. Thus, each coadjoint orbit is a symplectic manifold [7]. Moreover the group G acts on each orbit as a transitive group of canonical transformations,

$$\omega_{Ad^*_g(\rho)}(Ad_g(X), Ad_g(Y)) = \omega_\rho(X,Y), \tag{14}$$

where X, Y are tangent vectors at ρ and $Ad_g(X)$ and $Ad_g(Y)$ are tangent vectors at $Ad^*_g(\rho)$.

The mean-field Hamiltonian h may be defined now. Choose a coadjoint orbit $\mathcal{O}_\rho$ in the dual space and let $\mathcal{E}$ denote the restriction of the energy functional to this orbit. If X is a tangent vector at ρ, denote the derivative of the energy in the X direction by

$$d\mathcal{E}_\rho(X) = \frac{d}{dt}\mathcal{E}(Ad^*_g(\rho))|_{t=0}, \text{ where } g = \exp(\mathrm{i}tX). \tag{15}$$

The model Hamiltonian $h \in \mathfrak{g}/\mathfrak{g}_\rho$ is determined by

$$d\mathcal{E}_\rho(X) = \omega_\rho(h, X) \text{ for all } X \in \mathfrak{g}. \tag{16}$$

There is a unique solution to this equation for $h \in \mathfrak{g}/\mathfrak{g}_\rho$ because the symplectic form ω is non-degenerate on the tangent space. Note that the model Hamiltonian h depends on the point ρ in the coadjoint orbit.

In the special case when the Hamiltonian H is an element of the Lie algebra $\mathfrak{g}$, the energy functional is just $E[\rho] = \langle \rho, H\rangle$ and the model Hamiltonian h is equal to H. Otherwise the model Hamiltonian h is the optimal approximation to H relative to the Lie algebra $\mathfrak{g}$ in the sense that $\langle \phi \,|\, [H-h, X]\, \phi\rangle = 0$ for all $X \in \mathfrak{g}$ where $M(\phi) = \rho \in \mathcal{C}$.

The level surface points for which the Hamiltonian vector field h vanishes includes the Lie DFT density of the ground state. An integral curve of the Hamiltonian vector field on the coadjoint orbit is a time-dependent mean-field solution. The linearization of this Hamiltonian dynamical system in the neighborhood of an elliptic stationary density is a generalization of the random-phase approximation (RPA) for excited states.

5. SU(2) Lipkin-Menshkov-Glick Theory

The LMG model [21–23] is a useful test case for Lie DFT. The LMG Hamiltonian is a quadratic element in the enveloping algebra of $\mathfrak{su}(2)$, and the quantum Hilbert space is the irreducible unitary representation σ of dimension $2J+1$. The model is well-known to have a structural or quantum phase transition in the thermodynamic limit, $J \to \infty$.

The complexification of the matrix Lie algebra $\mathfrak{su}(2)$ is spanned by

$$S_0 = \begin{pmatrix} \frac{1}{2} & 0 \\ 0 & -\frac{1}{2} \end{pmatrix}, \quad S_+ = \begin{pmatrix} 0 & 1 \\ 0 & 0 \end{pmatrix}, \quad S_- = \begin{pmatrix} 0 & 0 \\ 1 & 0 \end{pmatrix}. \tag{17}$$

The image of the moment map is the density matrix ρ in the algebra's dual space

$$\rho = \begin{pmatrix} w & \bar{u} \\ u & -w \end{pmatrix}, \tag{18}$$

where the entries of ρ are the expectations of the $\mathfrak{su}(2)$ operators in the representation space, the real number $w = \langle \sigma(S_0) \rangle = \langle \rho, S_0 \rangle = \mathrm{Tr}(\rho S_0)$, $u = \langle \sigma(S_+) \rangle$, and its complex conjugate $\bar{u} = \langle \sigma(S_-) \rangle$.

The highest weight of an irreducible $\mathfrak{su}(2)$ representation of dimension $2J+1$ determines a diagonal density matrix, $\lambda = \mathrm{diag}(J, -J)$. One-half the positive root is the diagonal density $\delta = \mathrm{diag}(1/2, -1/2)$. The orbit representative for $\mathcal{O}_\varrho = \mathcal{O}_{\lambda+\delta}$ is $\varrho = \mathrm{diag}(\bar{J}, -\bar{J})$, $\bar{J} \equiv (J + 1/2)$, and the coadjoint orbit $\mathcal{O}_\varrho$ is the level surface of the Casimir function

$$\frac{1}{2}\mathrm{Tr}(\rho^2) = w^2 + |u|^2 = \bar{J}^2, \tag{19}$$

which is diffeomorphic to the two-sphere.

For any real $0 \leq x \leq 1$, the real-valued LMG energy function on the dual space is

$$\mathcal{E}(\rho) = (1-x)w + (x/2\bar{J})(u^2 + \bar{u}^2) \tag{20}$$

and its exterior derivative is

$$d\mathcal{E} = (1-x)dw + (x/\bar{J})\,(udu + \bar{u}d\bar{u})\,. \tag{21}$$

The construction of the mean-field Hamiltonian described in Sec. 4.3 requires the replacement of the differential dw by S_0 because $w = \langle\sigma(S_0)\rangle = \langle\rho, S_0\rangle$ and, similarly, du by S_+, $d\bar{u}$ by S_-:

$$h[\rho] = (1-x)S_0 + (x/\bar{J})\,(uS_+ + \bar{u}S_-)\,. \tag{22}$$

Note that the usual LMG energy replaces $\bar{J}$ by J in Eq. (20). To apply the coadjoint theory to the standard LMG model, simply replace $\bar{J}$ by J and, hence, replace $\mathcal{O}_{\lambda+\delta}$ by $\mathcal{O}_\lambda$ throughout this section.

The Lax equation

$$\mathrm{i}\frac{d}{dt}\rho = [h[\rho], \rho] = \begin{pmatrix} (x/\bar{J})\left(u^2 - \bar{u}^2\right) & (1-x)\bar{u} - (2x/\bar{J})wu \\ -(1-x)u + (2x/\bar{J})w\bar{u} & -(x/\bar{J})\left(u^2 - \bar{u}^2\right) \end{pmatrix} \tag{23}$$

determines a Hamiltonian dynamical system on a coadjoint orbit.

An equilibrium density commutes with the mean-field Hamiltonian, $[h[\rho], \rho] = 0$. This matrix equation may be solved directly as an algebraic system. There are three classes of equilibrium solutions:
Case 1. $u = 0$.

$$w = \pm\bar{J}, \rho = \pm\varrho, h[\rho] = \pm((1-x)/2\bar{J})\rho, \mathcal{E} = \pm(1-x)\bar{J}. \tag{24}$$

Case 2. u is real.

$$w = \frac{1-x}{2x}\bar{J},\ |u|^2 = \bar{J}^2 - w^2 \geq 0 \Leftrightarrow x \geq \frac{1}{3},$$
$$u = \pm v,\ h[\rho] = (x/\bar{J})\rho,\ \mathcal{E} = (5x^2 - 2x + 1)\bar{J}/4x. \tag{25}$$

Case 3. u is pure imaginary.

$$w = -\frac{1-x}{2x}\bar{J},\ |u|^2 = \bar{J}^2 - w^2 \geq 0 \Leftrightarrow x \geq \frac{1}{3},$$

$$]u = \pm iv,\ h[\rho] = -(x/\bar{J})\rho,\ \mathcal{E} = -(5x^2 - 2x + 1)\bar{J}/4x. \quad (26)$$

In Cases 2 and 3, $v = \frac{\bar{J}}{2x}\sqrt{(3x-1)(1+x)}$ and note that the mean-field solutions are two-fold degenerate in agreement with representation theory.

As a function of x, the energy of the ground-state density is

$$\mathcal{E}_{\text{g.s.}} = \begin{cases} -\bar{J}\,(1-x), & 0 \leq x \leq \frac{1}{3}\ (\text{Case1}) \\ -\bar{J}\,(5x^2 - 2x + 1)/4x, & \frac{1}{3} \leq x \leq 1\ (\text{Case 3}). \end{cases} \quad (27)$$

The SU(2) group element that effects the coadjoint transformation to the ground-state density is as follows:

$$\text{For } 0 \leq x \leq \frac{1}{3},\ g = \begin{pmatrix} 0 & 1 \\ -1 & 0 \end{pmatrix}; \quad (28)$$

$$\text{For } \frac{1}{3} \leq x \leq 1,\ g = \begin{pmatrix} \mp i\sin\theta & \cos\theta \\ -\cos\theta & \pm i\sin\theta \end{pmatrix},\ \tan\theta = \sqrt{\frac{3x-1}{1+x}}. \quad (29)$$

6. SU(3) Elliott Theory

The Elliott SU(3) theory models the rotation of a deformed nucleus [12] (cf. Chapter 3). The algebra is spanned by the angular momentum and the quadrupole moment (restricted to a single harmonic-oscillator shell) that characterizes the deformation. In nuclear physics, two representations corresponding to two different physical interpretations are relevant: (1) The original Elliott microscopic representation on fermion shell-model space [12] and (2) the interacting boson model representation on the model space of *ds* bosons [24] for systematic studies. SU(3) predicts rotational bands, which are a ubiquitous feature of the spectrum of open shell nuclei. The measured electromagnetic transition rates among the states of rotational bands indicate frequently that the two $\mathfrak{su}(3)$ Casimir invariants are constant.

We assume SU(3) is a dynamical group with a Hamiltonian which is a quartic polynomial in the $\mathfrak{su}(3)$ enveloping algebra. This energy choice is general enough to encompass most nuclear applications. Lie DFT finds approximate equilibrium densities and their energies, while normal mode solutions are discovered to describe precessions [25–27].

The rotation group SO(3) is a subgroup of SU(3) and the energy is SO(3)-invariant, i.e., the energy function is constant on each SO(3) coadjoint orbit in the dual space of $\mathfrak{su}(3)$-densities. Rotational symmetry can be exploited to simplify SU(3) DFT.

An SU(3) irreducible representation is labeled by a pair of non-negative integers, $(\lambda\,\mu)$. The density of the highest-weight vector is a diagonal matrix, $\frac{1}{3}\mathrm{diag}(-\lambda - 2\mu, -\lambda + \mu, 2\lambda + \mu)$. Because one-half the sum of the positive roots equals $\mathrm{diag}(-1, 0, 1)$, the orbit representative associated to the $(\lambda\,\mu)$ irrep is

$$\varrho = \frac{1}{3}\mathrm{diag}(-\bar{\lambda} - 2\bar{\mu}, -\bar{\lambda} + \bar{\mu}, 2\bar{\lambda} + \bar{\mu}), \tag{30}$$

where $\bar{\lambda} = \lambda + 1$ and $\bar{\mu} = \mu + 1$.

Each orbit space is a homogeneous space,

$$\mathcal{O}_\varrho = \begin{cases} \mathrm{SU}(3)/\mathrm{U}(1) \times \mathrm{U}(1), & \dim = 6,\ \bar{\lambda} \neq 0 \text{ and } \bar{\mu} \neq 0 \\ \mathrm{SU}(3)/\mathrm{U}(2), & \dim = 4,\ \bar{\lambda} = 0 \text{ or } \bar{\mu} = 0 \end{cases}. \tag{31}$$

When $\bar{\lambda} = \bar{\mu} = 0$ the trivial orbit is a point.

An orbit is a level surface of two Casimir functions with values

$$\begin{aligned} \mathcal{C}_2(\varrho) &= \frac{2}{3}\left(\bar{\lambda}^2 + \bar{\lambda}\bar{\mu} + \bar{\mu}^2\right) \\ &= \frac{2}{3}\left(\lambda^2 + \mu^2 + \lambda\mu + 3\lambda + 3\mu\right) + 2, \end{aligned} \tag{32}$$

$$\mathcal{C}_3(\varrho) = \frac{1}{9}(\bar{\lambda} - \bar{\mu})(2\bar{\lambda} + \bar{\mu})(\bar{\lambda} + 2\bar{\mu})$$
$$= \frac{1}{9}(\lambda - \mu)(2\lambda + \mu + 3)(\lambda + 2\mu + 3). \tag{33}$$

Note that the values of the Casimirs on the integral orbit differ merely by an additive constant from their eigenvalues in an irreducible $(\lambda\,\mu)$ representation (cf. Ref. [28] and Chapter 4).

Leschber, Draayer, and Castanos established a correspondence between the Bohr–Mottelson parameters (β, γ), which are measured experimentally and tabulated for deformed nuclei, and the $\mathfrak{su}(3)$ diagonal density labels $(\lambda\,\mu)$ [29,30]. The calculated $(\lambda\,\mu)$ values are real numbers and are not generally natural numbers.

6.1. SO(3) *Kinematical Symmetry*

SU(3) introduces an essential new feature, kinematical symmetry, to Lie DFT. To exploit rotational symmetry, the dual space is decomposed into a direct sum of irreducible subspaces with respect to the coadjoint action of the rotation group. The vector space $\mathfrak{su}(3)^*$ is the direct sum of the vector space of traceless real symmetric matrices (quadrupole irrep) and the vector space of pure imaginary antisymmetric matrices (vector irrep). A density is expressed uniquely as a sum

$$\rho = q - \frac{\mathrm{i}}{2}l, \tag{34}$$

where q is traceless real symmetric and l is real antisymmetric. An identical decomposition of the Lie algebra is compatible with the dual space decomposition,

$$q_{ij} = \left\langle \rho, \frac{E_{ij} + E_{ji}}{2} - \frac{1}{3}\delta_{ij} I \right\rangle,$$
$$l_{ij} = \langle \rho, -\mathrm{i}(E_{ij} - E_{ji}) \rangle. \tag{35}$$

where E_{ij} is the elementary 3×3 matrix whose row i and column j entry equals one and whose other entries equal zero and I is the identity matrix. The antisymmetric matrix l determines uniquely the pseudo-vector angular momemtum $\vec{l}$ where $l_{ij} = \varepsilon_{ijk} l_k$. The quantum operators for the single-shell quadrupole moment and the angular momentum in the representation space are

$$\hat{Q}_{ij} = \sigma \left(\frac{E_{ij} + E_{ji}}{2} - \frac{1}{3}\delta_{ij} I \right),$$

$$\hat{L}_{ij} = \sigma \left(-\mathrm{i}(E_{ij} - E_{ji}) \right). \tag{36}$$

6.2. *Reduction of the Dynamical System Due to Symmetry*

Without loss of generality, we align the direction of the angular momentum l with the 1-axis. Let $\mathcal{O}_\varrho^l$ denote the three-dimensional submanifold of the six-dimensional coadjoint orbit $\mathcal{O}_\varrho$ with fixed conserved angular momentum l,

$$\mathcal{O}_\varrho^l = \left\{ \rho = q - \frac{\mathrm{i}}{2} l \in \mathcal{O}_\varrho \right\}. \tag{37}$$

The trajectories of the rotationally invariant Hamiltonian dynamical system lie on the surface $\mathcal{O}_\varrho^l$. Let I^2 denote the squared length of the angular momentum vector, which is $l \cdot l = -(1/2)\mathrm{tr}(l^2) = \sum_{i<j} l_{ij}^2$.

The physical interpretation is simplified in the "body-fixed" or intrinsic frame. By definition, the real part of an intrinsic-frame density is a diagonal matrix. In the Elliott interpretation, these diagonal entries are proportional to the squared axis lengths of the inertia ellipsoid. Thus, an intrinsic-frame density provides a geometrical picture of the many-body quantum system.

The goal is to transfer the dynamical system on $\mathcal{O}_\varrho^l$ to an equivalent dynamical system on a surface of intrinsic densities. To achieve that goal, we construct a mapping from $\mathcal{O}_\varrho^l$ to a manifold of intrinsic densities. When $\rho = q - \frac{\mathrm{i}}{2} l \in \mathcal{O}_\varrho^l$, let $R \in \mathrm{SO}(3)$ denote a rotation that diagonalizes q, $\mathrm{Ad}_R^* \rho = \tilde{\rho} = \tilde{q} - \frac{\mathrm{i}}{2}\tilde{l}$. The diagonal matrix $\tilde{q}$ is unique when the eigenvalues are ordered, say $\tilde{q}_1 \geq \tilde{q}_2 \geq \tilde{q}_3$.

But the diagonalizing rotation R is not unique. If the eigenvalues are all distinct, then the rotation is unique modulo an element of the dihedral group, $Z_2 \times Z_2$, consisting of all diagonal SO(3) matrices $m = \mathrm{diag}(m_1, m_2, m_3)$, $m_k = \pm 1$, $m_1 m_2 m_3 = 1$. Thus, $\mathrm{Ad}^*_{mR}\,\rho = \mathrm{Ad}^*_m \tilde{\rho} = \tilde{q} - \frac{\mathrm{i}}{2} m\,\tilde{l}\,m$ for any m in the dihedral group.

Let $\vec{I}$ denote the pseudo-vector angular momentum corresponding to $\tilde{l}$. The pseudo-vector corresponding to $m\,\tilde{l}\,m$ has components $m_k I_k, k = 1, 2, 3$. We choose the element m such that the components I_1 and I_3 are both non-negative and say that $\tilde{l}$ is in *standard position*.

In the special case when two of the eigenvalues of q coalesce, say $\tilde{q}_1 = \tilde{q}_2 > \tilde{q}_3$ (oblate intrinsic density) or $\tilde{q}_1 > \tilde{q}_2 = \tilde{q}_3$ (prolate intrinsic density), there is more flexibility in the choice of the diagonalizing rotation R. In addition to the dihedral group, left multiplication of R by an element of the subgroup SO(2) of rotations about the symmetry axis fixes $\tilde{q}$ for an oblate or prolate density. We choose the SO(2) rotation to make $I_2 = 0$ and say that $\tilde{l}$ is in standard position.

Having made a unique choice for the rotation R, we have constructed a well-defined mapping ψ from $\mathcal{O}^l_\varrho$ to the space of intrinsic-frame densities,

$$\mathcal{M}^I_\varrho = \Big\{ \tilde{\rho} = \tilde{q} - \frac{\mathrm{i}}{2}\tilde{l} \in \mathcal{O}_\varrho \mid \tilde{q} = \mathrm{diag}(\tilde{q}_1, \tilde{q}_2, \tilde{q}_3),\ \tilde{q}_1 \geq \tilde{q}_2 \geq \tilde{q}_3, \\ \sum_i \tilde{q}_i = 0,\ \text{and } I^2 = \tilde{l} \cdot \tilde{l} \text{ with } \tilde{l} \text{ in standard position} \Big\}. \tag{38}$$

Because of rotational symmetry, the Hamiltonian dynamical system on the six-dimensional SU(3) coadjoint orbit $\mathcal{O}_\varrho$ can be reduced to a dynamical system on the two-dimensional space of intrinsic densities $\mathcal{M}^I_\varrho$. Define the time-dependent real antisymmetric matrix $\Omega(t) = \dot{R}\,R^{\mathrm{T}}$ which is called the angular velocity matrix. The Hamiltonian dynamical system (23) simplifies to,

$$\mathrm{i}\frac{d\tilde{\rho}}{dt} = [h_\Omega[\tilde{\rho}], \tilde{\rho}], \tag{39}$$

where $h_\Omega[\tilde{\rho}] = h[\tilde{\rho}] + \mathrm{i}\Omega$ is called the $\mathfrak{su}(3)$ Routhian.

6.3. *Rotational-scalar Energy Functions*

A rotational-scalar energy functional is a sum of integrity-basis functions. The trace of any polynomial in q and l is a rotational-scalar function. The quadrupole–quadrupole interaction is $q \cdot q = \operatorname{tr}(q^2) = \sum q_{ij}^2$. Because the quadratic Casimir function is $\mathcal{C}_2 = q \cdot q + \frac{1}{2} l \cdot l$, the squared length of the angular momentum vector and the quadrupole–quadrupole interaction are not functionally independent.

Using the results of Eq. (35) for the coordinate functions, q_{ij} and l_{ij}, the Hamiltonian vector field corresponding to any function is computed easily. For example, the exterior derivative of the squared length of the angular momentum is $d(l \cdot l) = 2\sum_{i>j} l_{ij}\, dl_{ij}$ and the corresponding vector field is $2\sum_{i>j} l_{ij}(-\mathrm{i})(E_{ij} - E_{ji}) = -2\mathrm{i}l$. The Hamiltonian vector fields associated with the Casimirs commute with ρ and, hence, are zero tangent vectors.

When the chosen energy function is

$$\mathcal{E}(\rho) = A(l \cdot l) + BX_3 + CX_4, \tag{40}$$

where A, B, C are real constants and the integrity-basis functions are given as (cf. Sec. 3.1 of Chapter 4)

$$X_3 = \operatorname{tr}(\tilde{l}\tilde{q}\tilde{l}), \tag{41}$$

$$X_4 = \operatorname{tr}(\tilde{l}\tilde{q}^2\tilde{l}) + (\tilde{l} \cdot \tilde{l})(\tilde{q} \cdot \tilde{q}), \tag{42}$$

the mean-field Hamiltonian is

$$\begin{aligned} h[\rho] = -2\mathrm{i}Al + 4B\left(q^2 - \frac{1}{3}(q \cdot q)I\right) \qquad (43)\\ + C\left(l^2 q + ql^2 - \frac{2}{3}X_3 I + 2(l \cdot l)q + 2\mathrm{i}(lq^2 + q^2 l - (q \cdot q)l)\right). \end{aligned}$$

6.4. $\mathfrak{su}(3)$ *Mean-field Dynamics*

The coadjoint rotation to the manifold of intrinsic densities maps the angular momentum l to $\tilde{l} = RlR^{\mathrm{T}}$. Because the angular momentum is a constant of the motion, the time-dependent intrinsic angular momentum obeys the dynamical equation,

$$\frac{d}{dt}\tilde{l} = [\Omega, \tilde{l}]. \tag{44}$$

When the antisymmetric matrices, $\tilde{l}$ and Ω, are expressed in terms of pseudovectors, $\vec{I}$ and $\vec{\omega}$, respectively, this dynamical equation becomes the equivalent familiar vector equation, $d\vec{I}/dt = -\vec{\omega} \times \vec{I}$ for the precession of the angular momentum vector in the body-fixed or intrinsic frame.

6.5. *Rotating Equilibrium Solutions*

When $\tilde{\rho}$ is a time-independent solution to the Routhian Lax equation (39), the system is rotating in equilibrium, i.e., the quadrupole deformation $\tilde{q}$ and the angular momentum $\tilde{l}$ in the rotating body-fixed frame are constants. For any rotationally invariant energy, Eq. (44) implies that

$$[\Omega, \tilde{l}] = 0 \Rightarrow \tilde{l} = \mathcal{I}\Omega \text{ or } \vec{I} = \mathcal{I}\vec{\omega}, \tag{45}$$

for some constant $\mathcal{I}$, which is called the moment of inertia. When the componenets I_k of the body-fixed angular momentum are non-zero, this equilibrium equation implies

$$\mathcal{I}^{-1} = \frac{\omega_1}{I_1} = \frac{\omega_2}{I_2} = \frac{\omega_3}{I_3}, \tag{46}$$

whose explicit form depends on the energy function.

We solve for equilibrium solutions in favorable cases analytically or, in more complex situations, numerically.

6.5.1. *Analytic solutions*

Because $\tilde{\rho}$ is a point on the orbit $\mathcal{O}_\varrho$, the components of $\tilde{q}$ and $\tilde{l}$ satisfy

$$\begin{aligned} 0 = \mathrm{tr}(\tilde{\rho}) &= \sum \tilde{q}_k, \\ \mathcal{C}_2(\varrho) = \mathrm{tr}(\tilde{\rho}^2) &= \sum \tilde{q}_k^2 + \frac{1}{2} I^2, \\ \mathcal{C}_3(\varrho) = \mathrm{tr}(\tilde{\rho}^3) &= \sum \tilde{q}_k^3 - \frac{3}{4} X_3, \end{aligned} \tag{47}$$

where X_3 is defined in Eq. (41) and $I^2 = -\frac{1}{2}\mathrm{tr}(\tilde{l}^2)$.

For the mean-field Hamiltonian of Eq. (43), constant $\tilde{q}$ implies

$$0 = \frac{d}{dt}\tilde{q}_k = C I_1 I_2 I_3 (\tilde{q}_i - \tilde{q}_j) \text{ for } i, j, k \text{ cyclic.} \tag{48}$$

Excluding spherical solutions, $\tilde{q} = 0$, at least one of the angular momentum components is zero.

Principal axis rotation. When two of the angular momentum components vanish, say $I_2 = I_3 = 0$, the body rotates about a principal axis. The three unknowns $(\tilde{q}_1, \tilde{q}_2, \tilde{q}_3)$ are solutions to the three equations (47) which depend parametrically on $I_1 = I$. There are three kinds of solutions in which the axis of rotation is short, long, or intermediate length:

$$\tilde{q}_1 = -\frac{\bar{\lambda} + 2\bar{\mu}}{3}, \quad 0 \le I \le I_{\max} = \bar{\lambda} \text{ (short)}, \tag{49}$$

$$\tilde{q}_1 = \frac{2\bar{\lambda} + \bar{\mu}}{3}, \quad 0 \le I \le I_{\max} = \bar{\mu} \text{ (long)}, \tag{50}$$

$$\tilde{q}_1 = -\frac{\bar{\lambda} - \bar{\mu}}{3}, \quad 0 \le I \le I_{\max} = \bar{\lambda} + \bar{\mu} \text{ (middle)} \tag{51}$$

and the deformations perpendicular to the rotation axis are

$$\tilde{q}_{2,3} = -\frac{\tilde{q}_1}{2} \pm \frac{1}{2}\sqrt{I_{\max}^2 - I^2}. \tag{52}$$

In the conventional notation of nuclear rotational bands when $\lambda > \mu$, the short axis solutions (49) form the $K = 0$ band, and the long axis solutions (50) are the $K = I$ sequence of bandheads, where K denotes the component of the intrinsic-frame angular momentum vector along the long axis.

The reciprocal of the moment of inertia is $\mathcal{I}^{-1} = 2(A+B\tilde{q}_1+C\tilde{q}_1^2)$ and the energy is $\mathcal{E} = I^2/(2\mathcal{I})$. The Routhian is $h_\Omega[\tilde{\rho}] = \mathrm{Ad}_g\epsilon$, where $\epsilon = \mathrm{diag}(\epsilon_1, \epsilon_2, \epsilon_3)$

$$\begin{aligned}
\epsilon_1 &= -\frac{2}{9}B(2\bar{\lambda}^2 - 4\bar{\mu}\bar{\lambda} - 4\bar{\mu}^2 - 3I^2) - \frac{4}{9}C(\bar{\lambda} + 2\bar{\mu})I^2, \\
\epsilon_2 &= \frac{1}{9}B(8\bar{\lambda}^2 + 8\bar{\mu}\bar{\lambda} - 4\bar{\mu}^2 - 3I^2) + \frac{2}{9}C(\bar{\lambda} + 2\bar{\mu})I^2, \\
\epsilon_3 &= -\frac{1}{9}B(4\bar{\lambda}^2 + 16\bar{\mu}\bar{\lambda} + 4\bar{\mu}^2 + 3I^2) + \frac{2}{9}C(\bar{\lambda} + 2\bar{\mu})I^2
\end{aligned} \tag{53}$$

or

$$\begin{aligned}
h_\Omega[\tilde{\rho}] = 2B &\begin{pmatrix} \frac{2}{3}(2\tilde{q}_1^2 - \tilde{q}_2^2 - \tilde{q}_3^2) & 0 & 0 \\ 0 & \frac{2}{3}(2\tilde{q}_2^2 - \tilde{q}_1^2 - \tilde{q}_3^2) & \mathrm{i}I\tilde{q}_1 \\ 0 & -\mathrm{i}I\tilde{q}_1 & \frac{2}{3}(2\tilde{q}_3^2 - \tilde{q}_1^2 - \tilde{q}_2^2) \end{pmatrix} \\
&+ \frac{2}{3}CI^2\tilde{q}_1\mathrm{diag}(2, -1, -1).
\end{aligned} \tag{54}$$

6.6. $\mathfrak{su}(3)$ *Normal Modes*

To find normal mode frequencies in the neighborhood of an elliptic equilibrium point $\tilde{\rho} \in \mathcal{M}_\varrho^I$, we may solve the linearized Lax equation.

A simple method is to use rotational invariance to reduce the normal mode system to a linear transformation of the two-dimensional tangent space $T_{\tilde{\rho}}\mathcal{M}_\varrho^I$. Suppose $\delta\tilde{\rho} = \delta\tilde{q} - \frac{\mathrm{i}}{2}\delta\tilde{l}$ is a tangent vector to $\mathcal{M}_\varrho^I$ at $\tilde{\rho} = \tilde{q} - \frac{\mathrm{i}}{2}\tilde{l}$. Because the vector is tangent to the intrinsic submanifold $\mathcal{M}_\varrho^I$, $\delta\tilde{q}$ is diagonal and $\mathrm{Tr}(\tilde{l}\,\delta\tilde{l}) = 0$. To be also tangent to the coadjoint orbit $\mathcal{O}_\varrho$, $\mathrm{Tr}(\tilde{\rho}\,\delta\tilde{\rho}) = \mathrm{Tr}(\tilde{\rho}^2\,\delta\tilde{\rho}) = 0$. Thus, for rotation about the principal 1-axis, the tangent space is the span of $\vec{v}_1$ and $\vec{v}_2$,

$$\vec{v}_1 = \begin{pmatrix} 0 & -\mathrm{i}/2 & 0 \\ \mathrm{i}/2 & 0 & 0 \\ 0 & 0 & 0 \end{pmatrix}, \quad \vec{v}_2 = \begin{pmatrix} 0 & 0 & \mathrm{i}/2 \\ 0 & 0 & 0 \\ -\mathrm{i}/2 & 0 & 0 \end{pmatrix}. \tag{55}$$

Denote a small displacement of the density near an equilibrium point by $\tilde{\rho}(t) = \tilde{\rho} + \delta\tilde{\rho}(t)$ and the perturbed angular velocity by $\Omega(t) = \Omega + \delta\Omega(t)$. The linearized Routhian equation is

$$\frac{d}{dt}\delta\tilde{\rho}(t) = -\mathrm{i}[h_\Omega[\tilde{\rho}], \delta\tilde{\rho}] - \mathrm{i}[\delta h[\tilde{\rho}; \delta\tilde{\rho}], \tilde{\rho}] + [\delta\Omega, \tilde{\rho}] \tag{56}$$

$$= L[\tilde{\rho}; \delta\tilde{\rho}]. \tag{57}$$

We choose $\delta\Omega$ such that $L[\tilde{\rho}, \delta\tilde{\rho}]$ is a tangent vector. $\delta\Omega$ depends linearly on $\delta\tilde{\rho}$, and, therefore, $L[\tilde{\rho}, \delta\tilde{\rho}]$ depends linearly on its second argument.

For rotation about the principal 1-axis, the linear operator evaluates to

$$L[\tilde{\rho}; \vec{v}_1] = L_{21}\vec{v}_2, \;\; L[\tilde{\rho}; \vec{v}_2] = -L_{12}\vec{v}_1, \tag{58}$$

where the equilibrium density $\tilde{\rho}$ is given in Sec. 6.5.1 and

$$\begin{aligned} L_{21} &= 2I(\tilde{q}_1 - \tilde{q}_3)(B - Cu_2), \\ L_{12} &= 2I(\tilde{q}_1 - \tilde{q}_2)(B - Cu_3), \\ u_k &= \tilde{q}_k - \frac{I^2}{4(\tilde{q}_1 - \tilde{q}_k)}. \end{aligned} \tag{59}$$

The non-zero frequency ω is an eigenvalue of the 2×2 matrix

$$\begin{pmatrix} 0 & -L_{12} \\ L_{21} & 0 \end{pmatrix}. \tag{60}$$

Hence, the square of the frequency is

$$\omega^2 = 4I^2(\tilde{q}_2 - \tilde{q}_1)(\tilde{q}_3 - \tilde{q}_1)\,(B - Cu_2)\,(B - Cu_3). \tag{61}$$

When $\tilde{q}_1$ is the long or short axis, ω^2 is positive and the equilibrium density is a stable elliptic point. But when $\tilde{q}_1$ is the middle length axis, ω^2 is negative and the equilibrium is unstable. The classical rigid rotor has a similar instability for middle axis rotation.

The stable solution to the linearized equation determines a precession with frequency ω of the angular momentum about the equilibrium rotation axis. In the rotational model, the $K = 0$ band of quantum states corresponds to rotations about the short axis with different angular momentum. The precession solutions with excitation energy ω determine the energies of $K \neq 0$ rotational bands.

7. Conclusion

This Chapter's paradigm examples illustrate the usefulness and practicality of Lie DFT. Applications require only elementary matrix computations. In $\mathfrak{su}(n)$ cases when n is small, such as $\mathfrak{su}(2)$ in Sec. 4 and $\mathfrak{su}(3)$ in Sec. 5, analytic results are attainable. Numerical calculations for $\mathfrak{su}(n)$ would be generally necessary when n is large or when the energy is a high-degree polynomial in the enveloping algebra. However, an $\mathfrak{su}(500)$ Lie DFT application is trivial for a modern laptop. When the dimension of an $\mathfrak{su}(n)$ algebra's irreducible representation space is astronomically large, Lie DFT computations on an integral coadjoint orbit still only involve $n \times n$ matrices.

The core mathematical ideas of Lie DFT are foundational components in the orbit method for the construction of irreducible representations and in the modern theory of classical mechanics. With respect to many-body physics, the Lie construction of mean-field dynamics and its linearization are generalizations of Hartree-Fock/RPA to any Lie algebra [31]. The deep connections among Lie DFT, the orbit method, and theoretical mechanics suggest several opportunities for interdisciplinary cross-fertilization. Do some integrable Lax systems from classical mechanics apply to Lie DFT quantum many-body systems of physical interest? Can the orbit method prove that, in a Lie group irrep, the densities of the eigenvectors of a Hamiltonian operator in the enveloping algebra equal (approximately) the critical points of an associated energy functional on the corresponding integral orbit? For a simple quadratic

Hamiltonian in the enveloping algebra, what is the associated Hohenberg–Kohn energy functional?

The Chapter's Lie DFT examples show that analytic results are achievable in special circumstances. In such favorable cases, it proves useful to characterize the coadjoint orbit as a level surface of the Casimirs. The Casimir identities impose constraints on the Lie algebra functions and enable the continued use of these mathematically and physically natural functions in the derivation of formulae. The Casimir method to characterize orbits has been applied to several cases, including Sp(3, $\mathbb{R}$) [32–34], GCM(3) [35], SO(5) [36], and SO(6) [37].

Lie DFT goes beyond the safe domain of the rigorous theorems of Kirillov integral orbit theory and makes estimates of the Hamiltonian operator's spectrum in the irrep space. Further mathematical developments in the Kirillov orbit method would provide a rigorous foundation for Lie DFT.

A.1. Appendix

The following table contrasts representation and coadjoint orbit models. It also provides a summary of main notations and references to the text.

Table A.1. Representation and coadjoint orbit models, where ϕ denotes an N-fermion antisymmetrized state vector in the Hilbert representation space $\mathcal{H}$ of a Lie algebra $\mathfrak{g}$ of observables X.

Conventional Lie group model	Mean-field model	Relevant discussion
1. $\mathfrak{g}$ = Lie algebra σ is a representation of $\mathfrak{g}$	$\mathfrak{g}^*$ = dual space of $\mathfrak{g}$	Sec. 2
2. Elements of $\mathfrak{g}$: quantum microscopic observables $\hat{X} = \sigma(X)$	Elements of $\mathfrak{g}^*$: densities ρ [moment maps, $M(\phi)$]	Sec. 2
3. $\phi \mapsto \sigma(g)\phi$, where g is an element of the Lie group G = exp($\mathfrak{g}$) and σ is the representation of G	Coadjoint transformation $\rho \mapsto Ad^\star_g(\rho)$	Sec. 4.1

(*Continued*)

Table A.1. (*Continued*)

Conventional Lie group model	Mean-field model	Relevant discussion
$\phi \mapsto \exp(\mathrm{i}t\hat{X})\phi$	$Ad_g^*(\rho) = \gamma_X(t)$, with $g = e^{\mathrm{i}tX}$ (see curves in Fig. 1(b); the coadjoint orbit $\mathcal{O}_\rho$ consists of all such curves as g ranges over the group G)	
4. Irreducible representation (irrep) space	Integral coadjoint orbit $\mathcal{O}_\rho$	Sec. 4.2
5. Dimension of irrep	Volume of coadjoint orbit	Sec. 4.2
6. Casimir operators $\hat{\mathcal{C}}_k$:	Casimir functions $\mathcal{C}_k$:	Sec. 4.2
— For an irrep of a compact Lie group, each vector in the irrep space is a simultaneous eigenvector for all the Casimir operators;	— A Casimir function is constant on each coadjoint orbit;	
— A simultaneous eigenspace of all the Casimir operators in a representation space is an irrep space.	— A common level surface of the Casimir functions for a compact Lie group is a single coadjoint orbit.	
7. Exact quasi-dynamical symmetry	Single non-integral coadjoint orbit	
8. Hamiltonian operator on representation space	Energy functional on coadjoint orbit $E[\rho]$: see Eq. (6)	Sec. 3
	Mean-field Hamiltonian $h[\rho]$ is determined by $d\mathcal{E}_\rho(X) = \mathrm{i}\langle\rho, [h[\rho], X]\rangle$ for all $X \in \mathfrak{g}$: see Eq. (16)	
9. Eigenstates of Hamiltonian	Critical points of energy functional	Sec. 3
	Densities are determined by the Lax equation $\mathrm{i}\frac{d}{dt}\rho = [h[\rho], \rho]$	Secs. 5 and 6

References

[1] P. Hohenberg and W. Kohn, *Phys. Rev. B.* **136**, 864 (1964).

[2] E. H. Lieb, *Int. J. Quant. Chem.* **24**, 243 (1983).

[3] R. G. Paar and W. Yang, *Density-Functional Theory of Atoms and Molecules.* Oxford University Press, Oxford (1989).

[4] J. Callaway and N. H. March, *Solid State Phys.* **38**, 135 (1984).

[5] D. Langreth and H. Suhl, *Many-Body Phenomena at Surfaces.* Academic Press, Orlando (1984).
[6] R. M. Dreizler and J. de Providencia, *Density Functional Methods in Physics.* Plenum Press, London (1985).
[7] A. A. Kirillov, *Lectures on the Orbit Method.* American Mathematical Society, Providence, RI (2004).
[8] J.-P. Souriau, *Structures Des Systemes Dynamiques.* Dunod, Paris (1969).
[9] B. Kostant, Orbits, symplectic structures and representation theory. In *Proc. US–Japan Seminar on Diff. Geom.*, Nippon Hyronsha, Tokyo (1966).
[10] D. Vogan, *Representations of Real Reductive Lie Groups.* Boston, Birkhäuser (1981).
[11] R. Abraham and J. E. Marsden, *Foundations of Mechanics.* Benjamin-Cummings, Reading, MA. (1978).
[12] J. P. Elliott, *Proc. Roy. Soc. A.* **245**, 128 (1958); *ibid.*, 562 (1958).
[13] C. Bahri, D. J. Rowe, and W. Wijesundera, *Phys. Rev. C.* **58**, 1539(1998).
[14] J. E. Marsden and T. S. Ratiu, *Introduction to Mechanics and Symmetry.* Springer-Verlag, New York (1994).
[15] M. Levy, *Proc. Natl. Acad. Sci. USA.* **76**, 6062 (1979).
[16] M. Levy, *Phys. Rev. A.* **26**, 1200 (1982).
[17] Ts. Dankova and G. Rosensteel, *Phys. Rev. C.* **63**, 054303 (2001).
[18] J. Dixmier, *Enveloping Algebras.* Sec. 2.4, North-Holland, Amsterdam (1977).
[19] J. E. Drut, R. J. Furnstahl, and L. Platter, *Prog. Part. Nucl. Phys.* **64**, 120 (2010).
[20] J. P. Perdew and S. Kurth. In C. Fiolhais, F. Nogueira, and M. Marques,(eds.), *A Primer in Density Functional Theory, Springer Lecture Notes in Physics.* Vol. 620, New York (2003), p. 1.
[21] H. J. Lipkin, N. Meshkov, and A. J. Glick, *Nucl. Phys. A.* **62**, 188 (1965).
[22] N. Meshkov, A. J. Glick, and H. J. Lipkin, *Nucl Phys A* **62**, 199 (1965).
[23] A. J. Glick, H. J. Lipkin, and N. Meshkov, *Nucl Phys A* **62**, 211 (1965).
[24] F. Iachello and A. Arima, *The Interacting Boson Model.* Cambridge Univ. Press, Cambridge (1987).
[25] G. Rosensteel and Ts. Dankova, *Phys. Rev. C.* **64**, 064303 (2001).
[26] G. Rosensteel and Ts. Dankova, *J Phys. A.* **35**, 1055 (2002).
[27] G. Rosensteel, *J Phys. A.* **37**, 6751 (2004).
[28] G. Rosensteel, *Nucl. Phys. A.* **341**, 397 (1980).
[29] Y. Leschber and J. P. Draayer, *Phys. Lett. B.* **190**, 1 (1987).
[30] O. Castanos, J. P. Draayer, and Y. Leschber, *Z. Phys. A.* **329**, 33 (1988).
[31] D. J. Rowe, A. Ryman, and G. Rosensteel, *Phys. Rev. A.* **22**, 2362 (1980).
[32] G. Rosensteel, *Phys. Rev. C.* **65**, 064321 (2002).
[33] G. Rosensteel and J. L. Graber, *J Phys. A.* **35**, L535 (2002).
[34] J. L. Graber and G. Rosensteel, *Phys. Rev. C.* **68**, 014301 (2003).
[35] G. Rosensteel, *J Phys. A.* **37**, 10967 (2004).
[36] G. Rosensteel, *J Phys. A.* **38**, 9221 (2005).
[37] G. Rosensteel, *J Phys. A.* **39**, 3113 (2006).

Chapter 11

Exactly Solvable Pairing in a Mean-field Framework: Models and Applications

Feng Pan[*,†,‡], Xin Guan[*] and Jerry P. Draayer[†]
[*] *Department of Physics, Liaoning Normal University*
Dalian 116029, P. R. China
[†] *Department of Physics and Astronomy, Louisiana State University*
Baton Rouge, LA 70803-4001, USA
[‡] *daipan@dlut.edu.cn*

Exact solutions of spherical or deformed mean field augmented by various types of pairing are briefly reviewed. Of particular interest are the standard pairing model, the nearest-level pairing model and the extended pairing model. The results for many pairing-driven quantities, such as odd–even mass differences and the moment of inertia, show that these models successfully describe the ground state of spherical nuclei or well-deformed heavy nuclei. Our recent results for level statistics of pair-excitation spectra in selected Ca isotopes described in the standard pairing model, together with the ground-state occupation probabilities of valence nucleon pairs with various angular momenta for even–even $^{156-162}$Yb and ground-state phase transitions in Sm isotopes described by the extended model, are also presented.

Contents

1. Introduction

Nuclear pairing correlation, as an important part of the residual interactions necessary to augment any nuclear mean-field theory, represents one of the main and longstanding pillars of current understanding of nuclear structure [1]. For example, the pairing interaction of the nuclear shell model plays a key role to reproduce low-energy spectroscopic properties of nuclei, such as binding energies, odd–even effects, single-particle energy-level occupancy, excitation spectra, and moments of inertia [2–4]. Bohr, Mottelson, Pines, and Belyaev were the first to introduce the Bardeen–Cooper–Schrieffer (BCS) theory of superconductivity in condensed matter [5] to descriptions of pairing phenomena in nuclei [2,6]. Though the BCS and the more refined Hartree–Fock–Bogolyubov (HFB) approximations provide simple and clear pictures in demonstrating pairing correlations in nuclei [2,7,8], tremendous efforts have been made in finding accurate solutions to the problem [9–14] to overcome serious drawbacks in the BCS and the HFB, such as spurious states and non-orthogonal solutions, resulting from particle-number nonconservation effects in these approximations [10,14–16].

Driven by the importance of having exact solutions of either spherical or deformed mean-field plus standard pairing Hamiltonian, considerable progress, building on Richardson and Gaudin's early work [17–21] and extensions to it based on the Bethe ansatz, has been made [22–31]. For all these algebraic Bethe ansatz approaches, the solutions are provided by a set of highly non-linear Bethe ansatz equations (BAEs). Though these applications demonstrate that the pairing problem is exactly solvable, solutions of these BAEs are not easy and normally require extensive numerical work, especially when the number of levels and valence pairs are large [27]. This limits the applicability of the methodology to relatively small systems. However, it has been shown recently that the set of Gaudin–Richardson equations for the standard pairing case can be solved relatively easily by using the extended Heine–Stieltjes (HS) polynomial approach [32, 33]. Since solutions of the standard pairing model can be obtained from zeros of the associated extended HS polynomials, the approach can be applied to study the model with more pairs over a larger number of single-particle levels.

Moreover, if nuclear pairing interaction is restricted to a nearest state in a deformed mean-filed theory, it leads to a simply solvable hard-core Bose–Hubbard model Hamiltonian, which may also be suitable to describe well-deformed nuclei [34–36]. More recently, the Nilsson mean-field plus extended pairing model has also been proposed to describe deformed nuclei [37], which includes pairing interactions among valence pairs in different orbits up to infinite order. It has been shown that the extended pairing model can be regarded as the standard pairing Hamiltonian at a first-order approximation, namely, when only the lowest energy eigenstate described by the Racah quasi-spin formalism of the standard pairing interaction is taken into consideration, and the results thus display similar pair structures as the ones found in the low-lying states of the standard pairing model [38]. The advantage of the model lies in the fact that it can also be solved exactly and more easily than the standard pairing model, especially when both the number of valence nucleon pairs and the number of single-particle orbits are large, which, therefore, is more suitable to be used for rare-earth nuclei [39].

In the following sections, we briefly review either spherical or deformed mean-field plus standard pairing, nearest-level pairing (NLP), and extended pairing models. Application of these models to describe odd–even mass differences, moment of inertia, level statistics, and possible phase transitions are also presented. A brief summary is given in the final section.

2. Exact Solution of the Standard Pairing Model

The Hamiltonian of the spherical mean-field plus pairing model within the proton or neutron sector is given by

$$\hat{H}_{\rm SP} = \sum_{j=1}^{n} \epsilon_j \hat{n}_j - G \sum_{jj'} S_j^+ S_{j'}^-, \tag{1}$$

where the sums run over all given j-levels of total number n, $G > 0$ is the overall pairing interaction strength, $\{\epsilon_j\}$ are non-degenerate single-particle energies obtained from any spherical mean-field theory, such as the independent-particle spherical shell model, $\hat{n}_j = \sum_m c_{jm}^\dagger c_{jm}$ is the number operator for valence particles in the jth level, and $S_j^+ = \sum_{m>0} (-)^{j-m} c_{jm}^\dagger c_{j\ -m}^\dagger$ $[S_j^- = (S_j^+)^\dagger]$ are pair creation [annihilation] operators.

According to the Richardson–Gaudin method [17, 21], k-pair eigenstates of Eq. (1) can be written as

$$|k; x; \nu_{j'}\rangle = S^+(x_1) S^+(x_2) \cdots S^+(x_k) |\nu_{j'}\rangle, \tag{2}$$

where $|\nu_{j'}\rangle$ is the pairing vacuum state with the seniority $\nu_{j'}$ that satisfies $S_j^- |\nu_{j'}\rangle = 0$ and $\hat{n}_j |\nu_{j'}\rangle = \delta_{jj'} \nu_j |\nu_{j'}\rangle$ for all j, and

$$S^+(x_i) = \sum_{j=1}^{n} \frac{1}{x_i - 2\epsilon_j} S_j^\dagger, \tag{3}$$

in which x_i $(i = 1, 2, \ldots, k)$ satisfy the following set of BAEs:

$$1 - 2G \sum_j \frac{\rho_j}{x_i - 2\epsilon_j} - 2G \sum_{i'=1(\neq i)}^{k} \frac{1}{x_i - x_{i'}} = 0, \tag{4}$$

where the first sum runs over all j-levels and $\rho_j = -\Omega_j/2$ for level degeneracy $\Omega_j = j + \frac{1}{2} - \delta_{jj'}\nu_{j'}$. For each $x^{(\xi)}$ solution, the corresponding eigenenergy is given by

$$E_k^{(\xi)} = \sum_{i=1}^{k} x_i^{(\xi)} + \nu_{j'}\epsilon_{j'}. \tag{5}$$

The standard pairing solution shown above was first derived by Richardson [17] and then discussed in a series of his papers [18–20]. Independently, similar solution to the spin-chain problem was considered by Gaudin [21]. The procedure used is now called Richardson–Gaudin method, which is also equivalent to the algebraic Bethe ansatz method, which is an effective tool in deriving exact solutions of standard pairing and other integrable models [23–25, 28].

It was observed that BAEs similar to those shown in Eq. (4) can be solved from zeros of the corresponding extended HS polynomials [32, 33, 40]. Through the HS correspondence for the BAEs (4), one may consider the second-order Fuchsian equation

$$A(x)y''(x) + B(x)y'(x) - V(x)y(x) = 0. \tag{6}$$

Here, $A(x) = \prod_j (x - 2\epsilon_j)$ is a polynomial of degree n, the polynomial $B(x)$ is given as

$$B(x)/A(x) = \sum_j \frac{2\rho_j}{x - 2\epsilon_j} - \frac{1}{G}, \tag{7}$$

where the sum runs over all j-levels and $V(x)$ are called Van Vleck polynomials [32] of degree $n-1$, which are determined according to Eq. (6). In the original electrostatic analogue considered by Heine and Stieltjes [41], the parameters $\{\rho_j\}$ that specify fixed charges should all be positive for no external electrostatic field, $1/G \to 0$. Therefore, the polynomials $y(x)$ with negative $\{\rho_j\}$ and $1/G \neq 0$ are called extended HS polynomials. They approach the original HS polynomials with negative $\{\rho_j\}$ in the $G \to \infty$ limit.

In search for polynomial solutions of Eq. (6), we write

$$y(x) = \sum_{j=0}^{k} a_j x^j, \quad V(x) = \sum_{j=0}^{n-1} b_j x^j, \tag{8}$$

where $\{a_j\}$ and $\{b_j\}$ are the expansion coefficients to be determined. Substitution of (8) into Eq. (6) yields two matrix equations. Namely, the condition that the coefficients in front of x^i $(i = 0, \ldots, k)$ must be zero yields a $(k+1) \times (k+1)$ matrix $\mathbf{F}$ with $\mathbf{F}\mathbf{v} = b_0\mathbf{v}$, where the eigenvector $\mathbf{v}$ of $\mathbf{F}$ is simply given by the expansion coefficients $\mathbf{v} = \{a_0, \ldots, a_k\}$. In addition, the condition that the coefficients in front of x^i $(i = k+1, \ldots, n+k-1)$ must be zero yields another $(n-1) \times (k+1)$ upper-triangular matrix $\mathbf{P}$ with $\mathbf{P}\mathbf{v} = 0$, which provides a unique solution for b_i $(i = 1, \ldots, n-1)$ in terms of $\{a_j\}$. Entries of the two matrices are all linear in the coefficients $\{b_1, b_2, \ldots, b_{n-1}\}$. Matrices $\mathbf{F}$ and $\mathbf{P}$ can be easily constructed, and solutions can be computed efficiently using a MATHEMATICA computer code [33]. Furthermore, if we set $a_k = 1$ in $y(x)$, the coefficient a_{k-1} becomes equal to the negative sum of the $y(x)$ zeros, $a_{k-1} = -\sum_{i=1}^{k} x_i$, and hence, yields the corresponding eigenenergy $E_k = -a_{k-1} + \nu_{j'}\epsilon_{j'}$ according to Eq. (5).

2.1. *Pairing Gaps in* Ni *and* Sm *Isotopes*

Using the HS correspondence and the polynomial approach to the standard pairing model (1), hereafter referred to as "HS Pairing", one can evaluate various ground-state quantities of the model and compare with experimental results, even when valence space is relatively larger. For example, pairing gaps for Ni and Sm isotopes are calculated in the model [32], and are compared with the corresponding experimental data [42] and results obtained by using the BCS approximation [5, 15] (Fig. 1). In the calculation, the empirical like-particle pairing gap is estimated by the third derivative of binding energy, E_B, with respect to the number of valence like-particles [1], which for neutrons is,

$$P(N) \equiv \frac{1}{4}(E_B(Z, N-2) - 3E_B(Z, N-1) + 3E_B(Z, N) - E_B(Z, N+1)). \tag{9}$$

We find that a paring strength of $G = 23/A$ yields a close reproduction of experimental results for Ni isotopes (Fig. 1(a)).

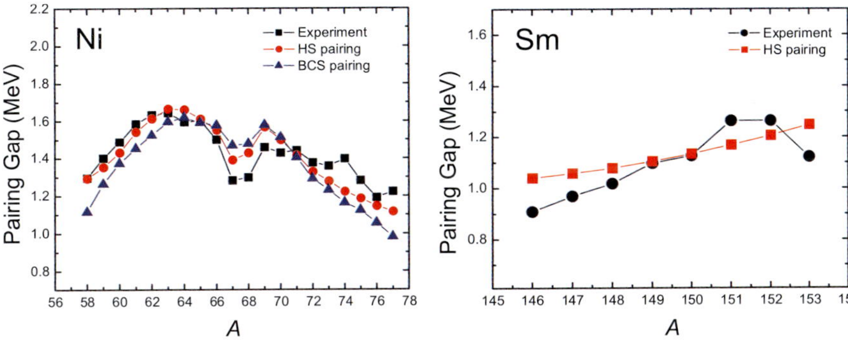

(a) (b)

Fig. 1. Pairing gaps in MeV (a) for $^{58-77}$Ni calculated in the standard pairing model of Eq. (1) using the HS correspondence ("HS pairing") and in the conventional BCS approach ("BCS pairing") [5, 15] with four j-shells, $f_{5/2}, p_{1/2,3/2}, g_{9/2}$, and $G = 23/A$ MeV, and compared to experimental data, and (b) for $^{146-153}$Sm calculated in the HS pairing with six j-levels in the $pfhi$ shell and $G = 24/A$ MeV, and compared to the corresponding experimental data. The figures are taken from Ref. [32].

For Sm isotopes, the calculations are performed for valence neutrons in the sixth HO shell with a model space of 6 j-levels, $f_{5/2,7/2}, p_{1/2,3/2}, h_{9/2}, i_{13/2}$ (Fig. 1(b)). It can be observed that the theoretical paring gaps with empirical pairing strength $G = 24/A$ MeV also closely reproduce the experimental pattern.

2.2. *Level Statistics in* Ca *Isotopes*

Level statistics for the HS pairing model can also be studied by using the polynomial approach, in which two typical statistical measures [43, 45, 46], $p(S)$ and $\Delta_3(L)$, defined below, are used to analyze the level fluctuation properties of a k-pair even- or odd-mass system. The distribution of the nearest-neighbor level spacing $p(S)$ is defined as the probability of two nearest-neighbor energy levels to have a spacing S, which is calculated from the unfolded spectrum[a]

[a]The unfolded levels $\{\tilde{E}_k^{(\xi)}\}$ are obtained from the mapping $N(E_k^{(\xi)}) \mapsto \tilde{E}_k^{(\xi)}$, where $\{E_k^{(\xi)}\}$ are given by Eq. (5) with ξ labeling the ξth excited level, and the mapping is

$\{\tilde{E}_k^{(\xi)}\}$ as

$$S_\xi = \tilde{E}_k^{(\xi+1)} - \tilde{E}_k^{(\xi)}. \tag{10}$$

As is commonly accepted, Poisson statistics with

$$p(S) = e^{-S} \tag{11}$$

characterizes a regular system, while the Wigner distribution, which is almost identical to the Gaussian orthogonal ensemble (GOE) prediction,

$$p(S) = (\pi/2)Se^{-\pi S^2/4}, \tag{12}$$

characterizes a chaotic system. The spectral rigidity is defined as

$$\Delta_3(\alpha, L) = \frac{1}{L}\min_{A,B}\int_\alpha^{\alpha+L}[N(\tilde{E}) - (A\tilde{E} + B)]^2 d\tilde{E}, \tag{13}$$

which is the average of the least-square deviations between the number staircase function $N(\tilde{E})$ for the unfolded spectrum and its best linear fit $(A\tilde{E} + B)$ over the energy interval $[\alpha, \alpha + L]$. The average of $\Delta_3(\alpha, L)$ over n_α intervals $(\alpha, \alpha + L)$, which overlap by $L/2$ successively, yields a smoother $\overline{\Delta}_3(L)$ measure,

$$\overline{\Delta}_3(L) = \frac{1}{n_\alpha}\sum_\alpha \Delta_3(\alpha, L). \tag{14}$$

It is shown that

$$\overline{\Delta}_3(L) = L/15, \tag{15}$$

for a regular spectrum, while, in the large-L limit,

$$\overline{\Delta}_3(L) \approx \frac{1}{\pi^2}(\ln L - 0.0687), \tag{16}$$

for a chaotic system with the GOE statistics.

Level statistics in Ca isotopes described by the HS pairing model has been studied in Ref. [47], in which single-particle energies and

---

determined by fitting a smooth six-degree polynomial function to the number staircase function, $N(E_k^{(\xi)})$, that counts the number of levels below $E_k^{(\xi)}$ [43, 44].

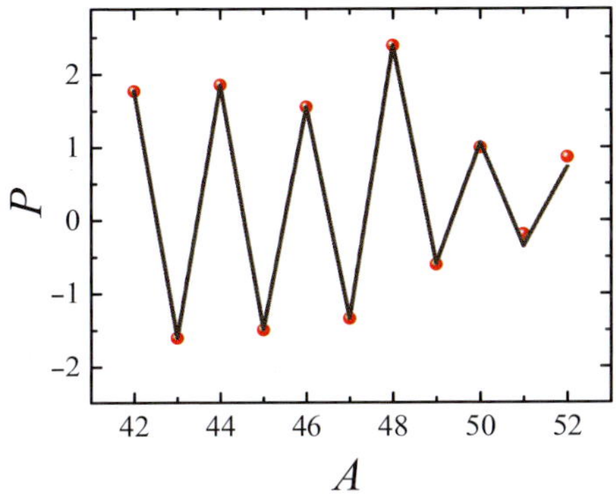

Fig. 2. Theoretical values connected by solid lines and the corresponding experimental values [42] (circles) of the pairing gap $P(A)$ (in units of MeV) of $^{42-52}$Ca as a function of the mass number A. The figure is taken from Ref. [47].

pairing strengths G deduced from experimental data [42, 48] are adopted. In particular, in the calculation, the proton energy is treated as a constant. The pairing interaction strength G for neutrons is determined from pairing gaps deduced from experimental binding energies [42]. The pairing gaps of $^{42-53}$Ca are thus calculated by

$$P(A) = E_B(A+1) + E_B(A-1) - 2E_B(A), \tag{17}$$

where $E_B(A)$ is the binding energy of a nucleus with a mass number A, which is either taken from experiment or calculated as the ground-state energy of the pairing Hamiltonian (1) (Fig. 2).

The spectral statistical properties for each Ca isotope are expected to be strongly influenced by the interplay of the mean field and the pairing interaction. The level spacing distribution $p(S)$ and spectral rigidity $\overline{\Delta}_3(L)$ for $^{48-53}$Ca are shown in Fig. 3. In most cases, such as ^{48}Ca and $^{51-53}$Ca, the $p(S)$ and $\overline{\Delta}_3(L)$ are of the Poisson type. The two exceptions, that are found to obey the GOE statistics, are ^{49}Ca and ^{50}Ca, as revealed by the $p(S)$ and $\overline{\Delta}_3(L)$ quantities. Since these two nuclei are close to a subshell closure (neutron number 28) and hence, an effective ^{48}Ca core may be adopted, their unusual statistical behavior may be predominantly dictated by the shell structure. This, together with the GOE-like spectral statistics shown in Fig. 3, suggests that ^{49}Ca and ^{50}Ca realize

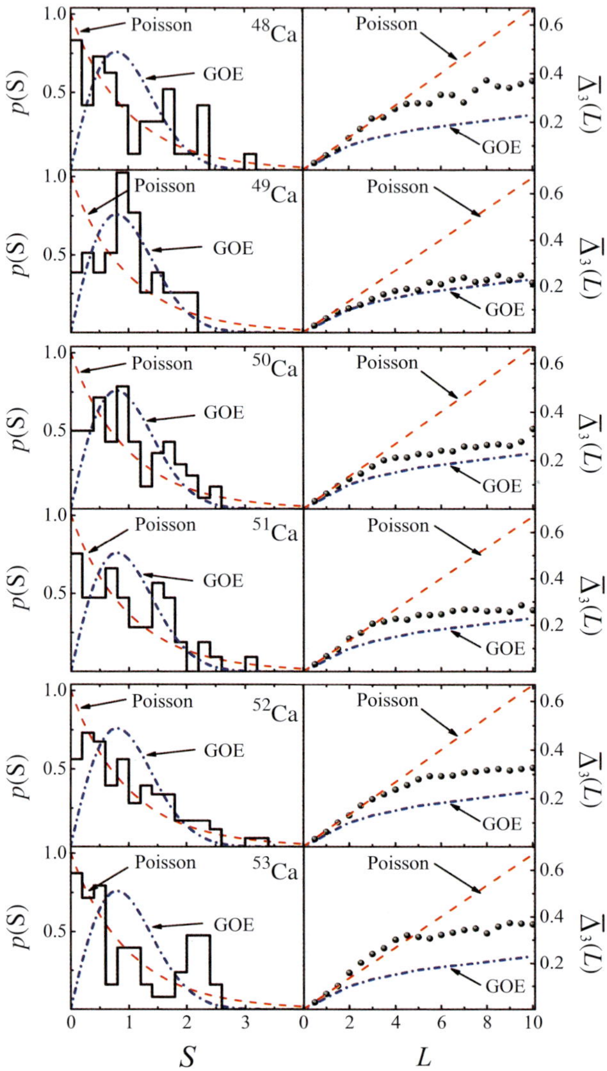

Fig. 3. Level spacing distribution $p(S)$ as a function of the spacing S, and spectral rigidity $\overline{\Delta}_3(L)$ as a function of the energy interval L for the pair-excitation levels in $^{48-53}$Ca. Results are shown for the HS pairing model (1) (black), for the Poisson statistics (red), see Eqs. (11) and (15), and for the GOE statistics (blue), see Eqs. (12) and (16). These figures are taken from Ref. [47].

a realistic example of nuclei in the phase transitional region from the localized to the superconducting phase. The GOE-like behavior of the level statistics at the exceptional G points in the model may be due to the fact that these points are related locally to the behavior of the potential, representing generically the corresponding classical instability points, as conjectured in Ref. [49]. The results indicate that pairing-driven nuclei can indeed manifest a chaotic behavior, even though the regular behavior is more common.

3. Deformed Mean-field Plus Nearest-level Pairing Model

Within the deformed mean field, there are several types of pairing interactions that can be solved exactly. The NLP interaction is inspired by the level-dependent Gaussian-type pairing interaction [11] with

$$G_{ii'} = -A_{ii'} e^{-B(\varepsilon_i - \varepsilon_{i'})^2}, \tag{18}$$

where i and i' represent doubly occupied Nilsson levels with single-particle energies ε_i and $\varepsilon_{i'}$, generated from the Nilsson model, the parameters $A_{ii'} > 0$ and $B > 0$ are adjusted in such a way that the location of the first excited eigensolution lies approximately at the same energy as for the standard pairing case. Of course, there is some freedom in adjusting the parameters, allowing one to control in a phenomenological way the interaction among the levels. The expression (18) implies that scattering between particle pairs occupying levels with single-particle energies that lie close to each other are favored; scattering between particle pairs in levels with distant single-particle energies are not favored. As an approximation, this pairing interaction is further simplified to the nearest-level coupling [34–36], namely, $G_{ii'}$ is given by

$$G_{ii} = -A, \; G_{ii+1} = G_{i+1i} = -Ae^{-B(\varepsilon_i - \varepsilon_{i+1})^2}, \tag{19}$$

where $A > 0$ are real parameters, while $G_{ii'}$ are taken to be 0 otherwise. Hence, the deformed mean-field plus NLP Hamiltonian can be expressed as

$$\hat{H}_{\mathrm{NLP}} = \sum_{\rho=1}^{r} \varepsilon_{q_\rho} \hat{n}_{q_\rho} + \sum_{i,i'}{}' t_{ii'} b_i^\dagger b_{i'}, \tag{20}$$

where the first sum runs over the levels occupied by a single particle which occurs in the description of odd-A nuclei or several unpaired particles in broken-pair cases, and the second primed sum runs only over levels that are occupied by pairs of particles. For the NLP interaction, the t-matrix is tridiagonal given by $t_{ii} = 2\varepsilon_i + G_{ii} = 2\varepsilon_i - A$ and $t_{ii+1} = t_{i+1i} = G_{ii+1}$, with $t_{ii'} = 0$ otherwise. The fermion pair operators in this expression are given by

$$b_i^\dagger = a_i^\dagger a_{\bar{i}}^\dagger, \quad b_i = a_{\bar{i}} a_i, \tag{21}$$

where $a_i^\dagger$ is the creation operator for the ith single-particle level in the Nilsson model and $a_{\bar{i}}^\dagger$ is the corresponding time-reversed state. The $b_i^\dagger$ and b_i satisfy the following commutation relations:

$$[b_i, b_j^\dagger] = \delta_{ij}(1 - 2N_i), \quad [N_i, b_j^\dagger] = \delta_{ij} b_j^\dagger, \quad [N_i, b_j] = -\delta_{ij} b_j, \tag{22}$$

where $N_i = \frac{1}{2}(a_i^\dagger a_i + a_{\bar{i}}^\dagger a_{\bar{i}})$ is the pair number operator in the ith level for even–even nuclei.

The eigenstates of Eq. (20) for k-pair excitation can be expressed as

$$|k; \eta, n_{q_1}, n_{q_2}, \ldots, n_{q_r}\rangle$$
$$= \sum\nolimits'_{i_1 < i_2 \cdots < i_k} C^{(\eta)}_{i_1 i_2 \cdots i_k} b_{i_1}^\dagger b_{i_2}^\dagger \cdots b_{i_k}^\dagger |n_{q_1}, n_{q_2}, \cdots, n_{q_r}\rangle, \tag{23}$$

where the total number of particles $n = 2k + \sum_{\rho=1}^{r} n_{q_\rho}$, $q_1, q_2, \ldots, q_r$ are the levels occupied by r single particles, the prime indicates that $i_1,\ i_2, \ldots, i_k$ can not be taken to be $q_1,\ q_2, \ldots, q_r$ in the summation, and $n_{q_\rho} = 1$ $(\rho = 1, 2, \ldots, r)$. If only even–even and odd-A nuclei are treated without including broken pairs, r is taken to be 1 for odd-A nuclei, and 0 for even–even nuclei. In Eq. (23), $C^{(\eta)}_{i_1 i_2 \ldots i_k}$ is a determinant given by

$$C^{(\eta)}_{i_1 i_2 \cdots i_k} = \begin{vmatrix} g^{\eta_1}_{i_i} & g^{\eta_1}_{i_2} & \cdots & g^{\eta_1}_{i_k} \\ g^{\eta_2}_{i_i} & g^{\eta_2}_{i_2} & \cdots & g^{\eta_2}_{i_k} \\ \cdots & \cdots & \cdots & \cdots \\ g^{\eta_k}_{i_i} & g^{\eta_k}_{i_2} & \cdots & g^{\eta_k}_{i_k} \end{vmatrix}, \tag{24}$$

where η is a shorthand notation for a selected set of k eigenvalues of the $\tilde{t}$ matrix (which denotes the t matrix without the corresponding r rows and columns) that can be used to distinguish the eigenstates with the same number of pairs, k; g^{η_p} is the pth eigenvector of the $\tilde{t}$ matrix with eigenvalue $E^{(\eta_p)}$, that is

$$\sum_{i'} \tilde{t}_{ii'} g^{\eta_p}_{i'} = E^{(\eta_p)} g^{\eta_p}_i. \tag{25}$$

The excitation energies corresponding to Eq. (23) can be expressed as

$$E^{(\eta)}_k = \sum_{\rho=1}^{r} \varepsilon_{q_\rho} + \sum_{i=1}^{k} E^{(\eta_i)}, \tag{26}$$

where the first sum runs over r Nilsson levels each occupied by a single valence nucleon (for odd-A nuclei or in broken-pair cases), the second one is a sum of k different eigenvalues of the $\tilde{t}$ matrix. Obviously, $\tilde{t}$ is an $(N - r) \times (N - r)$ matrix, since those levels occupied by single valence nucleons are excluded as a result of the Pauli blocking.

3.1. *Odd–even Mass Difference: Actinide and Rare-earth Nuclei*

For even–even nuclei, the k-pair excitation energies are determined by the sum of k different eigenvalues chosen from the N eigenvalues of the $\tilde{t}$ matrix with $r = 0$, and hence the total number of excited levels is $N!/k!(N - k)!$. While for odd-A nuclei, the Nilsson level that is occupied by the single valence nucleon should be excluded in the original t matrix. In the latter case, the eigenvalue problem of Eq. (20) can be solved simply by diagonalizing the corresponding $\tilde{t}$ matrix as shown in Eq. (25) with the eigenstates and the corresponding eigenenergy given by Eqs. (23) and (26), respectively. When proton–proton and neutron–neutron pairing interactions are all considered, the model Hamiltonian includes both neutron and proton parts according to Eq. (20). We also consider the case with no proton–proton pair excitations, and in this case, we take the proton mean-field and proton pairing-energy contribution to the total energy as a constant in the fit of the theoretical binding energies to experimental data. Actually, the proton mean field and the proton pairing interaction mainly contribute to the energy and moment of inertia of low-lying excited states, and less affect odd–even mass difference as shown in Fig. 4.

3.2. *Moments of Inertia of Actinide Nuclei*

Moment of inertia of the first 2^+ state defined by $\Im_{\rm exp} = 3\hbar^2/E(2_1^+)$, where $E(2_1^+)$ is the experimental energy of the first 2^+ state, can also be calculated according to the Inglis formula from the cranking model [52]

$$\Im_{\rm th} = 2\hbar^2 \sum_n \frac{|\langle n|J_{x'}|{\rm g}\rangle|^2}{E_n - E_{\rm g}}, \tag{27}$$

where $J_{x'}$ is total angular momentum projected onto the x' axis of the intrinsic frame, $|{\rm g}\rangle$ and $|n\rangle$ stand for ground and excited states calculated in the model, and the sum runs over all excited states with broken pairs. Since excited states with more than one broken pair lie higher in energy and contribute much less to the moment of inertia according to Eq. (27), only one broken-pair states are considered in our calculation. Fig. 5 shows the moment of inertia of the first 2^+ state for even–even nuclei of $^{226-233}$Th, $^{230-239}$U, and $^{236-243}$Pu, calculated in the model (20) with both proton–proton and neutron–neutron pairing and compared with the corresponding experimental values, the standard pairing with the BCS approximation [53], and

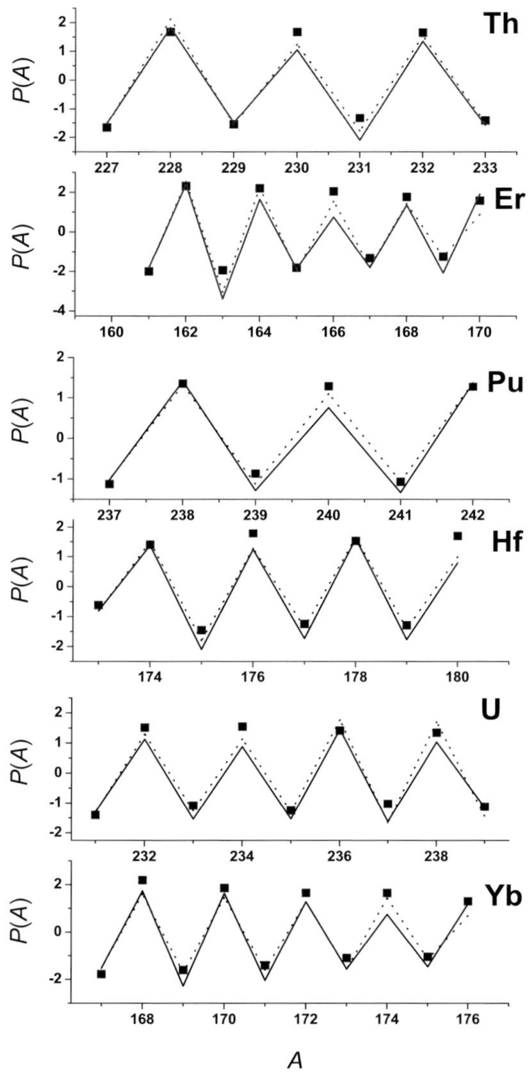

Fig. 4. Odd–even mass differences $P(A)$ in MeV of $^{226-234}$Th, $^{160-170}$Er, $^{236-243}$Pu, $^{172-180}$Hf, $^{230-239}$U, and $^{166-176}$Yb nuclei, where the filled squares correspond to the experimental values [42, 50, 51]. The results calculated in the model (20) with both proton–proton and neutron–neutron pairing are connected by solid lines, while those calculated in the model with only the neutron part considered are connected by dashed lines. The figures are taken from Ref. [36].

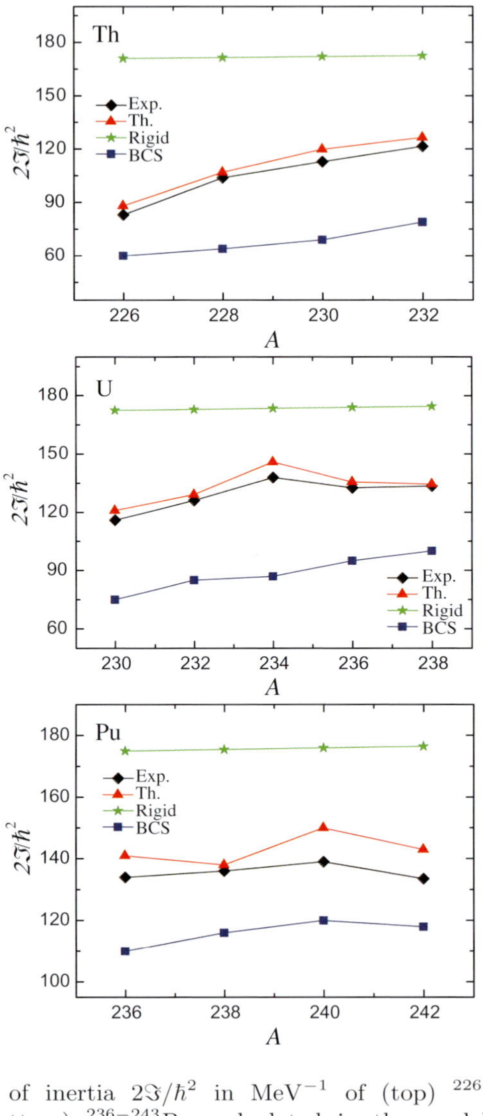

Fig. 5. Moment of inertia $2\Im/\hbar^2$ in MeV^{-1} of (top) $^{226-233}$Th, (middle) $^{230-239}$U, and (bottom) $^{236-243}$Pu, calculated in the model (20) with both proton–proton and neutron–neutron pairing (denoted as "Th.", red triangles) and compared to the corresponding experimental values (black diamonds) [42, 50, 51], the standard pairing with the BCS approximation (blue squares) [53], and results from the rigid-body formula (green stars). These figures are taken from Ref. [36].

results from the rigid-body formula,

$$\Im_{\text{rigid}} \simeq \frac{2}{5} M R_0^2 (1 + 0.31\beta), \tag{28}$$

where $R_0 \approx 1.2A^{\frac{1}{3}} \times 10^{-13}$cm is the empirical nuclear radius, β and M are the quadrupole deformation parameter and nuclear mass, respectively. It can be seen from Fig. 5 that the rigid-body results are too large if no pairing interaction is included, while the reduced moments of inertia in the BCS approximation indicate that this model may overestimate the pairing interaction in realistic nuclear systems, as was pointed out by Nilsson and Prior in their early work [53]. In our calculation, however, only the nearest-level pairing is considered, for which the results are much better than those from the standard pairing with the BCS approximation as far as these actinide nuclei are concerned.

4. Deformed Mean-field Plus Extended Pairing Model

Based on the standard pairing, one can construct an extended pairing model [37], which is given by

$$\begin{aligned} \hat{H} &= \sum_{j=1}^{p} \epsilon_j n_j - G_{\text{ex}} \sum_{i,j=1}^{p} b_i^\dagger b_j \\ &- G_{\text{ex}} \sum_{\mu=2}^{\infty} \frac{1}{(\mu!)^2} \sum_{i_1 \neq i_2 \neq \cdots \neq i_{2\mu}} b_{i_1}^\dagger b_{i_2}^\dagger \cdots b_{i_\mu}^\dagger b_{i_{\mu+1}} b_{i_{\mu+2}} \cdots b_{i_{2\mu}}, \end{aligned} \tag{29}$$

where $G_{\text{ex}} > 0$ is the overall pairing strength. Besides the Nilsson mean field and the standard pairing interaction, the Hamiltonian (29) also includes many-pair hopping terms that allow nucleon pairs to simultaneously scatter (hop) among different Nilsson levels, which is thus exactly solvable. Due to the Pauli principle and the particle number conservation, the infinite sum in Eq. (29) naturally truncates, namely, $\mu \leq [p/2]$, where $[x]$ denotes the integer part of x. It is also clear that each term of the form $b_i^\dagger \cdots b_j^\dagger$ that enters into the eigenstates of Eq. (29) should have different indices $i \neq \cdots \neq j$. Let

$|j_1, \ldots, j_m\rangle$ be the pairing vacuum state that satisfies

$$b_i|j_1, \ldots, j_m\rangle = 0 \tag{30}$$

for $1 \le i \le p$, where each of the m levels, $j_1, j_2, \ldots, j_m$, is occupied by a single nucleon. With this extension in place, we have shown that the model is exactly solvable. The k-pair eigenstates of Eq. (29) can be written as

$$|k; \zeta; j_1, \ldots, j_m\rangle = \sum_{1 \le i_1 < i_2 < \cdots < i_k \le p} \Lambda^{(\zeta)}_{i_1 i_2 \cdots i_k} b^+_{i_1} b^+_{i_2} \cdots b^+_{i_k} |j_1, \ldots, j_m\rangle, \tag{31}$$

where $\Lambda^{(\zeta)}_{i_1 i_2 \cdots i_k}$ is the expansion coefficient with

$$\Lambda^{(\zeta)}_{i_1 i_2 \cdots i_k} = \frac{1}{1 - x^{(\zeta)} \sum_{\mu=1}^{k} \epsilon_{i_\mu}}, \tag{32}$$

where, similar to the results given in the Bethe ansatz approach, $x^{(\zeta)}$ should satisfy

$$\frac{2}{x^{(\zeta)}} + \sum_{1 \le i_1 < i_2 < \cdots < i_k \le p} \frac{G_{\text{ex}}}{1 - x^{(\zeta)} \sum_{\mu=1}^{k} \epsilon_{i_\mu}} = 0. \tag{33}$$

The additional quantum number ζ labels the ζth solution of Eq. (33). The corresponding k-pair excitation energies of Eq. (31) are given by

$$E^{(\zeta)}_k = \frac{2}{x^{(\zeta)}} - G_{\text{ex}}(k-1). \tag{34}$$

Similar results for odd–even systems can also be derived by using this approach except that the index j of the level occupied by the single nucleon should be excluded from the summation in Eq. (31) and the single-particle energy ϵ_j contributing to the eigenenergy from the first term of Eq. (29) should be included in Eq. (34). Extensions to many broken-pair cases are thus straightforward.

It has been shown that, if only the first few eigenstates are considered, the pair structure of these states in the extended pairing model and the standard pairing model are similar, especially in the ground state [38]. It is not difficult to show that the expressions for the eigenstates of the extended pairing Hamiltonian and those of the standard pairing Hamiltonian (1) in the first-order approximation are the same. For k-pair excitations, $G_{\rm ex}$ in the extended pairing model and the parameter G in the standard pairing Hamiltonian (1) satisfy the relation

$$G_{\rm ex} = \frac{(p-k)!k!}{p!}(p-k+1)kG. \tag{35}$$

Once the overall pairing strength G is fixed, and the parameter $G_{\rm ex}$ is chosen according to Eq. (35), it is easy to show that the difference between the ground-state energy of the extended model, $E_k^{(\rm gs)}(\rm ex)$, and that of the standard pairing model in the first-order approximation, $E_k^{(\rm gs)}$, is given by

$$E_k^{(\rm gs)}(\rm ex) - E_k^{(\rm gs)} = -(k-1)G_{\rm ex}, \tag{36}$$

which shows that the extended pairing interaction among valence pairs is slightly more attractive than the standard pairing interaction in the first-order approximation. Since $G_{\rm ex}$ decreases drastically with an increasing number of pairs k, filling up through the middle of a major shell, the ground-sate energy difference for the two Hamiltonians becomes negligible for many pairs k. In addition, since the extended pairing model can be solved exactly with a single one-variable equation (33), which is simpler than the Richardson–Gaudin equations with k variables shown in (4) for the standard pairing Hamiltonian, the extended pairing model can be applied to relatively large systems, especially when only ground state and a few low-lying eigenstates, together with the corresponding eigenenergies, need to be considered.

4.1. *Odd–even Mass Differences and Moments of Inertia of Rare-earth Nuclei*

Odd–even mass differences and the first pairing-excitation state of $^{153-163}$Er, $^{155-165}$Yb, and $^{157-167}$Hf have been calculated in the extended pairing model (29) and fitted to the experimental values [42,51]. As shown in Fig. 6, the odd–even mass differences for

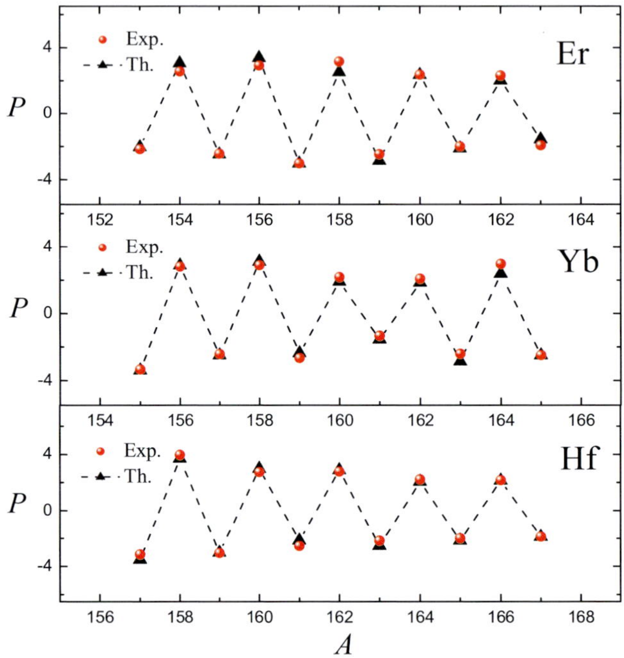

Fig. 6. Theoretical and experimental odd–even mass differences (in MeV) for $^{153-163}$Er, $^{155-165}$Yb, and $^{157-167}$Hf, where experimental values are denoted as "Exp." and theoretical values calculated in the extended pairing model (29) are denoted as "Th.".

these isotopes are very close to the corresponding experimental data. The moments of inertia and the odd–even differences of the moments of inertia of $^{157-164}$Yb are also calculated in the framework of the extended pairing model, according to the Inglis cranking formula (27). Similar to the definition of the odd–even mass difference, the relative odd–even difference of the moments of inertia may be defined as [56]

$$P_{\Im} = \frac{\delta\Im}{\Im}\bigg|_{\text{av}} = \frac{\Im(A) - \frac{1}{2}[\Im(A+1) + \Im(A-1)]}{\frac{1}{2}[\Im(A+1) + \Im(A-1)]}, \tag{37}$$

where A is the mass number and $\frac{1}{2}[\Im(A+1) + \Im(A-1)]$ is the average of the ground-state band moments of inertia of the neighboring nuclei. The theoretical and experimental values of the

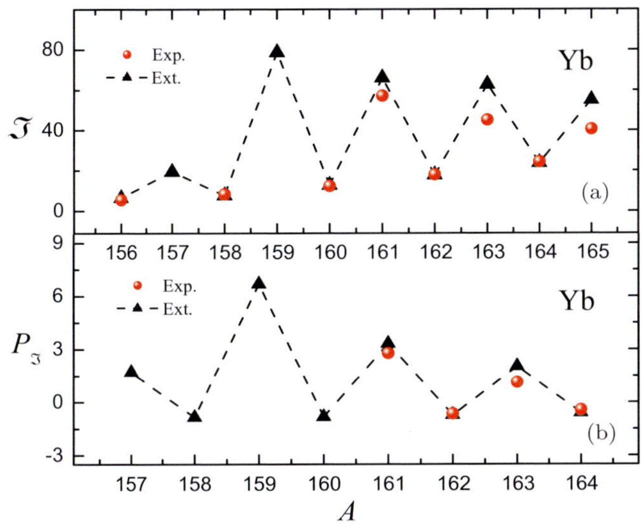

Fig. 7. (a) Theoretical and experimentally deduced moments of inertia in $\hbar^2(\mathrm{MeV})^{-1}$ for $^{156-165}$Yb. (b) Relative odd–even differences of the moments of inertia of $^{157-164}$Yb calculated by Eq. (37). Theoretical results are calculated in the extended pairing model (29) and "Exp." denotes the corresponding values extracted from the experimental spectra of these nuclei [54,55].

moment of inertia $\Im$ for both even–even and odd-A Yb nuclei are shown in Fig. 7a, while the relative odd–even differences of the moments of inertia $P_\Im$ for the Yb isotopes are shown in Fig. 7(b). Clearly, the theoretical values of the moment of inertia are in a good agreement with the corresponding experimentally deduced values for even–even nuclei, while there are small deviations between the theoretical and experimental moments of inertia for odd-A nuclei.

4.2. *Occupation Probabilities in* Yb *Isotopes*

Ground-state occupation probabilities of valence nucleon pairs with J angular momentum (for various J quantum numbers) for even–even $^{156-162}$Yb can also be calculated in the extended pairing model (29). For the ith Nilsson level, the pair creation operators $b_i^\dagger$ can be expressed in terms of the single-particle creation operators of the spherical harmonic-oscillator shell model,

$$b_i^\dagger = \sum_{j_i j_i'} W_{j_i}^i W_{j_i'}^i (-)^{j_i'-\Omega_i} c_{j_i\Omega_i}^\dagger c_{j_i'\,\overline{\Omega}_i}^\dagger, \tag{38}$$

where Ω_i is the projection of j_i onto the third axis of the intrinsic frame, and $W_{j_i}^i$ is a normalized expansion coefficient of the ith Nilsson state expanded in terms of a set of spherical shell-model states. In addition,

$$c_{j_i\Omega_i}^\dagger c_{j_i'\overline{\Omega}_i}^\dagger = \sum_{J_i} \langle j_i\Omega_i j_i'\overline{\Omega}_i | J_i 0\rangle B_{j_i j_i' J_i 0}^\dagger, \tag{39}$$

where $\langle j_i\Omega_i j_i'\overline{\Omega}_i | J_i 0\rangle$ is a Clebsch–Gordan coefficient, and $B_{j_i j_i' J_i 0}^\dagger$ is the pairing operator with total angular momentum quantum number J_i. By substituting Eq. (39) into Eq. (38), we have

$$b_i^\dagger = \sum_{j_i j_i'} W_{j_i}^i W_{j_i'}^i (-)^{j_i'-\Omega_i} \sum_{J_i} \langle j_i\Omega_i j_i'\overline{\Omega}_i | J_i 0\rangle B_{j_i j_i' J_i 0}^\dagger. \tag{40}$$

By substituting Eq. (40) into Eq. (31), the k^ρ-pair eigenstates of the model can be expressed as

$$\begin{aligned} |k^\rho;\zeta;0\rangle = & \sum_{1\le i_1<\cdots<i_k\le p} C_{i_1 i_2\cdots i_k}^{(\zeta)} \prod_{i=1}^{k^\rho} \sum_{j_i j_i'} W_{j_i}^i W_{j_i'}^i (-)^{j_i'-\Omega_i} \\ & \times \sum_{J_i} \langle j_i\Omega_i j_i'\overline{\Omega}_i | J_i 0\rangle B_{j_i j_i' J_i 0}^\dagger(\rho)|0\rangle, \end{aligned} \tag{41}$$

where $\rho = \pi$ for protons or $\rho = \nu$ for neutrons.

The number of like-nucleon pairs with angular momentum J in the ground state ($\zeta = 1$) can then be calculated by

$$n_J^\rho = \left\langle k^\rho;\zeta=1;0 \left| \sum_{jj'} B_{jj'J0}^\dagger(\rho) \frac{\partial}{\partial B_{jj'J0}^\dagger(\rho)} \right| k^\rho;\zeta=1;0 \right\rangle. \tag{42}$$

Since both the proton and neutron sectors are considered, the ground-state occupation probability of valence nucleon pairs with

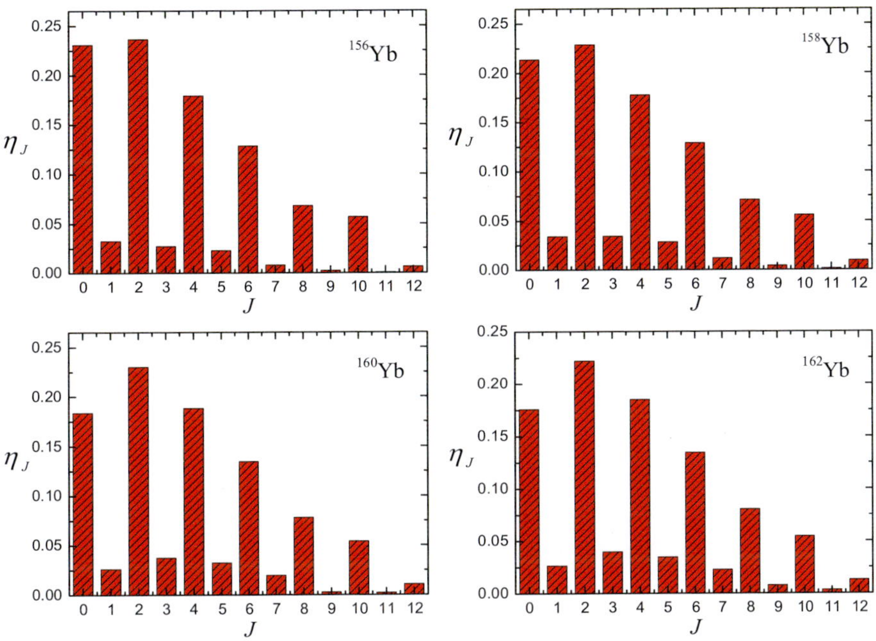

Fig. 8. Ground-state occupation probability of valence nucleon pairs with angular momentum quantum number J in $^{156-162}$Yb calculated in the Nilsson mean-field plus extended pairing model (29).

angular momentum J can be expressed as

$$\eta_J = \frac{n_J^{\pi} + n_J^{\nu}}{k}, \tag{43}$$

where the total number of pairs is $k = k^{\pi} + k^{\nu}$.

Figure 8 displays the calculated results for the ground-state occupation probabilities of valence nucleon pairs with angular momentum $J = 0$ to $J = 12$ for $^{156-162}$Yb. It is obvious that the pair occupation probability decreases with increasing J and is much higher for even-J pairs as compared to odd-J pairs. Figure 8 also shows that these conclusions are independent of the number of valence nucleon pairs k and, therefore, hold for any k cases. Moreover, the S ($J = 0$), D ($J = 2$), and G ($J = 4$) valence nucleon pairs dominate in the ground state of these nuclei. Specifically, their total occupation probability is higher than 60%. However, the occupation probabilities of $J = 6$

and $J = 8$ pairs, even those of $J = 10$ and $J = 12$ pairs, are non-negligible, while odd-J pairs also contribute to the ground state noticeably. Hence, the IBM with s-, d-, and g-bosons seems to provide a reasonable simplified description of the collective motion of these deformed nuclei [57].

4.3. *Phase Transitions in* Sm *Isotopes*

The odd–even differences of various separation energies may provide signals of a ground-state phase transition. It is well known that the odd–even effects, such as those shown in the odd-even mass differences, are the most important evidence of pairing interactions in nuclei [3, 58]. Therefore, a possible microscopic origin of the phase transition may be revealed from a shell-model study. As an illustrative example, Sm isotopes are considered in the extended pairing model of Eq. (29) [59]. Specifically, the odd–even mass difference for the isotopes calculated in the extended pairing model (29) is fitted to the experimental data to determine the neutron pairing strength G, for which the results are shown in Fig. 9. As shown in Fig. 9(a), the odd–even mass difference for the Sm isotopes can be well described by the extended pairing model. Particularly, the phase transitional character around $N = 90$ may be explicitly reproduced by the theoretical results (Figs. 9(b) and 9(c)). In addition, a sudden flattening around $N = 90$ is clearly presented by the pairing strength difference defined as $\Delta G = G(Z, N + 2) - G(Z, N)$, as shown in the inset of Fig. 9(b). A similar situation also appears in ΔG for the odd-mass Sm nuclei but with the sudden flattening occurring around $N = 91$, as shown in Fig. 9(c). As is well-known, the phase transition occurring around $N = 90$ in Sm isotopes corresponds to a transition from a vibrational spectrum to a rotational one [60]. Indeed, the sudden flattening in the pairing strength difference ΔG (Figs. 9(b) and (c)) occurs exactly around the vibrational–rotational transitional point (the critical point). Further investigation indicates that a similar behavior of the pairing strength G also emerges in Nd and Gd isotopes. One can deduce from the above-mentioned analysis that the phase transitional phenomena relevant to pairing in these

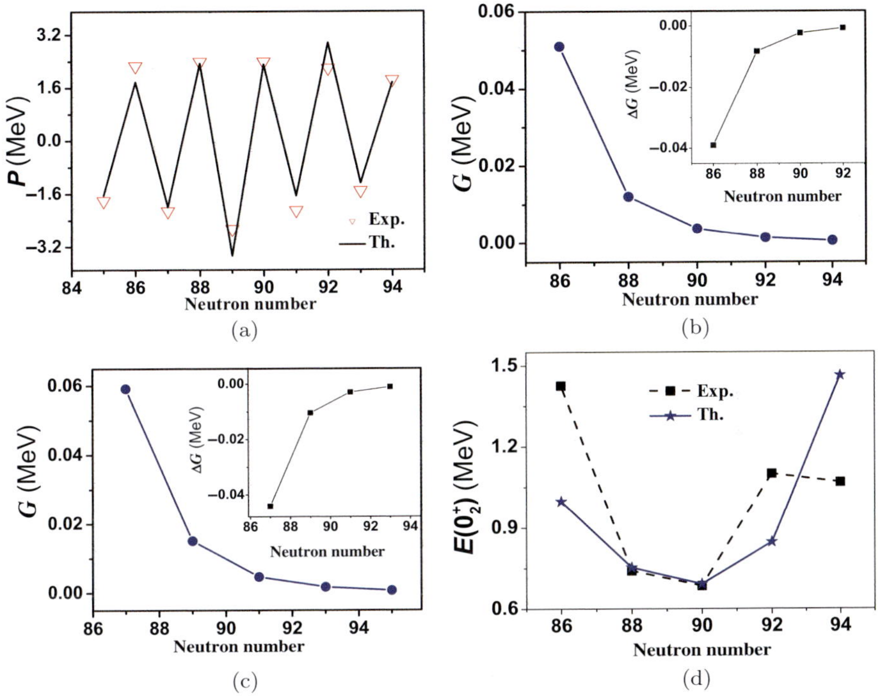

Fig. 9. (a) Odd–even mass difference $P(A)$ (in MeV) for Sm isotopes, calculated in the extended pairing model (29) and fitted to the experimental values. (b) Neutron pairing strength G (in MeV) together with its difference (in MeV) $\Delta G = G(Z, N + 2) - G(Z, N)$ (inset) for the even-mass Sm nuclei. (c) Same as (b) but for the odd-mass Sm nuclei. (d) Pair-excitation (PE) energy (in MeV) calculated in the extended pairing model (stars) and the corresponding $E_{0_2^+}$ experimental data [61] for the even-mass Sm nuclei (squares). The figure is taken from Ref. [59].

isotopes may be driven by decreasing of the pairing strength with a sudden flattening in ΔG around the critical point.

In addition, the pair-excitation (PE) energy, which is also a quantity that is sensitive to the pairing strength G, is analyzed for even–even Sm isotopes. Since the angular momentum projection along the third axis in the intrinsic frame is considered to be a conserved quantity in the model, the PE states determined by the Hamiltonian (29) have the same spin and parity as those of the ground state of a nucleus and can reasonably describe the corresponding excited states. The results of the first PE energy calculated in the model are compared with the experimental energy of the first 0^+ excited state, $E_{0_2^+}$, of even-mass Sm nuclei (Fig. 9(d)). It should be noted that the calculated first PE energy here corresponds to the first neutron pair excitation because the calculated energy for the first excited proton pair in the Sm nuclei is always greater than that of the first excited neutron pair. As shown in Fig. 9(d), a clear evidence for a phase transition is revealed by the PE energy around $N = 90$, where the PE energy reaches its minimum. This coincides with the region where the sudden flattening of ΔG occurs, as discussed above. More importantly, the phase transitional behavior shown by the theoretical PE energies is indeed confirmed by the corresponding experimental data, which can be systematically reproduced by the extended pairing model (29). Clearly, the results shown in Fig. 9 indicate that the transitional behavior governed by the pairing interaction in Sm isotopes can be qualitatively described by the extended pairing model reviewed here.

5. Summary

In summary, exact solutions of either the spherical or deformed mean field plus the standard pairing, and deformed mean field plus the nearest-level pairing or extended pairing are briefly reviewed. It is shown that odd–even mass differences for Ni and Sm isotopes can be fitted well by the spherical mean-field plus the standard pairing model. The analysis on the level statistics of Ca isotopes shows that the model can indeed manifest the pairing-driven chaotic behavior in ^{49,50}Ca, though the regular behavior is more common. Some odd–even mass differences and moments of inertia of low-lying states for selected rare-earth and actinide nuclei show that the NLP or the extended pairing model can equally be used to describe pairing-driven phenomena in these well-deformed nuclei. The results for the ground-state occupation probability of valence nucleon pairs with J angular momentum for even–even $^{156-162}$Yb calculated in the extended pairing model show that the S, D, and G pairs (that is, $J = 0, 2,$ and 4) dominate in the ground state, which thus provides a fermion shell-model reasoning for approximations in the interacting boson model, while additional contribution of about 40% of higher-J pairs and, less important, odd-J pairs is revealed. Pairing-driven ground-state phase transitions in Sm nuclei are examined based on the extended pairing model, which shows that the critical phenomena relevant to pairing may be microscopically driven by a continuous decrease in the pairing strength with a sudden flattening of the pairing strength difference around the critical point. Thus, it is demonstrated in this Chapter that the NLP model and the extended pairing model work well for the ground state of well-deformed heavy nuclei, and that the standard pairing model works well for the ground state of spherical nuclei. However, these models still need to be extended in order to elucidate the detailed structure of nuclei, especially for excited states. For the spherical mean field, besides the pairing interaction, the quadrupole–quadrupole interaction as well as isovector and isoscalar pairing interactions (that is, isospin-1 and isospin-0 proton–neutron pairs) should also be considered in some cases, which are neglected or simplified in the standard pairing model. For the deformed case, in order to describe excited states, angular momentum projection must be carried out for both the nearest-level and extended pairing models. The relevant work is now in progress.

Acknowledgements

Support from the U. S. National Science Foundation (OCI-0904874, ACI-1516338), U.S. Department of Energy (DE-SC0005248), the Southeastern Universities Research Association, the China–U.S. Theory Institute for Physics with Exotic Nuclei (CUSTIPEN) (DE-FG02-13ER42025), the NSFC (11405080, 11175078, and 11375080), and the LSU–LNNU joint research program (9961) is acknowledged.

References

[1] A. Bohr and B. R. Mottelson, *Nuclear Structure*. Vol. 1, Benjamin, New York (1969).
[2] S. T. Belyaev, *Mat. Fys. Medd. Dan. Vid. Selsk.* **31**(11) (1959).

[3] P. Ring and P. Schuck, *The Nuclear Many-Body Problem*. Springer Verleg, Berlin (1980).
[4] M. Hasegawa and S. Tazaki, *Phys. Rev. C.* **47**, 188(1993).
[5] J. Bardeen, L. N. Cooper, and J. R. Schrieffer, *Phys. Rev.* **108**, 1175 (1957).
[6] A. Bohr, B. R. Mottelson and D. Pines, *Phys. Rev.* **110**, 936 (1958).
[7] H. C. Pradhan, Y. Nogami, and J. Law, *Nucl. Phys. A.* **201**, 357 (1973).
[8] H. J. Mang, *Phys. Rep.* **18**, 325 (1975).
[9] M. Bishari, I. Unna, and A. Mann, *Phys. Rev. C.* **3**, 1715 (1971).
[10] J. Y. Zeng, C. S. Cheng, *Nucl. Phys. A.* **405**, 1 (1983); **411**, 49 (1984).
[11] H. Molique, J. Dudek, *Phys. Rev. C.* **56**, 1795 (1997).
[12] G. D. Dans and A. Klein, *Phys. Rev.* **143**, 735 (1966).
[13] A. Covello and E. Salusti, *Phys. Rev.* **162**, 859 (1967).
[14] A. Volya, B. A. Brown, and V. Zelevinsky, *Phys. Lett. B.* **509**, 37 (2001).
[15] A. K. Kerman, R. D. Lawson, and M. H. MaeFartane, *Phys. Rev.* **124**, 162 (1961).
[16] V. Zelevinsky and A. Volya, *Phys. At. Nucl.* **66**, 1781 (2003).
[17] R. W. Richardson, *Phys. Lett.* **3**, 277 (1963).
[18] R. W. Richardson, *Phys. Lett.* **5**, 82 (1963).
[19] R. W. Richardson and N. Sherman, *Nucl. Phys.* **52**, 221 (1964).
[20] R. W. Richardson and N. Sherman, *Nucl. Phys.* **52**, 253 (1964).
[21] M. Gaudin, J. Phys. **37**, 1087 (1976).
[22] F. Pan, J. P. Draayer, and W. E. Ormand, *Phys. Lett. B.* **422**, 1 (1998).
[23] J. Dukelsky, C. Esebbag, and P. Schuck, *Phys. Rev.* Lett. **87**, 066403 (2001).
[24] J. Dukelsky, C. Esebbag, and S. Pittel, *Phys. Rev.* Lett. **88**, 062501 (2002).
[25] H.-Q. Zhou, J. Links, R. H. McKenzie, and M. D. Gould, *Phys. Rev. B.* **65**, 060502(R) (2002).
[26] F. Pan and J. P. Draayer, *Phys. Rev. C.* **66**, 044314 (2002).
[27] S. Rombouts, D. Van Neck, and J. Dukelsky, *Phys. Rev. C.* **69**, 061303(R) (2004).
[28] G. Ortiz, R. Somma, J. Dukelsky, and S. Rombouts, *Nucl. Phys. B.* **707**, 421 (2005).
[29] J. Dukelsky, V. G. Gueorguiev, P. Van Isacker, S. Dimitrova, B. Errea, and S. Lerma H., *Phys. Rev.* Lett. **96**, 072503 (2006).
[30] A. B. Balantekin, J. H. de Jesus, and Y. Pehlivan, *Phys. Rev. C.* **75**, 064304 (2007).
[31] A. B. Balantekin and Y. Pehlivan, *Phys. Rev. C.* **76**, 051001(R) (2007).
[32] X. Guan, K. D. Launey, M. Xie, L. Bao, F. Pan, and J. P.Draayer, *Phys. Rev. C.* **86**, 024313 (2012).
[33] X. Guan, K. D. Launey, M. Xie, L. Bao, F. Pan, J.P. Draayer, *Comp. Phys. Commun.* **185**, 2714 (2014).
[34] F. Pan and J. P. Draayer, *J. Phys. A.* **33**, 9095 (2000).
[35] Y. Chen, F. Pan, G. S. Stoitcheva, J. P. Draayer, *Int. J. Mod. Phys. B.* **16**, 2071 (2002).
[36] F. Pan, Y. Wang, X. Guan, J. Lu, X.-R. Chen, and J. P. Draayer, *AIP Conf. Proc.* **1342**, 63 (2011).

[37] F. Pan, V. G. Gueorguiev, and J. P. Draayer, *Phys. Rev.* Lett. **92**, 112503 (2004).
[38] F. Pan, M.-X Xie, X. Guan, L.-R Dai, and J. P. Draayer, *Phys. Rev. C.* **80**, 044306 (2009).
[39] X. Guan, H. Li, Q. Tan, F. Pan, and J. P. Draayer, *Chin. Phys. C.* **35**, 747 (2011).
[40] F. Pan, L. Bao, L. Zhai, X. Cui, and J. P. Draayer, *J. Phys. A.* **44**, 395305 (2011).
[41] G. Szegö, *Orthogonal Polynomials*, 4th edn. Vol. 23, *Amer. Math. Soc. Colloq. Publ.* Providence, RI (1975).
[42] G. Audi, A. H. Wapstra, and C. Thibault, *Nucl. Phys. A.* **729**, 337 (2003).
[43] T. A. Brody, J. Flores, J. B. French, P. A. Mello, A. Pandey, and S. S. M. Wong, *Rev. Mod. Phys.* **53**, 385 (1981).
[44] Y. Alhassid and A. Novoselsky, *Phys. Rev. C.* **45**, 1677 (1992).
[45] F. J. Dyson, J. Math. Phys. **3**, 140 (1962); *ibid.*, 157 (1962); *ibid.*, 166 (1962); *ibid.*, 1191 (1962); *ibid.*, 1199 (1962).
[46] M. L. Metha, *Random Matrices*, 2nd edn. Academic Press, Boston (1991).
[47] X. Guan, K. D. Launey, Jianzhong Gu, Feng Pan, and J. P. Draayer, *Phys. Rev. C.* **88**, 044325 (2013).
[48] J. A. Cameron and B. Singh, *Nucl. Data Sheets.* **94**, 429 (2001).
[49] W. D. Heiss, R. G. Nazmitdinov, and S. Radu, *Phys. Rev.* Lett. **72**, 2351 (1994).
[50] P. Möller, J. R. Nix, and K.-L. Kratz, *At. Data Nucl. Data Tables.* **66**, 131 (1997).
[51] P. Moller, J. R. Nix, W. D. Myers, and W. J. Swiatecki, *At. Data Nucl. Data Tables.* **59**, 185 (1995).
[52] D. R. Inglis, *Phys. Rev.* **96**, 1059 (1954).
[53] S. G. Nilsson and O. Prior, *Mat. Fys. Medd. K. Dan. Vidensk. Selsk.* **32**(16) (1961).
[54] P. C. Sood, D. M. Headly, and R. K. Sheline, *At. Data Nucl. Data Tables* **47**, 89 (1991).
[55] A. K. Jain, R. K. Sheline, P. C. Sood, and K. Jain, *Rev. Mod. Phys.* **62**, 393 (1990).
[56] J. Y. Zeng, Y. A. Lei, T. H. Jin, and Z. J. Zhao, *Phys. Rev. C.* **50**, 746 (1994).
[57] F. Iachello and I. Talmi, *Rev. Mod. Phys.* **59**, 339 (1987).
[58] K. Heyde, *Basic Ideas and Concepts in Nuclear Physics*, 2nd edn. Institute of Physics Publishing (1999).
[59] Y. Zhang, L. Bao, X. Guan, F. Pan and J. P. Draayer, *Phys. Rev. C.* **88**, 064305 (2013).
[60] F. Iachello and A. Arima, *The Interacting Boson Model.* Cambridge University Press, New York (1987).
[61] G. Audi, A. H. Wapstra and C.Thibault, *Nucl. Phys. A.* **729**, 337 (2003). (LBNL Isotopes Project Nuclear Data Dissemination Home Page. Retrieved September, 2015, from http://ie.lbl.gov/toi.html.)

Author Index

Subject Index